T0314297

High-Energy Astrophysics

High-Energy Astrophysics

Fulvio Melia

PRINCETON UNIVERSITY PRESS

PRINCETON AND OXFORD

Copyright ©2009 by Princeton University Press

Published by Princeton University Press, 41 William Street, Princeton, New Jersey 08540

In the United Kingdom: Princeton University Press, 6 Oxford Street, Woodstock, Oxfordshire 0X20 1TW

All Rights Reserved

Library of Congress Cataloging-in-Publication Data

Melia, Fulvio.
High-energy astrophysics/Fulvio Melia.
        p.   cm. — (Princeton series in astrophysics)
    Includes bibliographical references and index.
    ISBN 978-0-691-13543-4 (hardcover : alk. paper) — ISBN 978-0-691-14029-2 (pbk. : alk. paper)
1. Nuclear astrophysics.   2. Astrophysics.   I. Title.
    QB464.M45 2009
    523.01'97–dc22   2008022287

British Library Cataloging-in-Publication Data is available

This book has been composed in Sabon and Swiss

Printed on acid-free paper. ∞

press.princeton.edu

Printed in the United States of America

10 9 8 7 6

TO THE MEMORY OF
DAVID CHU

Contents

Preface

Designed to detect X-rays from the moon, a sounding-rocket experiment in 1962 instead ended up discovering a pervasive and isotropic X-ray background and the brightest source (Sco X-1) of high-energy radiation outside the solar system. Within a mere thirty years, the German-US-UK ROSAT satellite had increased this census dramatically by uncovering as many as 150,000 distinct objects emitting photons at 0.1–2.5 keV.

The growth of high-energy astrophysics as an experimental space science has been breathtaking. And theorists have keenly kept up the pace, interpreting the wide array of exotic phenomena revealed during this exploration, as dramatic discovery has followed dramatic discovery.

Very soon after ROSAT, the *Compton* Gamma Ray Observatory opened a new γ-ray window stretching our sensing capability to 30 GeV, and the *Chandra* X-ray Observatory peered deeply at supermassive black holes near the beginning of time. Most recently, stereoscopic Cerenkov telescopes in Namibia imaged the sky at TeV energies with unprecedented spatial clarity, all the while creating a flurry of theoretical activity with the promise of extending the boundaries of our physical laws to uncharted territory. Amazingly, neutron stars and black holes had not yet even entered our lexicon at the dawn of high-energy astrophysics. Today, we are preparing an experiment that will actually image the event horizon of a supermassive object at the heart of the Milky Way.

This subject generates excitement with little input from its practitioners. The purview of high-energy astrophysics is the realm of physical conditions and interactions that we simply cannot replicate here on Earth. So it fosters within us a natural curiosity and desire to probe farther and deeper, to test theories of strong gravity, and to bring the early universe to a focus so that we may see how it all began.

Several attempts have already been made at synthesizing the developments of this field into the form of a book. But it is difficult to balance the needs and aspirations of experimental high-energy astrophysics with a formal treatment and theoretical discussion of the objects uncovered by the observations.

The present volume is the outgrowth of a graduate course in high-energy astrophysics that I have now taught at the University of Arizona for over 15 years. Though by no means exhaustive, the topics covered by this material constitute an attempt to seamlessly blend together a discussion of the instrumental techniques and initiatives of the various theoretical ideas and of the methods astronomers employ to study the many classes of high-energy objects that grace this field. Its intended use is not only for instructional purposes at the graduate and advanced undergraduate levels, but also for professional scientists who may wish to consult a compact, self-contained monograph covering the essential aspects—in both theory and experiment—of this rapidly growing discipline.

I have learned much about this topic directly from my own work, and for generously supporting this research for more than a decade and a half, I gratefully acknowledge the National Science Foundation, the National Aeronautics and Space Administration, and the Alfred P. Sloan Foundation.

I thank the hundreds of students with whom I have had the pleasure of sharing my knowledge of high-energy astrophysics, for their interest and warmth, and for helping me refine this material into a streamlined—yet nonetheless comprehensive—subject.

To Patricia, Marcus, Eliana, and Adrian and to my parents, whose guidance has been priceless, I extend my enduring love and gratitude.

Fulvio Melia
Tucson, Arizona

High-Energy Astrophysics

Chapter One

Introduction and Motivation

1.1 THE FIELD OF HIGH-ENERGY ASTROPHYSICS

Compared with optical astronomy, which traces its foundations to prehistoric times,[1] the field of high-energy astrophysics is a relatively new science, dealing with astronomical sources and phenomena largely discovered since the advent of space-based instrumentation.[2] Ironically, however, the earliest signs of high-energy activity from space appeared in laboratory equipment on the ground, and were not recognized as extraterrestrial for many years.

Physicists had noted since about 1900 that some unknown ionizing radiation had to be present near Earth's surface, from the manner in which the leaves of an electroscope always came together, presumably as a result of a gradual discharging. By 1912, Victor Hess (1883–1964) had made several manned balloon ascents to measure the ionization of the atmosphere as a function of altitude, making the surprising discovery[3] that the average ionization above $\sim$1.5 km increased relative to that at sea level. By demonstrating in this manner that the source of the ionizing radiation—named *cosmic rays* by Robert Millikan (1868–1953) in 1925—must therefore be extraterrestrial, Hess opened up the frontier of high-energy astrophysics and was eventually awarded the Nobel prize in physics in 1936.

Not long after World War II, the ready availability of sounding rockets facilitated the first major discovery of the high-energy sources themselves. The name *sounding rocket* comes from the nautical term "to take a sounding," meaning to make a measurement. Typically consisting of a solid-fuel rocket motor and a payload, these devices are commonly used to carry instruments and to take readings from about 50 to 1500 km above Earth's surface, between the regions sampled by weather balloons and orbiting satellites. The sounding rocket consumes its fuel during the rising portion of the flight, then the payload separates to complete the arc and land on the ground with a parachute. Data are collected for up to 15 min at a time while the payload is flying ballistically above Earth's atmosphere.

The first scientific flight of a sounding rocket in 1949 carried a photon (Geiger) counter and showed unambiguously that the Sun emits X-rays.[4] Not long after

---

[1] See, e.g., Ruggles and Saunders (1993).

[2] We will learn shortly in this chapter why high-energy observations must be made above Earth's atmosphere.

[3] Hess's balloon flights reached as high as 5 km above sea level. One of his followers, Kolhörster, ascended to 9 km a few years later. See Hess (1912).

[4] See Friedman, Lichtman, and Byram (1951).

that, in 1962, Riccardo Giacconi and colleagues carried out another sounding rocket experiment that many consider to be the birth of X-ray astronomy.[5] Though flown to investigate X-rays from the moon, the 1962 experiment instead discovered Sco X-1, the brightest X-ray source in the sky, and with it, a completely unexpected diffuse glow of X-rays coming from all directions—the cosmic X-ray background.

Experimental and observational astronomy have grown rapidly since then, with the numbers and size of detection systems increasing steadily with satellites in the 1970s, to a broadening international effort in the 1980s, to even larger and more imposing spacecraft in the 1990s and beyond. Today, high-energy instrumentation is being developed and deployed not only by the United States, but also by countries in Europe, member states of the former Soviet Union, Asia, and South America.

But though much of what we have learned about the ultraviolet (UV), X-ray, and γ-ray sky has occurred in space, we should not come away with the thought that high-energy astrophysics is *exclusively* a space-based discipline. From the beginning, high-altitude balloons could carry heavy payloads of larger aperture for flights lasting $\sim$10 h or more, at altitudes of $\sim$40 km. As we shall see, only X-rays exceeding $\sim$20 keV in energy can penetrate down this far in the atmosphere, but to this day, balloon experiments continue to provide important observational results in the 20- to 60-keV portion of the spectrum.

At even higher energies, the atmosphere itself becomes part of the detector, providing a target for the highest energy photons as they descend from space, thereby producing a cascade of particles and radiation that reach Earth's surface. Thus, at TeV energies and higher (where 1 TeV $= 10^{12}$ eV), observational high-energy astronomy is actually best conducted from the ground. Early in the 21st century, some of the most spectacular gains in this field have been made by air Cerenkov telescopes, as they are known, and we will consider these in detail, along with other detector techniques, later in this chapter.

Regardless of the photon energy, however, the principal feature that distinguishes high-energy astrophysics from other branches of astronomy is the relative paucity of photons in the X-ray and γ-ray bands. Instrument builders must factor this into their designs. Several years ago, the National Aeronautics and Space Administration (NASA) in the United States was seriously considering funding the development of a detector, to be attached to the international space station, that would collect a scant 300–500 photons over its entire two-year lifetime.

There are at least two reasons why this occurs. First, the energy of the photons precludes large numbers of them being emitted concurrently. A 1-MeV γ-ray carries as much energy as one million optical photons, so for a given (limited) energy budget, it is a lot more difficult to produce high-energy γ-rays than photons detected by optical means. Second, the sky is far less crowded in X-rays and γ-rays than

---

[5]Giacconi's extensive work in high-energy astrophysics was recognized with the awarding of the Nobel prize in physics in 2002, "for pioneering contributions to astrophysics, which have led to the discovery of cosmic X-ray sources." The results from the 1962 sounding-rocket flight were reported in Giacconi et al. (1962).

in other spectral regimes. One can actually count the high-energy point sources. Indeed, until very recently, all of them could be assigned individual names.[6]

Given the complexity of building and deploying these detectors, experimental high-energy astrophysics now tends to be what physicists would call a "big science." As in particle physics, nothing meaningful can be done anymore without the strong participation of a large group of collaborators, often at several different universities and government laboratories. In high-energy astrophysics, this effort must also be supplemented by the key participation of one or more space agencies, such as NASA in the United States and the European Space Agency (ESA).

High-energy astrophysics is also an important theoretical discipline, encompassing many other sub-branches of physics. Broadly speaking, high-energy astrophysics involves the study of (1) large quantities of energy, usually coupled to relativistic matter, (2) the rapid release of this energy in events of extreme violence, sometimes completely destroying the underlying source, (3) the interaction of matter and radiation under the extreme conditions of superstrong gravity and magnetic fields, and (4) the emission of large fluxes of X-rays, γ-rays, and sometimes also UV radiation.

Although interesting as an instrumental and observational science, high-energy astrophysics is therefore also particularly attractive to theoreticians because it provides new physical problems and tests of fundamental theories, such as general relativity, under conditions that are totally inaccessible in the laboratory, or even within the solar system. Consider this: one teaspoonful of neutron-star material weighs as much as all of humanity combined. And a marshmallow dropped onto the surface of that neutron star releases enough gravitational energy to produce an explosion equal to that of a medium-sized atomic bomb on Earth. This is the realm of physical reality we will be exploring in this book.

1.2 ENERGIES, LUMINOSITIES, AND TIMESCALES

Before we move on to consider the hurdles one must face in first building high-energy telescopes and then using them effectively to interrogate the electromagnetic signals reaching Earth from distant high-energy sources, let us orient ourselves by considering several characteristic scales. For example, it matters when building a device to measure variability in the source whether the emitter changes its profile over a period of several microseconds or much more slowly over an interval of many months.

As we develop the theoretical tools used in high-energy astrophysics throughout this book, we will encounter a variety of particle acceleration mechanisms, each scaled by its own critical energy and fiducial acceleration time. The simplest to consider is gravity, so we will here estimate certain critical numbers characterizing the high-energy emission by compact objects accreting from their environment.

The release of gravitational potential energy by matter falling onto a compact object is believed to be the principal source of power in sources as diverse as the

---

[6]As we shall see later in this chapter, the most dramatic change in our view of the X-ray sky occurred when the satellite ROSAT was deployed in the early 1990s. Within the first six months of operation, ROSAT produced a catalog exceeding 100,000 X-ray sources.

accreting star in a low-mass X-ray binary and the significantly more powerful and distant supermassive black hole in active galactic nuclei (AGNs) and quasars.

A mass m falling onto an object of (bigger) mass M and radius R releases a quantity of energy

$$\Delta E_{\rm acc} = \frac{GMm}{R}. \tag{1.1}$$

Thus, for a neutron star with mass $M_{\rm ns} \sim 1 M_\odot$ and radius $R_{\rm ns} \approx 10\,{\rm km}$ (the size of a typical city),

$$\Delta E_{\rm acc}({\rm ns}) \approx 10^{20}\,{\rm ergs\,g}^{-1}. \tag{1.2}$$

As we shall see, on a neutron star this energy eventually emerges in the form of X-rays and γ-rays. By comparison, hydrogen burning into helium releases a quantity of energy

$$\Delta E_{\rm nuc}({\rm H} \to {\rm He}) \approx 6 \times 10^{18}\,{\rm ergs\,g}^{-1}, \tag{1.3}$$

so the accretion yield for a neutron star is about 20 times bigger than the process we commonly associate with the release of energy in main sequence stars (hence the "doomsday" marshmallow anecdote from before).

These numbers, however, are not universal since $\Delta E_{\rm acc}$ depends on the compactness M/R of the accreting object. For example, a white dwarf also has a mass $M_{\rm wd} \sim 1 M_\odot$, but a radius $R_{\rm wd}$ ($\sim 10^9$ cm) about equal to that of Earth. For such an object, M/R is so small that

$$\Delta E_{\rm acc}({\rm wd}) \approx \tfrac{1}{50} \Delta E_{\rm nuc}({\rm H} \to {\rm He}). \tag{1.4}$$

On the other hand, black holes have a characteristic radius $R_S = 2GM/c^2$, the event (or Schwarzschild) horizon to be defined in chapter 3. For them, $M/R \sim c^2/2G$, and

$$\Delta E_{\rm acc}({\rm bh}) \approx 5 \times 10^{20}\,{\rm ergs\,g}^{-1}. \tag{1.5}$$

Black holes do not have a hard surface (like a neutron star or a white dwarf) so we do not expect all of the energy released gravitationally to be observable as electromagnetic radiation, since some of it is presumably advected through the horizon.

There are at least two additional questions we should ask about the energy released by gravity. First, what is the total luminosity (or power) associated with such a process? Second, how quickly can the emission region vary? Knowing the luminosity of a typical high-energy source permits us to infer its energy flux arriving at Earth, and hence the required instrument sensitivity to measure it. These days, analyzing the timing behavior of an astronomical source can be just as informative as studying its spectral characteristics, so, as we have already indicated, we must also have an idea of the source's variability.

To address the first of these, let us suppose that we have a fully ionized hydrogen gas accreting isotropically onto a compact star. Aside from the effects of a magnetic (or perhaps even an electric) field, there are generally two forces on the plasma, one arising from the inward pull of gravity, and a second (an outward force) due to Thomson scattering between the charged particles in the gas and the radiated photons diffusing outward. It is the balance between these two influences that sets the luminosity scale.

Although both electrons and protons are present in the infalling plasma, the outward force is dominated by scatterings between the photons and electrons. Thomson scattering by protons is much less important due to their significantly smaller cross section. To understand this in a heuristic manner, let us think of the scattering classically as a one-dimensional process, in which the charge behaves as an oscillator responding to an electric wave $\mathbf{E}(z, t)$ passing through the medium in the $\hat{\mathbf{z}}$-direction. From Newton's law

$$\frac{d\mathbf{p}}{dt} = q\mathbf{E}, \tag{1.6}$$

where $\mathbf{p}$ is the particle's momentum and q is its charge. For a harmonic wave

$$\mathbf{E}(z, t) = E_0(z)e^{i\omega t}\,\hat{\mathbf{x}}, \tag{1.7}$$

the particle's (one-dimensional) equation of motion reads

$$m\frac{d^2x}{dt^2} = qE_0(z)e^{i\omega t}. \tag{1.8}$$

The solution to this differential equation is trivial, and may be written in the form

$$x(t) = -Ae^{i\omega t}, \tag{1.9}$$

where A is the amplitude of the motion, given by the expression

$$mA\omega^2 = qE_0(z), \tag{1.10}$$

or more simply as

$$A = \frac{qE_0(z)}{m\omega^2}. \tag{1.11}$$

Evidently, the amplitude of the charge's oscillation is inversely proportional to its mass, and the cross-sectional area, which is roughly proportional to A^2, will therefore go as $\sim m^{-2}$. So labeling the proton and electron Thomson cross sections as σ_{Tp} and σ_{Te}, respectively, we infer that

$$\frac{\sigma_{Tp}}{\sigma_{Te}} = \left(\frac{m_e}{m_p}\right)^2 \approx 3 \times 10^{-7}. \tag{1.12}$$

The actual value of this ratio, here deduced classically, should not be taken too literally. However, we will use the large disparity between σ_{Tp} and σ_{Te} to justify our assumption that only electron Thomson scattering needs to be considered in calculating the outward radiation force on the infalling plasma.

Within the plasma, therefore, the gravitational force is stronger for the protons, these being the more massive particles, whereas the outward radiative force is greater for the electrons, these having the bigger cross section. But the two particle species do not separate spatially since the Coulomb attraction between them prevents this from happening. For the purpose of finding the force balance, it suits us to think of the ionized gas as comprising loosely coupled electron–proton pairs, even though in the ionized state, the two oppositely charged sets of particles are more like two fluids passing through each other. The gravitational force on each couplet is then simply

$$f_{grav} = -\frac{GM(m_p + m_e)}{r^2}. \tag{1.13}$$

To find the corresponding force f_{rad} due to Thomson scattering, we must consider the rate of momentum transfer from the outwardly streaming radiation to the electrons in the plasma. As we shall see in chapter 3, a photon of energy ϵ carries a momentum ϵ/c, so radiation with energy flux F (in units of energy per unit area per unit time) also carries a momentum flux

$$\Pi = \frac{F}{c}. \tag{1.14}$$

The radiative force on the electron in each couplet is the momentum per unit time transferred to it from Π. Henceforth denoting σ_{Te} as simply σ_T, we therefore have

$$f_{rad} = \frac{F}{c}\sigma_T. \tag{1.15}$$

For an isotropic point source of radiation with luminosity L, this can also be written

$$f_{rad} = \frac{\sigma_T L}{4\pi c r^2}, \tag{1.16}$$

and so the net force on the particle couplet is

$$f_{grav} + f_{rad} = \left[-GM(m_p + m_e) + \frac{\sigma_T L}{4\pi c}\right]\frac{1}{r^2}. \tag{1.17}$$

When the two forces appearing on the left-hand side of equation (1.17) are equal (and opposite), the right-hand side of this equation must, of course, vanish. The critical luminosity for which this occurs is known as the Eddington limit (or luminosity). Since $m_p \gg m_e$, one simply puts $m_p + m_e \approx m_p$, which gives

$$L_{edd} \equiv \frac{4\pi c G M m_p}{\sigma_T}. \tag{1.18}$$

Numerically, we have

$$L_{edd} \approx 1.3 \times 10^{38} \left(\frac{M}{M_\odot}\right) \text{ergs s}^{-1}. \tag{1.19}$$

Thus, the Eddington luminosity for a neutron star with mass $M_{ns} = 1.4\,M_\odot$ is $\sim 10^4\,L_\odot$ (all in X-rays, as we shall see shortly), whereas for an AGN black hole with mass $M_{AGN} = 10^7\,M_\odot$, the Eddington luminosity is $\sim 10^{45}$ ergs s^{-1}, roughly 10 times the luminosity of the entire galaxy.

The Eddington limit is an important characteristic luminosity because sources with $L > L_{edd}$ expel nearby matter and quench the process of accretion. For this reason, the Eddington luminosity is a reliable measure of the power expected from a very active source of mass M, deriving its energy by absorbing matter from its environment.

For compact objects, this power is usually manifested as UV photons or X-rays, though sometimes secondary processes may recycle this radiation to yet higher energies. The effective temperature T_{eff} of a blackbody radiating a total power L_{edd} is given by

$$4\pi R^2 \sigma_B T_{\text{eff}}^4 = L_{\text{edd}}, \tag{1.20}$$

where σ_B is the Boltzmann constant. With equation (1.18), this gives

$$T_{\text{eff}} = \left(\frac{cGMm_p}{R^2 \sigma_T \sigma_B} \right)^{1/4}. \tag{1.21}$$

Thus, for a neutron star, $T_{\text{eff}}(\text{ns}) \approx 2 \times 10^7$ K, corresponding to a characteristic photon energy $h\nu = kT_{\text{eff}} \approx 1.6\,\text{keV}$ (X-rays). White dwarfs, on the other hand, emit with an effective temperature $T_{\text{eff}}(\text{wd}) \approx 6 \times 10^5$ K, with characteristic photon energy $\approx 50\,\text{eV}$ (UV). In the case of a black hole, the spectrum cannot be interpreted without understanding the accretion process more fully, since the absence of a hard surface means that most of the gravitational energy is released over an extended region of space.

We are now in a position to estimate the photon counting rate required of our detectors in order to measure, say, the X-ray flux reaching us from a typical neutron star near the center of our Galaxy. The photon number flux impinging on Earth's atmosphere from one of these sources is roughly $F_{\text{ph}} \equiv L_{\text{edd}}/4\pi D^2 \langle \epsilon \rangle$, where D ($\approx 8.5\,\text{kpc}$) is the distance to the galactic center and $\langle \epsilon \rangle$ ($\sim 1.6\,\text{keV}$) is the average photon energy. Thus, $F_{\text{ph}} \sim 10\,\text{ph}\,\text{cm}^{-2}\,\text{s}^{-1}$. By comparison, the very first X-ray orbital mission in 1970, named *Uhuru*, achieved a limiting sensitivity of $\sim 6 \times 10^{-3}\,\text{ph}\,\text{cm}^{-2}\,\text{s}^{-1}$, rendering such neutron-star sources easily detectable over the mission lifetime. The *Chandra* observatory (see section 1.5), one of the most recently deployed X-ray satellites, achieves a sensitivity of about $2 \times 10^{-6}\,\text{ph}\,\text{cm}^{-2}\,\text{s}^{-1}$ in 10^5 seconds of integration, greatly enhancing our capability of discovering very faint sources, many of which accrete well below the Eddington limit.

The second question we posed with regard to the transfer of energy from gravity to the radiating plasma concerns the variability timescale t_{var} in the emitting region. For gravity, t_{var} is heavily influenced by the dynamical timescale t_{dyn}, which we estimate as follows.

Consider an element of mass m falling under the influence of gravity above the surface of a star. An upper limit to t_{var} is the time it takes for m to fall a distance R, the star's radius. Let us put

$$\tfrac{1}{2} a t_{\text{dyn}}^2 \equiv R, \tag{1.22}$$

where $a = GM/R^2$ is the gravitational acceleration. Rewriting this in the form

$$t_{\text{dyn}} = \left(\frac{2R^3}{GM} \right)^{1/2}, \tag{1.23}$$

we find that $t_{\text{dyn}}(\text{ns}) = 10^{-4}$ s for a neutron star. Similarly, for a white dwarf we estimate that $t_{\text{dyn}}(\text{wd}) = 3$ s. For a black hole, the situation is again unclear without a more complete understanding of the emitter's geometry, particularly since jets are often important—if not dominant—contributors to the overall emission from these objects. Doppler shifts and light-travel time effects, which we will develop later in this book, can seriously modify the variability timescale inferred by a distant observer relative to what is actually taking place in the emitter's frame. Observationally, we already know that some AGNs, like the BL Lac object H 0323+022, display variability in all wavebands, including the X-ray,[7] where the source varies on a timescale as short as $\sim$30 s. That is surprising, considering the fact that the light-travel time ($\sim 2R_S/c$) around a black hole with mass $\sim 10^8 \, M_\odot$ is about half an hour, which is the shortest interval one would think of associating with the local dynamical time. So clearly some other effect, probably due to relativistic length contraction, is at play in transforming the variability timescale in the emitter's frame into that measured by the observer.

Our instruments will detect changes in the emitter's configuration only if they can accumulate reasonable photon counting statistics to measure the incoming flux from typical high-energy sources in several seconds; the integration time needs to be even shorter—only a fraction of a millisecond—in the case of neutron stars. As we shall see later in this chapter, the high-energy observatories built thus far have tended to specialize in either imaging capability or spectral energy resolution, though only with a modest ability to resolve the source temporally. The one exception has been the *Rossi* X-ray Timing Explorer, launched in December 1995, whose remarkable temporal resolution has allowed astronomers to resolve changes occurring over only a fraction of a millisecond in the brightest objects.

1.3 ATMOSPHERIC ABSORPTION

We next familiarize ourselves with the difficulties faced by experimental high-energy astrophysicists in constructing X-ray and γ-ray detectors that, with appropriate deployment, can sense the radiation reaching us from astronomical sources. Their triumph in overcoming nature's hurdles has been recognized by the Nobel prize committee, which has thus far awarded two prizes in this field.[8]

There is a very good reason why our eyes have developed the greatest sensitivity to light in the $\sim$400- to 700-nm range of wavelengths—the aptly labeled "visible" portion of the spectrum. Without this capability, the long development of optical astronomy since prehistoric times would have been impossible. This waveband not only encompasses most of the Sun's spectral output—roughly a blackbody peaking at a wavelength of $\sim$600 nm, right in the middle of this range—but also represents one of the few regions in the spectrum where atmospheric absorption is virtually nonexistent. These two factors mean that most of the light reaching us from our environment has a wavelength between 400 and 700 nm. Conversely, the intensity of light at many

---

[7] See Feigelson et al. (1986).

[8] As noted earlier in this chapter, Victor Hess was the first recipient, in 1936, for discovering cosmic rays. Riccardo Giacconi received his award in 2002 for his life's work in essentially creating the field of X-ray astronomy.

other wavelengths is severely attenuated near Earth's surface. In a way, this is actually a good thing as far as our survival is concerned, given that UV, X-ray, and γ-ray photons tend to seriously damage our cells. Unfortunately, the atmosphere's effective absorption of radiation at wavelengths shorter than the visible also strongly mitigates any possibility of viewing cosmic high-energy radiation directly from the ground, even up to the TeV range, where what we see are the secondary particles produced by collisions between the extraterrestrial photons and the upper atmosphere.

Photons can interact with atmospheric particles in several ways. Below about 60 keV (i.e., for wavelengths $\lambda \gtrsim 2 \times 10^{-9}$ cm, or 0.2Å), the dominant interaction is the photoelectric effect, in which an atomic electron completely absorbs the incident photon and is ejected. As such, the cross section for this process increases sharply at the so-called absorption edges, where the photon energy is sufficient to knock out an electron in the next (deeper) energy level. In the simplest description of radiation transfer through a medium with density ρ, the attenuation suffered by a ray with intensity I (in units of ergs cm^{-2} s^{-1} ster^{-1}) is described according to the differential equation

$$\frac{dI}{ds} = -\mu\rho I, \tag{1.24}$$

where the absorption coefficient μ is defined as the cross section per unit mass of the material, and s is the pathlength. If the medium is uniform, equation (1.24) integrates easily and the solution is

$$I(d) = I_0 \exp(-\mu\rho d), \tag{1.25}$$

where d is the distance into the medium, or simply its thickness if the ray has penetrated all the way through. Figure 1.1 shows the mass absorption coefficient μ as a function of photon energy for three elements—silicon, chlorine, and iron—in which the absorption edges are clearly visible. The larger the value of μ is, the shorter is the distance d required for a photon at this energy to be absorbed.

Above $\sim$10 MeV, photons passing through the atmosphere begin to produce electron–positron pairs. An individual photon traveling through vacuum cannot spontaneously materialize in this fashion because, as we shall see in chapter 3, its energy–momentum relation is not the same as that of particles with a nonzero rest mass. A single photon splitting into an electron–positron pair therefore cannot simultaneously conserve both energy and momentum. Within the Coulomb field of another charge, however, such as the nucleus of an atom or ion, this process is physically permitted because the incoming photon can interact with a virtual photon while their center of momentum moves subluminally in the observer's frame.[9] The absorption coefficient κ for this process is also straightforward to write down from QED, though we will here only show the results graphically (see figure 1.2 below).

Between these two energies (i.e., between $\sim$60 keV and $\sim$10 MeV), the dominant interaction suffered by extraterrestrial radiation in Earth's atmosphere is inelastic

---

[9]The first individuals to consider this process, Landau and Lifshitz, have written extensively on this subject, including the book *Quantum Electrodynamics*, a volume in their highly regarded series on theoretical physics. See Landau and Lifshitz (1982).

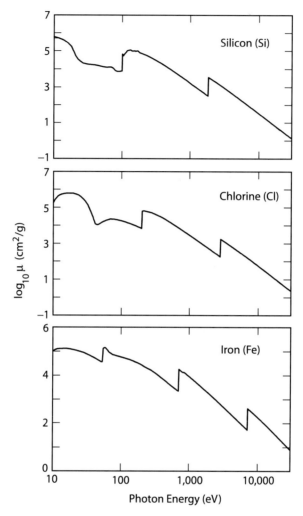

Figure 1.1 Mass absorption coefficients for silicon, chlorine, and iron as functions of energy. The absorption coefficient is dominated by photoabsorption over most of the energy range 10 eV to 30 keV. As discussed in the text, μ is related to the intensity I transmitted through a material density ρ and thickness d by $I = I_0 \exp(-\mu\rho d)$, where I_0 is the entering intensity. (Based on data from Henke, Gullikson, and Davis 1993)

Compton scattering, where the incoming photon scatters with a free electron and transfers a fraction of its energy to it. In cosmic-ray astronomy, it is common to separate the Compton scattering coefficient σ_c into two parts, the coefficient σ_a representing the energy absorbed by the particles, and σ_s representing the scattered photon energy. What really counts as far as attenuation is concerned, however, is the total coefficient

$$\sigma_c = \sigma_a + \sigma_s,$$

(1.26)

where

$$\sigma_a \equiv \sigma_c \frac{\langle \Delta \epsilon \rangle}{\epsilon} \qquad (1.27)$$

and

$$\sigma_s = \sigma_c \left(1 - \frac{\langle \Delta \epsilon \rangle}{\epsilon} \right). \qquad (1.28)$$

In these expressions, $\langle \Delta \epsilon \rangle$ is the average energy lost by an incoming photon with energy ϵ once it scatters with the medium.[10]

Figure 1.2 shows the total attenuation coefficient μ_0 for photons propagating through air, given as

$$\mu_0 = \mu_a + \sigma_s, \qquad (1.29)$$

where μ_a is the total absorption coefficient defined as

$$\mu_a \equiv \tau + \sigma_a + \kappa. \qquad (1.30)$$

The coefficient for Rayleigh scattering, labeled σ_r in this figure, drops precipitously with increasing energy and may be ignored in high-energy astronomy.

Returning now to equation (1.25), we see that for a significant fraction of the incoming intensity I to survive down to a certain distance d in the atmosphere, the "optical depth" $\mu \rho d$ must be of order one or less. Therefore, the main conclusion we draw from figure 1.2 is that for photon energies $\epsilon \gtrsim 0.5$ MeV, astronomical measurements must be made at column densities $\rho d \lesssim 1/\mu_0 \sim 10$ g cm^{-2} into the atmosphere, since $\mu_0 \lesssim 0.1$ cm^2 g^{-1} at these energies. On Earth, this atmospheric column density corresponds to a height of approximately 30 km above sea level (see figure 1.3), and for this reason, high-energy astrophysics is primarily a high-altitude and space-based scientific discipline.

The exercise we have just carried out for photon energies $\epsilon \gtrsim 0.5$ MeV can be extended over the entire spectrum, from radio to ultra high-energy γ-rays, and the result can be represented graphically as shown in figure 1.4. This diagram shows the penetration depth of extraterrestrial photons descending to a level above the ground where the attenuation factor in equation (1.25) drops below $1/e$. Radio waves have no trouble reaching Earth's surface, though absorption by the humidity in the atmosphere for wavelengths shorter than ~ 0.5 mm is as efficient as any of the other absorptive processes we discussed earlier. The visible band is quite clean, but at UV energies and higher, virtually all of the incoming intensity is absorbed well above the ground. We will see shortly how these natural constraints can be overcome to provide astronomers with the tools they need to observe extraterrestrial high-energy sources.

---

[10]Be aware that the symbol σ is more commonly used to represent a cross section whose units are simply area. The quantities σ_c, σ_a, and σ_s, with units of area per unit mass, are employed in cosmic-ray astronomy. We will emphasize the difference whenever any possible confusion may arise.

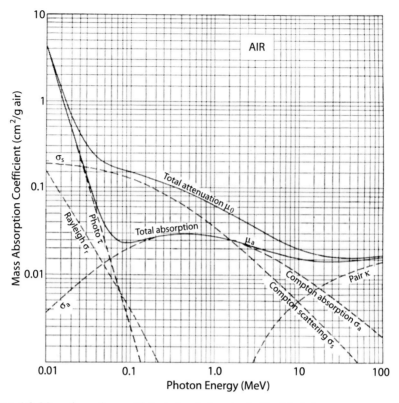

Figure 1.2 Mass absorption coefficients for photons in air. The labeled curves correspond to Compton absorption (σ_a), Compton scattering (σ_s), photoelectric absorption (τ), pair production (κ), and Rayleigh scattering (σ_r), which is elastic and confined to small angles, and can usually be ignored at the energies shown here. The curve marked "total absorption" shows the sum $\tau + \sigma_a + \kappa$. Adding Compton scattering to this yields the "total attenuation" coefficient μ_0. The composition for air in this diagram is 78.04 volume percent nitrogen, 21.02 volume percent oxygen, and 0.94 volume percent argon. (From Evans 1955)

1.4 EXPERIMENTAL TOOLS OF HIGH-ENERGY ASTROPHYSICS

Besides deploying our detector judiciously above most of Earth's atmosphere, we must also deal with at least two other challenges that loom in high-energy astronomy, requiring technological developments beyond the extensive instrumentation already available at longer wavelengths. One of these is the development of methods to focus high-energy radiation onto the detector, and the second is the design of the detectors themselves. High-energy photons penetrate much deeper into matter than their longer wavelength counterparts and are therefore difficult to reflect (and refract). At the highest energies, γ-rays are also very destructive, both to themselves and to the material they encounter. Before we discuss high-energy telescopes, we therefore need to first consider the physics of collecting and analyzing photons in the UV to TeV energy range.

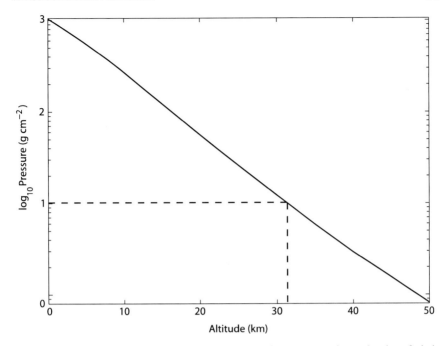

Figure 1.3 Atmospheric pressure (actually represented here as a column density of air in units of g cm^{-2}) as a function of altitude (in km) above Earth's surface. The pressure drops to a value below $\sim$10 g cm^{-2} only for heights above $\sim$30 km. (Based on the atmospheric model of Arguado and Burt 1999)

Modern UV and soft X-ray observatories use charge-coupled devices (CCDs) to capture incoming photons. A CCD is an array of (usually) silicon or germanium pixel elements with bias voltages that make each of them a small potential well. The incident photon ejects an electron that remains trapped in the well until the array is read out by a single low-noise amplifier.

When atoms are bound together in a metal or crystal, their electrons become bound to the material as a whole, and their energy levels cease to be precise. Instead, the electrons find themselves confined to energy bands; they are permitted to reside anywhere in the bands, but are forbidden to occupy energy states between them. Since the density of electrons in an energy band is limited by the Pauli exclusion principle, once a particular band is filled, any additional electrons must occupy a state in the next-higher band.

Metals are conductors because their highest energy band is only partially filled. Thus, when an electrical potential is placed across a sample, the electrons jump to higher energies within the same band and create a current by flowing down the potential. For insulators, however, all of the energy bands with electrons are filled, and no current is produced by an applied potential drop because the electrons are not permitted to change their energy and move down the potential, unless they acquire sufficient energy to jump to a higher band which contains no electrons.

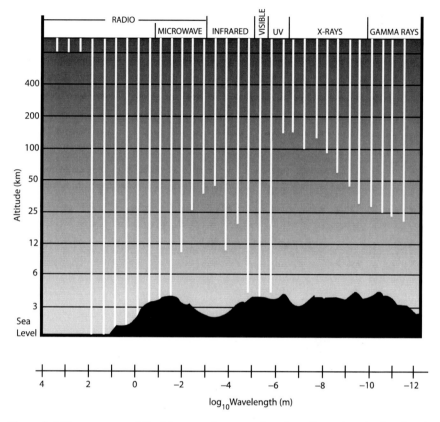

Figure 1.4 Transparency of Earth's atmosphere as a function of wavelength, showing the depth (in km) to which photons of a given energy will penetrate before being absorbed or scattered. The surface of the Earth is completely shielded from that portion of the spectrum (beyond the UV) of particular relevance to high-energy astrophysics. (Image courtesy of Jerry Woodfill and NASA)

The electrons in a CCD absorb UV photons or X-rays and jump to a higher energy band (leaving an electron "hole" behind), where they are collected and counted. The number of electrons is proportional to the X-radiation absorbed by the pixels (see figure 1.5). Depending on the type of material used in the detector elements, the energy required to produce an electron–hole pair is only a few eV (e.g., 3.5 eV per pair in silicon and 2.94 eV per pair in germanium), so these detectors are ideally suited to UV observations.

CCDs such as this have allowed astronomers to study one of the least explored windows of the electromagnetic spectrum—the extreme UV (EUV). Until recently, astronomers thought that the mixture of hydrogen gas and other less abundant gases filling the interstellar medium would absorb virtually all EUV radiation before reaching Earth. This region of the spectrum therefore became known as the "unobservable ultraviolet." But the high quantum efficiency of these devices has facilitated the detection of very faint objects. For example, later in this book, we will learn

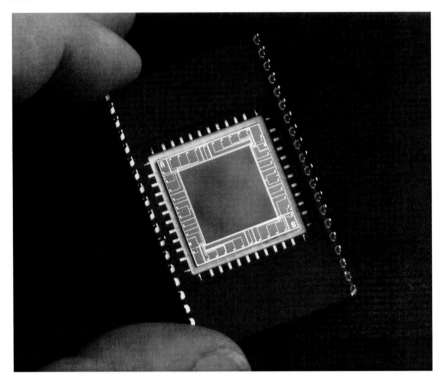

Figure 1.5 Photograph of a specially developed charge-coupled device (CCD) for use in ultraviolet imaging. A CCD is a sensor consisting of an integrated circuit containing an array of coupled capacitors (i.e., pixels). Typically several cm in size, each such detector contains thousands of pixels no bigger than ~15 μm across. A UV or soft X-ray photon entering one of these pixels liberates an electron, which remains trapped until an external circuit transfers the collected electric charge from one capacitor to the next, or to a receptor outside the CCD. Because of the high quantum efficiency of these devices, photons can be counted very accurately, allowing even very faint objects to be observed. (Image courtesy of the Imaging Technology Laboratory at the University of Arizona)

about Cataclysmic variables—binary star systems in which the mass of one star is transferred to a white dwarf accretor, causing dramatic changes in EUV brightness.

The idea of the incident photon knocking out an electron within the detector is central to most high-energy devices, though the actual structure of the instrument depends on whether the dominant electron–photon interaction is photoelectric, Compton scattering, or pair creation. In our discussion of Earth's atmosphere, we already introduced the idea that the photoelectric effect dominates at energies $\epsilon \lesssim$ 60–80 keV, the X-ray and soft γ-ray portion of the spectrum. During the pioneering days of high-energy astronomy, a commonly used device for X-rays ($\epsilon \lesssim 20$ keV) was the proportional counter, basically a gas-filled discharge tube with a voltage drop across the gas. An X-ray photon entering the tube produced a high-energy electron, which then initiated a cascade of electron–ion pairs as it accelerated across the tube. In these devices, the current pulse measured by the counter was proportional to the

Figure 1.6 A microcalorimeter is a small device for measuring energy absorbed in the form of heat. It consists of an absorber, coupled thermally to a constant-temperature heat sink and to a thermometer that can measure changes in its temperature. The energy of an absorbed X-ray photon is dissipated into heat, causing a minuscule change in temperature. The photon's incoming energy is proportional to this change. Shown here is a scanning electron microscope image of 240-μm cantilevered Bi/Cu X-ray absorbers on a fully integrated array. The imaging spectrometer on the Constellation-X incorporates a 32 × 32-microcalorimeter array, and has high quantum efficiency, with 4-eV resolution near 6 keV, 2-eV resolution near 1 keV, and the ability to count at rates of up to 1000 events per second per pixel. (Image courtesy of NASA)

deposited energy, which then yielded the information necessary to reconstruct the X-ray spectrum.

Unfortunately, during this era the choice between wavelength dispersive devices (such as Bragg crystals or grazing incidence diffraction gratings) and nondispersive spectrometers (solid-state detectors or proportional counters) presented somewhat of a dilemma. Dispersive spectrometers offered very good energy resolution, but only at low throughput (the rate at which photons can enter the device and are detected). Nondispersive spectrometers, on the other hand, had very high efficiency, but relatively poor resolution.

Today, the detector of choice for X-ray instrumentation is the microcalorimeter, which provides several advantages over these older methods (see figure 1.6). The energy deposited by the incoming high-energy photon is measured after it has been converted into heat, without the need to worry about the charge transport properties of the absorber. Calorimeters can therefore incorporate a wide variety of materials.[11]

---

[11] The first microcalorimeters for X-ray astronomy were developed by the University of Wisconsin and the NASA Goddard Space Flight Center. They were first employed on a sounding-rocket experiment (the X-ray quantum calorimeter XQC) launched in late 1995. This device had an array of microcalorimeters to study the X-ray background in the energy range 30–1000 eV.

A cryogenic microcalorimeter is composed of three parts, an absorber, a sensor that detects temperature variation in the absorber, and a weak thermal link between the detector and a heat sink. When an X-ray photon hits the absorber, its energy is eventually converted into thermal phonons and the temperature of the detector first rises and then returns to its original value set by the weak thermal link to the heat sink. This detector is also a type of proportional counter, in that the temperature change is proportional to the energy of the incident X-ray. The sensor is a resistor whose resistance is a strong function of temperature. The requirements for a good microcalorimeter are therefore an absorber with a small heat capacity able to convert the incident radiation into thermal phonons quickly and with high efficiency, and a sensor with low heat capacity and high sensitivity to temperature variations.

The development of microcalorimeters has been so promising that the Constellation-X Observatory, a key mission in NASA's Beyond Einstein Program, will feature microcalorimeters in its detection infrastructure. This mission will be a combination of several X-ray telescopes flying in unison to generate the equivalent observing power of one giant telescope, with a projected launch date of 2016. The combination of four telescopes in the array will provide a sensitivity 100 times greater than any other X-ray instrument ever flown. To meet the design requirements of efficiency and spectral resolution, Constellation-X will employ the X-ray microcalorimeter spectrometer, consisting of a 32×32 array of microcalorimeters with superconducting transition-edge sensor thermometers (see figure 1.6). One of the most challenging aspects of this design will be to meet the cooling power necessary to reduce the ambient temperature in the device to ~ 50 mK.

All X-ray telescopes, including Constellation-X (see section 1.5), must also be able to focus X-rays onto the detector. Unlike optical photons, X-rays reflect from the surface of conducting materials only at large angles of incidence i. For example, at energies $\epsilon \sim 1$ keV, reflection occurs for $i \gtrsim 87°$, and i increases rapidly toward 90° with increasing photon energy. X-ray telescopes must therefore incorporate a mirror assembly shaped like nested "barrels" (see figure 1.7), in which incoming photons approach along essentially parallel paths from great distances and reflect at grazing angles off a combination of hyperbolic and parabolic metallic surfaces to arrive at a focal point several meters beyond the telescope's front end. This type of structure therefore produces telescopes with relatively long focal lengths, as one can see with a quick inspection of plates 1 and 2, showing artists' views of two recently flown X-ray observatories (*Chandra* and XMM-*Newton*).

The efficiency of X-ray detectors drops off quickly above ~ 20 keV, either because the ionizing cross section depends strongly on the incoming photon's energy ($\sim \epsilon^{-3}$), or because photons become progressively more penetrative and difficult to capture by the microcalorimeter pixels. For energies between ~ 20 keV and several MeV, the most common device for detecting radiation is the scintillation counter.

The basic element of this technique is a crystal (such as CsI or NaI) in which the entering γ-ray Compton scatters several times before photoelectric absorption occurs. At these energies, the scattering is inelastic, and the photon loses energy with each interaction. The crystal subsequently converts the ionization energy lost by the excited electrons and crystal atoms into visible light, which in turn is converted into an electrical signal by a photomultiplier tube. It is important to stress that

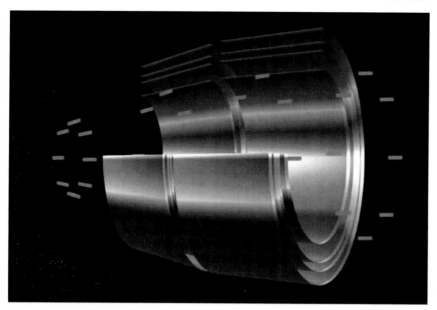

Figure 1.7 Unlike visible light, which reflects off mirrors at all angles of incidence, X-rays have such high energy that they penetrate (and get absorbed) below the surface of the reflector, unless they strike at grazing angles. X-ray telescope mirrors must therefore be shaped like "barrels" to reflect the incoming photons (from right to left in the image) and focus them to a point at the detector. Fortunately, this configuration allows many such mirrors (tens or more) to be nested together about the common symmetry axis, thereby greatly enhancing the flux of X-ray photons that may be brought to a focus. (Image courtesy of NASA/CXC/SAO)

because of the multiple Compton scatterings, it is difficult to use the excited electrons within the crystal to indicate directionality. As such, a hard X-ray/soft γ-ray detector utilizing this technology must rely on other telescope components (e.g., a coded mask aperture, as illustrated in figure 1.10) to provide information on the incoming photon's trajectory.

In the 1990s, one of the major instruments aboard the *Compton* Gamma Ray Observatory (GRO), known as COMPTEL, was a standout example of how successful this technique can be. COMPTEL consisted of two detector arrays, as shown in figure 1.8. A liquid scintillator, NE 213A, was used in the upper one, and NaI crystals appeared in the lower. Incoming γ-rays were detected by two successive interactions: the first was a Compton scattering event in the upper detector; the second was total absorption in the lower one. By measuring the locations of the interactions and energy losses in both cases, one could then determine the energy and direction of the incoming photon.

As shown in figure 1.8, the two detector planes were separated by a distance of 1.5 m. The 7 cylindrical D1 modules in the upper plane were each 27.6 cm in diameter, 8.5 cm thick, and the whole assembly was surrounded by eight photomultiplier tubes. The total area of the upper detector was approximately 4188 cm^2. By comparison, the 14 D2 modules in the lower plane consisted of cylindrical NaI blocks,

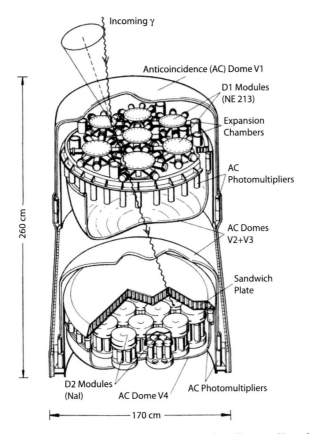

Figure 1.8 The Compton telescope approach to γ-ray detection, illustrated here for the particular implementation in the COMPTEL telescope on the *Compton*-GRO. (From Schoenfelder et al. 1993)

each 7.5 cm thick and 28 cm in diameter. Each block of NaI was viewed from below by seven photomultiplier tubes. The total geometrical area of the lower detector was 8620 cm^2.

As one can imagine, a chief source of error in measuring the incoming photon's energy is the possible loss of information along the sequence of interactions within the detector assembly. The photon's energy is estimated by summing the energies deposited in the upper and lower detectors, assuming total absorption of the scattered photon. Then, from the energy losses and the interaction locations, one determines the arrival direction of the incoming γ-ray. In practice, one must also fold into the analysis a detailed understanding of the properties of the scintillation detectors, and of the response of the telescope to background, multiple interactions, and partial absorption events, to arrive at a final determination of the photon's energy and direction of flight. A cutaway image of the COMPTEL telescope is shown in figure 1.9.

Figure 1.9 A cutaway image of the COMPTEL telescope on the *Compton-GRO*. COMPTEL consists of two detector arrays. A liquid scintillator, NE 213A, is used in the upper seven cylindrical modules, each of which is 27.6 cm in diameter, 8.5 cm thick, and viewed by eight photomultiplier tubes. In the lower array, 14 cylindrical blocks of 7.5 cm thickness and 28 cm diameter contain NaI crystals, each of which is viewed from below by seven photomultiplier tubes. The two detectors are separated by 1.5 m, and each is surrounded by a thin anticoincidence shield of plastic scintillator that rejects charged particles. The incoming γ-rays are detected by two successive interactions: an incident photon is first Compton scattered in the upper detector and is then totally absorbed in the lower one. The interaction sites are measured and the energy losses are determined. The overall energy and angular resolution of the telescope depends on the accuracy of these measurements. (Image courtesy of COMPTEL/NASA)

Today's hard X-ray/soft γ-ray detector designs often call for the inclusion of another component to help with directionality. The most common approach is the so-called coded aperture imaging to localize the incident photons. A coded aperture is a mask (see figure 1.10) positioned in front of the actual γ-ray detectors. The coded mask aperture sits about one meter above the detector plane and is made from ~50,000 lead tiles arranged in a random half-open/half-closed pattern. Each tile

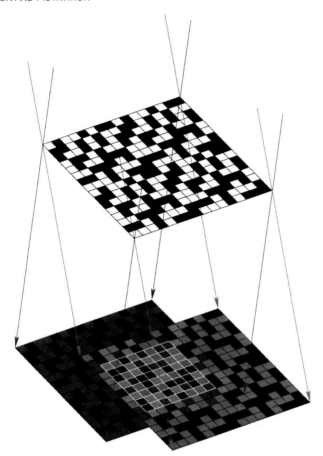

Figure 1.10 Since radiation above ~20 keV cannot be focused with current technology, present day high-energy detectors use a technique known as coded aperture imaging, in which an absorbing mask (such as lead) is positioned in front of the position-sensitive γ-ray detector. The lead tiles are arranged in a random half-open/half-closed pattern. Some of the incoming γ-rays are stopped by the mask; others pass on through to the detectors. Analysis of the resulting shadow determines the direction of incoming photons. If more than one source exists in the field of view, one may find the direction to each source by finding the multiple shadows cast on the detector plane. Several-arcminute spatial resolution is feasible with this technique. (Image courtesy of IBIS/INTEGRAL and ESA)

might be as small as $5 \times 5 \times 1$ mm^3. The idea behind this approach is that the lead tiles stop some incoming photons, while the apertures let in the rest. In so doing, the mask casts a shadow on the detector plane, and using the measured position of this shadow, and the known locations of the lead tiles, one can then determine the direction of the γ-ray source.

The coded mask aperture yields a less precise determination of the incoming photon's direction than X-ray mirrors do (see figure 1.7), but at energies beyond ~20 keV there is currently no better choice. Note that when more than one source is present in the field of view, one can still find the direction to each object by finding the multiple shadows cast on the detector. Telescopes designed on this principle can locate a source with an accuracy of several arcminutes. This method only works for locating point sources, however; diffuse emission cannot be imaged with this technique.

At energies $\epsilon \gtrsim 3\,\mathrm{MeV}$, photons become energetic enough to create electron–positron pairs, as we learned in our earlier discussion on atmospheric absorption. This interaction takes place within the CsI or NaI crystal, and one can, in principle, also use the direction of motion of the leptons (which tend to move in a forward direction relative to the incoming γ-ray) to infer the photon's angle of incidence into the detector.

But there's a limit to how far up in energy we can go with this technique because very high-energy γ-rays do not stop with the creation of just one pair of leptons; they produce cascades of particles and secondary radiation that cannot all be trapped within the detector. In this domain, the atmosphere actually becomes a help rather than a hindrance, because incoming photons strike the mesosphere, some 30–100 km above Earth's surface, and produce a glow of light detectable with air Cerenkov detectors on the ground (see figure 1.11). Of course, this process occurs for many different types of energetic particles entering the atmosphere, though our principal interest here is γ-rays.

This technology has evolved significantly since the early days. In the late 1920s, the French scientist Pierre Auger discovered the phenomenon of extensive air showers using ionization chambers, Geiger counters, and cloud chambers. He determined that very energetic cosmic rays (including γ-rays) could produce showers of secondary particles that spread out over several hundred meters. Atmospheric fluorescence was incorporated into the detector design of the 1980s, using the idea that charged particles passing by molecules in the atmosphere transfer some of their energy to the target particles. This has the effect of "shaking up" the electrons inside the molecules, thereby emitting secondary (or fluorescent) radiation as the electrons return to their normal configuration. Nitrogen molecules, the main constituents of air, produce blue fluorescent light, which can be sensed by photomultipliers on the ground. But this light is so faint that it can be seen only on moonless nights without clouds. This technique has successfully been used by several facilities, including the Fly's Eye experiment in Utah and, most recently, the Pierre Auger Observatory in Argentina.

The cascade of particles produced by the incident γ-ray (or, more generally, by the incident cosmic ray) also produces a glow of radiation by another mechanism, known as the Cerenkov effect. The speed of light in transparent materials is less than its value in vacuum ($c = 3 \times 10^{10}\,\mathrm{cm\ s^{-1}}$). In water, for example, light travels at $0.7c$. Particles with mass, however, can travel at speeds greater than this, creating a shock front of light that spreads out in a cone around the particle. Photomultiplier tubes placed within this medium sense the Cerenkov light, and one can then use the intensity and pattern of this radiation to infer the energy of the incident high-energy γ-ray. The use of several optical telescopes on the ground can even provide information on the source direction (to within a degree or so) with reconstruction of the various images.

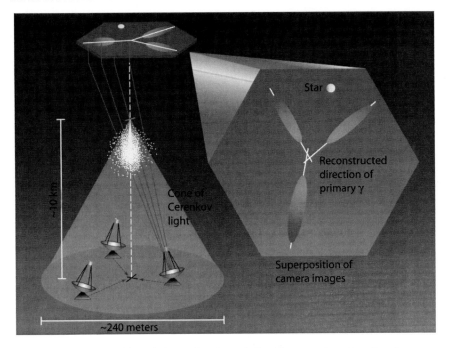

Figure 1.11 An energetic particle passing through Earth's atmosphere transfers its energy to charges at an impact point tens of kilometers above the ground. The cascade ensuing from this process includes particles moving through the medium at speeds greater than the speed of light in that region, producing a shock front of light, known as Cerenkov radiation, spreading out in a cone toward the ground. One or more Earth-based telescopes image this light, providing a measure of the incoming cosmic ray's direction and energy. (Image from Clery 2004. Illustration by K. Buckheit. Reprinted with permission from AAAS)

1.5 HIGH-ENERGY TELESCOPES

The development of high-energy astrophysics as an observational science has been so rapid since the rocket experiments of the early 1960s that it is no longer possible to list all of the missions within a reasonably limited space. The best we can do now is to catalog the principal missions since 1990, and to highlight the capabilities and scientific motivation for some of the more prominent satellites. But though the vast majority of high-energy astrophysics experiments have been conducted in space, we should not forget that observations may also be carried out at high altitudes within Earth's atmosphere, and we begin with a "status" report of where things stand with regard to NASA's ballooning program.

High-altitude balloons have for many years offered the capability of inexpensively lofting high-energy instruments to altitudes of 30 km or more (above most of Earth's atmosphere) for long-duration flights lasting several months (see figure 1.12). While the basics of ballooning remain unchanged over time, the balloons themselves have increased in size, and have become more reliable with each succeeding generation.

Figure 1.12 An artist's concept of NASA's Ultra Long Duration Balloon (ULDB), a structure composed of a lightweight polyethylene film about the thickness of ordinary plastic food wrap. Designed to lift 6000 pounds, it can rise to an altitude of about 33 km above 99% of Earth's atmosphere, for durations approaching 100 days. (Image courtesy of NASA)

The Goddard Space Flight Center's Wallops Flight Facility, which manages the Balloon Program, launches missions from permanent launch sites in Palestine, Texas, and Ft. Sumner, New Mexico, and from remote sites in the United States, Canada, Australia, New Zealand, and Antarctica.

NASA's balloons are made of a thin polyethylene material with a thickness like that of an ordinary sandwich wrap. When fully inflated, these balloons range up to 200 m in diameter and are taller than a 60-story building. The complete facility includes the balloon, a parachute, and the instruments themselves. Given their enormous size, these structures have a powerful lift, and can carry a payload with the weight of three small cars.

One of the most notable successes of this program was the balloon mission in 1998, named the Balloon Observations of Millimetric Extragalactic Radiation and Geophysics experiment (more commonly known as BOOMERANG), which returned what were then the highest-ever-resolution images of the cosmic microwave background. Perhaps just as important, all of the devices that flew on the Cosmic Background Explorer (COBE), the pioneering satellite that revealed fluctuations in the background radiation in the early 1990s, had been flight-tested on earlier balloon experiments.

In fact, the ballooning program continues to serve this very useful purpose of flight testing exploratory (or new concept) instrumentation for possible use in future orbital missions. It also provides the capability for carrying out small-scale missions that would not be feasible with a full launch into space. For example, in late 2006, the Wallops Facility launched the High Altitude Student Platform, carrying science experiments developed by student groups from across the United States, to an altitude of approximately 40 km for a total flight time of about 18 h. The scientific goals included the study of the cosmic ray flux, tests of various rocket nozzle designs, tests of an accelerometer based inertial guidance system, and remote sensing imaging.

In space-based X-ray astronomy, the first Earth-orbiting explorers were launched in the early 1970s (*Uhuru*, SAS-3, and Ariel-5), followed later in that decade by larger missions, such as HEAO-1, *Einstein*, EXOSAT, and Ginga. Their success, particularly in providing the first X-ray images of the sky (discussed at greater length in chapter 2), moved X-ray astronomy into the mainstream of scientific research. In the 1990s (see figure 1.13), the US–European ROSAT survey detected over 100,000 X-ray objects, the Japanese ASCA mission obtained the first detailed X-ray spectra of these sources, and the *Rossi*-X-ray Timing Explorer (RXTE) studied their timing behavior. Near the end of this decade, the launch of the *Chandra* and XMM-*Newton* observatories introduced high-resolution imaging and a high-throughput capability to our repertoire.

Gamma-ray astronomy has also advanced significantly since the middle of the 20th century, first with the Small Astronomy Satellite 2 (SAS-2) and Cos-B in the early 1970s, which made the first surveys of the γ-ray sky, and then most impressively with the *Compton* Gamma Ray Observatory in 1990, and more recently with the US–European INTEGRAL mission in 2002. But one cannot fail to notice that the number of γ-ray missions is significantly lower than their X-ray counterparts. As we noted earlier in this chapter, very high-energy photons are more difficult to corral and measure, so progress in improving spatial and energy resolution is slower the higher the energy of the radiation. For this reason, new γ-ray missions enter the pipeline more slowly than experiments in the UV and X-ray domain.

In the remainder of this chapter, we describe a handful of the principal missions featured in figure 1.13, not to exclude the rest, but simply to provide an overview of the various capabilities required for a comprehensive high-energy astrophysics program. The very beginning of the 1990s saw the successful launch and deployment of the Roentgen Satellite (ROSAT), a German–US–UK collaboration. This mission lasted almost 9 yrs and produced, within the first 6 months of operation, one of the most dramatic all-sky surveys ever made. (We will see a sample image from this survey in plate 5.)

By way of comparison, ROSAT's sensitivity was about a factor 1000 better than that of *Uhuru*, the first Earth-orbiting explorer. ROSAT's energy coverage was 0.1–2.5 keV in X-rays, and 62–206 eV in the Extreme UV (EUV). One of the principal instruments aboard the satellite was a pair of position-sensitive proportional counters (based on technology that preceded the microcalorimeter development discussed in section 1.4) with an energy resolution $\Delta E / E \sim 0.5$. A second instrument was the High Resolution Imager (HRI) with a spatial resolution of $\sim 2''$. The scientific return from this mission was enormous, encompassing not only the highly detailed all-sky surveys at X-ray and EUV energies, but also extensive pointed observations of supernova remnants and clusters of galaxies, studies of unusual objects such as

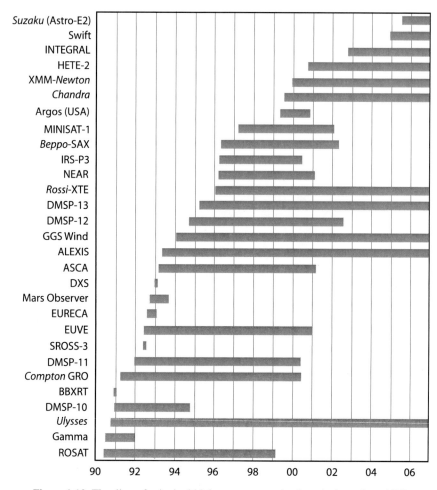

Figure 1.13 Timeline of principal high-energy astrophysics missions since 1990.

Geminga (for which X-ray pulsations were finally detected), and the discovery and observation of isolated neutron stars.

The *Compton* Gamma Ray Observatory (CGRO) was launched the following year and made breathtaking discoveries in the γ-ray sky until its gyroscopes began to fail, forcing NASA to bring it down from orbit in June 2000. Unlike most other satellites, CGRO was too big (about the size of a school bus) to burn up entirely during reentry in the atmosphere and so its descent had to be controlled by NASA, forcing a termination to what would otherwise have been a longer mission. The second of NASA's great observatories, CGRO had four instruments covering six orders of magnitude in energy, from 30 keV to 30 GeV. Its payload included the Burst and Transient Source Experiment (BATSE), an all-sky monitor covering the energy range 30 keV to 1 MeV, whose main role was to study the mysterious nature of γ-ray bursts (GRBs). Among its many exciting discoveries, BATSE produced the catalog of

GRBs shown in plate 8, demonstrating without any lingering doubt that these events must occur at cosmological distances, given their isotropic distribution in the sky.

The second instrument aboard CGRO was the Oriented Scintillation Spectrometer Experiment (OSSE), covering the energy range 50 keV to 10 MeV, which produced, among other things, a map of the Milky Way in the ^{26}Al γ-ray line. The third detector was the Compton Telescope (COMPTEL), capable of $\sim$1-steradian imaging in the energy range 0.8–30 MeV. This facility was featured earlier in figures 1.8 and 1.9, and the all-sky survey it produced is shown in plate 8. The fourth (and final) instrument on CGRO was the Energetic Gamma Ray Experiment Telescope (EGRET), spanning the range 30 MeV to 30 GeV. EGRET discovered the powerful γ-ray luminosity of superluminal AGNs, and produced the all-sky survey shown in plate 11.

The *Rossi* X-ray Timing Explorer (RXTE) broke new ground by being the first major mission designed for timing studies, as opposed to pure imaging or spectral measurements. Launched in December 1995, RXTE is capable of only moderate spectral resolution, but can discern source variability on timescales of months down to microseconds, covered in a broad spectral range from 2 to 250 keV. As discussed in section 1.2, this coverage in timing resolution is sufficient to detect even the highly dynamic variability on the surface of a neutron star. Its strength is the very large collecting area—6500 cm^2 for its proportional counter array (PCA), and 1600 cm^2 on the High-Energy X-ray Timing Experiment—which accumulates good photon statistics quickly to allow for fine temporal resolution. Among its scientific achievements, RXTE discovered kilohertz quasi-periodic oscillations (QPOs) in compact binaries (see section 10.4), X-ray afterglows in some GRBs, state transitions in the black-hole binary Cygnus X-1, and accretion instabilities in accreting neutron-star systems.

As we approach the current era in figure 1.13, we encounter one of the finest high-energy missions ever flown. The *Chandra* X-ray Telescope (shown schematically in plate 1) has made some truly unprecedented observations in the energy range 0.1–10 keV. With its unmatched $\sim 0.''5$ angular resolution, eight times better than any other X-ray telescope ever flown, *Chandra* has provided us with detailed images of the full impact of supernova blast waves, brown-dwarf flares, the cannibalization of one small galaxy by another, and the rich high-energy phenomena pervading the central region of our Galaxy. We will be featuring observations made with *Chandra* throughout this book, including the Deep Field view of the X-ray cosmos in plate 6, and the highly detailed view of the galactic center in plates 20 and 21.

Launched in July 1999, *Chandra* became the third in NASA's family of great observatories, following the *Hubble* Space Telescope and the *Compton* Gamma Ray Observatory. One of its greatest contributions to X-ray astronomy has been the resolution of the X-ray background, made possible not only because of *Chandra's* resolution, but also by its sensitivity to detect sources more than 20 times fainter than before.

XMM-*Newton* is a contemporary of *Chandra*, and is suitably designed to complement the latter's capabilities (see plate 2). The European Space Agency's (ESA) X-ray Multi-Mirror Mission was launched by Ariane 504 in December 1999, the second cornerstone satellite in ESA's Horizon 2000 Science Program. Whereas *Chandra's* imaging power is unmatched, XMM-*Newton's* strength is its unprecedented effective area ($\sim$1500 cm^2 for each of three co-aligned telescopes), which, together with its orbit that permits long, uninterrupted exposures, provides high sensitivity for detecting faint sources with excellent spectral resolution (with $E/\Delta E$ as high as 200–800 between 0.35 and 2.5 keV on the Reflection Grating Spectrometer). This satellite also carries an optical monitor, the first ever flown on an X-ray observatory, permitting simultaneous X-ray and optical observations of the same source.

Two examples that highlight XMM-*Newton's* strengths are provided in figures 9.14 and 12.11. These illustrations span the range of source size, from a compact binary containing an accreting white dwarf in the former, to the supermassive black hole at the center of our Galaxy in the latter. The X-ray flare seen in figure 12.11 is particularly important because it is the first event ever to be compellingly associated with phenomena occurring just outside a supermassive black hole's event horizon. As we shall see in chapter 12, the quasi-periodic modulations in this lightcurve correspond to an emitter orbiting at $\sim$3 Schwarzschild radii, right around the marginally stable orbit where the accretion disk is thought to end before its material plummets toward oblivion.

The most recent X-ray mission, now known as *Suzaku*, was actually part of *Chandra* in its original design. More accurately, one should say that *Chandra* and *Suzaku* were both part of the same mission concept called the Advanced X-ray Astrophysics Facility (AXAF). This was to be a combination X-ray imager (which became *Chandra*) and spectrometer (now *Suzaku*). Such a comprehensive design became prohibitively expensive and unmanageable, forcing NASA to abandon the simultaneous mission idea. But Astro-E2 (*Suzaku's* previous appellation) survived thanks to the Japanese Institute of Space and Astronautical Science (ISAS), which continued its development with US participation. *Suzaku* was launched in July 2005 from the Uchinoura Space Center in Japan, the first ever orbital mission to carry an X-ray microcalorimeter (see figure 1.6) for unparalleled energy resolution: at 6 keV, the X-ray Imaging Spectrometer (XRS) aboard *Suzaku* was designed to achieve an energy resolution of $\sim$6.5 eV. Sadly, the XRS failed shortly after launch and its potential scientific return has been lost to the mission. But having survived and operated flawlessly for 3 weeks after the satellite was deployed, the XRS has proven the point that this microcalorimeter technology can indeed work in space, and its promise will eventually be realized as new flight opportunities (e.g., with Constellation-X) become available.

The last two missions we will feature here are both γ-ray experiments, one already in orbit, the other to be launched in the very near future. The International Gamma-ray Astrophysics Laboratory (INTEGRAL) was launched into orbit by ESA in October, 2002 (see plate 3). Lifted into space on a Russian Proton rocket from Baikonur in Kazakhstan, INTEGRAL now revolves on a 72-hour elliptical orbit, ranging from 9000 km up to 155,000 km above Earth's surface. As the successor to CGRO, it is producing the next generation map of the sky in soft γ-rays, and is capable of performing high spectral and spatial observations of point sources. An example of what can be done with it is shown in figure 13.3, featuring the electron–positron annihilation-line distribution at the galactic center. Its energy coverage (3 keV to 10 MeV) is provided by a complement of several instruments, including the SPI spectrometer and the IBIS imager, both of which utilize a coded mask aperture for directionality (see figure 1.10).

As we look to the future, two principal high-energy missions loom on the horizon: Constellation-X, which we have already discussed in connection with the microcalorimeter technology, and the Gamma-ray Large Area Space Telescope (GLAST) (see plate 4). The former is in the pipeline for launch around 2016; the latter will be in orbit by the end of 2007. GLAST is the next generation high-energy γ-ray observatory designed for making observations of celestial sources in the energy band extending from 10 MeV to 100 GeV. Though it shares many features in common with EGRET on CGRO, its capabilities extend γ-ray astronomy well beyond what was achieved in the 1990s. Its detector is made of segmented 20-cm CsI bars, arranged to give both longitudinal and transverse information about the energy deposition pattern once a γ-ray comes in and causes a scintillation reaction. These bars function as small calorimeters and the light flash is photoelectrically converted into a voltage for readout.

The science mission of GLAST is rather ambitious, beginning with a one-year all-sky survey that will improve our view of the γ-ray sky from what we have now (courtesy of EGRET) in plate 11, to those simulated in plates 12 and 13. With a field of view (~ 2.5 steradians) about twice that of EGRET, and a sensitivity about 50 times greater at 100 MeV, its two-year limit for source detection is expected to be $\sim 1.6 \times 10^{-9}$ photons cm^{-2} s^{-1}, with a point-source positional accuracy of 30$''$ to 5 arcminutes. Its targets will include AGNs, whose jets are still poorly understood, the γ-ray background sky, which is currently blending the emission from the interstellar medium and a large number of as yet unidentified sources, GRBs, and possibly even sites of supersymmetric dark matter annihilation, which may produce monoenergetic γ-ray lines above 30 GeV.

The high-energy telescopes we have surveyed here only provide a brief overview of the general concepts. Looking back over the past 40 yrs, it is truly amazing to see how quickly technology has developed in this field, to the point where some missions are actually driving the development of instrumentation (e.g., microcalorimeter X-ray detection) that will find use in many other fields. Sadly, however, one cannot ignore a rather unsettling trend in figure 1.13: the second-order derivative of the launch times for these missions is negative. Rising costs and an ever-tightening budget in the United States are making it increasingly difficult to sustain a high-paced program. The growing participation by the European Space Agency, and countries such as Japan, however, is partially mitigating this effect. Let us all hope that high-energy astrophysics, a truly space-based science, continues to flourish in the decades to come and beyond.

SUGGESTED READING

A good collection of scholarly papers on the subject of archaeastronomy and the co-evolution of astronomy and culture may be found in the edited volume by Ruggles and Saunders (1993).

The beginning of high-energy astrophysics may rightly be traced to the earliest exploration for the source of ionization seen in laboratory experiments. An early paper on this topic was that by Hess (1912).

A complete history of high-energy astrophysics does not yet exist in book form. However, NASA's Goddard Space Flight Center has put together a reliable website attempting such a compilation, which the reader may find at http://heasarc.gsfc.nasa.gov/docs/heasarc/headates/heahistory.html.

X-ray astronomy began in 1962 with the first discovery of a high-energy source outside the solar system. Read about this observation in Giacconi et al. (1962).

In designing high-energy instruments, one must know the energy of the photons being detected, their flux, and how quickly the detection must be made in order to track the source variability. We have therefore included in this introductory chapter a brief discussion of a typical source luminosity and the associated dynamical timescale, even though we will not address the theory behind these concepts until later in the book. The reader will find a more complete presentation of issues dealing with radiative processes, including all the pertinent references, in chapter 5. In the meantime, it will also be useful to read chapter 1 in Rybicki and Lightman (1985).

In discussing the various interactions between incoming photons and Earth's atmosphere, we have limited our description to the essential results from quantum electrodynamics, without necessarily probing too far into the details. Many books on this subject now exist in the literature, but one of most insightful is still Landau and Lifshitz (1982).

To find a more detailed (and technical) description of the various detector technologies in high-energy astrophysics, the reader should consult articles in the proceedings of conferences on this topic. A good starting point is Siegmund and Flanagan (1999).

Throughout this book, but particularly in this chapter, we often refer to individual high-energy instruments and their capabilities. The National Aeronautics and Space Administration (NASA) in the United States, the European Space Agency (ESA) in Europe, and the Aerospace Exploration Agency (JAXA) in Japan, all maintain excellent instrument-specific websites, many of which continue to be updated as new data are collected and/or analyzed. The reader can find these at http://www.nasa.gov/missions/index.html, http://www.esa.int/esaSC/index.html, and http://www.jaxa.jp/index_e.html.

Chapter Two

The High-Energy Sky

In our overview of experimental high-energy astrophysics, we have been emphasizing how quickly this field has grown over the past 40 years. This type of expansion does not occur without the essential feedback provided by new discoveries. It is safe to say that every time a new energy window has opened up, or the detector sensitivity has improved, dramatic new celestial vistas have emerged to deepen our understanding of the cosmos and, at the same time, to whet our appetite with the promise of additional discoveries to be made with a deeper search. For much of the early history of this field, each mission has seemingly functioned as an explorer preparing the scene for the next device, rather than as an experiment designed to provide us with definitive scientific answers. Indeed, many high-energy satellites (and balloon flights) have incorporated the term "explorer" in their name.

What this means, of course, is that the proliferation of high-energy instrumentation has produced a parallel growth in the discovery of new sources and phenomena associated with them. The 1962 sounding-rocket mission revealed one dominant X-ray source, Sco X-1, and a diffuse (i.e., unresolved) background. After ROSAT in the early 1990s, our high-energy source catalog has grown to 150,000 individual objects, many of which were resolved out of the erstwhile background. No doubt, this process will continue, though much of the effort now is to study the known objects deeper and to gain a better insight into their origin and evolution, rather than to merely add more entries to the source catalog.

In this chapter, we begin our discussion of the newly identified sources by partially mimicking the pattern of discovery followed by a typical observatory. We will examine the all-sky surveys they produce to familiarize ourselves with a broad view of the cosmos, and then begin to introduce and describe many of the disparate objects we glean from the corresponding catalogs according to their principal defining characteristics. Of course, we cannot completely divorce this analysis from the theoretical underpinnings that provide us with an interpretation of their behavior, for otherwise their classification would simply become an exercise in collating seemingly unrelated facts. However, we will leave the main discussion on the interpretation of what we see to chapters 9–13, following our development of relativity theory and the language of four-dimensional spacetime, and other tools we will need to understand the full suite of interactions between matter and high-energy radiation.

2.1 X-RAY MAPS UP TO 10 keV

One of the most comprehensive all-sky imaging X-ray surveys ever made was that by ROSAT during its first six months of operation in the early 1990s.[1] Plate 5 shows these sources in galactic coordinates, with north at the top and longitude increasing to the left (i.e., to the east) of center.

The ROSAT All-Sky Survey, as it is formally known, covers the entire celestial sphere in the $\sim$0.5- to 2.5-keV range down to a typical limiting sensitivity of 5×10^{-13} ergs cm^{-2} s^{-1}. (For photons with a typical energy of $\sim$1 keV, this corresponds to a photon flux of $\sim$2 $\times$ 10^{-4} ph cm^{-2} s^{-1}; see discussion in section 1.2.) Note, however, that the exposure times (and hence sensitivity limits) vary markedly from the ecliptic pole to equator due to the scanning protocol. Some 80,000 X-ray sources are included in this image, with positional accuracies of $\sim$20$''$.

Once high-energy sources have been discovered in this manner, it is often necessary to conduct follow-up observations at other wavelengths to extract information, such as redshift, for a complete identification. In this case, however, many important conclusions may be drawn from the survey data directly. For example, this catalog of point sources immediately yields an accurate determination of the number counts versus flux relationship (usually called the log N–log S diagram, representing the cumulative number N of sources down to a flux level S), and therefore a direct measurement of the contribution of faint point sources to the diffuse X-ray background. One also infers directly from the X-ray data that this survey detects previously cataloged, well-studied bright stars within $\sim$10 pc of the Sun in great numbers, indicating the prevalence of X-ray emission by their hot coronae.

Still, the very faint end of the ROSAT catalog constitutes a rather heterogeneous mix of astrophysical objects, from the closest M dwarfs to very high redshift quasars and AGNs, and the X-ray data alone cannot indicate whether a given object is galactic or extragalactic. With such ambiguity, the intrinsic X-ray luminosity of a given ROSAT source may be uncertain by as much as a factor of 10^{14}. For this reason, the identification of infrared or optical counterparts to the X-ray catalog is essential for a proper study of these sources, particularly at the detection limit.

This correlative work is painstaking and long, consuming many years of effort and the utilization of various ground-based telescopes and instruments. An X-ray satellite (*Einstein*, also known as HEAO-2) from the previous generation flew from 1978 until 1981, producing its own X-ray survey, sampling about 10% of the celestial sphere with a spatial resolution of several arcseconds in the energy range 0.2–20 keV. The ambitious program to identify the sources in the *Einstein* catalog has required more than twenty years of patient work,[2] yielding about 800 optical counterparts, including normal stellar coronae, pre-main-sequence stars, interacting binaries, supernova remnants, nearby galaxies, clusters of galaxies, and AGNs with a broad range of luminosities.

---

[1]These data and their analysis appeared in several publications authored by the principal investigator J. Trümper and the ROSAT team.
[2]See Stocke et al. (1991).

Fortunately, we now have access to much more efficient means of acquiring optical counterparts to X-ray sources. The Sloan Digital Sky Survey (SDSS)[3] is very well matched to the ROSAT catalog for a variety of reasons. For example, the SDSS coverage corresponds to the region of greatest ROSAT sensitivity, because it lies in regions of high ecliptic latitude (where ROSAT could integrate for long times) and high galactic latitude (where the interstellar photoelectric opacity is so low that absorption of the low-energy X-rays is minimized).

Given the known range of X-ray to optical luminosity for all common classes of object,[4] both galactic and extragalactic, even the faintest counterparts to the ROSAT sources are accessible to the SDSS photometric and spectroscopic surveys. The SDSS is therefore providing very accurate colors and magnitudes for all of the ROSAT survey counterparts, facilitating a solid identification of the tens of thousands of X-ray sources in this rich database.

The extraordinary imaging power of *Chandra* (with an angular resolution of $\sim0.''5$) allows us to complement ROSAT's all-sky survey with a very deep exposure of selected regions of the sky. This extends the faint source catalog to much lower flux limits, revealing objects (as we shall see) emitting X-rays at cosmological redshifts of 6 or greater.

Plate 6 shows one such field, known as the Chandra Deep Field North, produced by integrating an area of the sky only three-fifths the size of the Moon for 23 days. This image, together with its counterpart in the southern hemisphere (called the Chandra Deep Field South), constitutes the most sensitive (or "deepest") X-ray exposure ever made. To put it in colloquial terms, the faintest sources in this image transmitted only one X-ray photon into *Chandra's* detector every 4 days.

One can count over 500 individual sources in this sample, and because the pointing was chosen to minimize any contamination from local objects within our own Galaxy, the vast majority of them are extragalactic. We are witnessing here the X-ray emission produced by supermassive black holes in the early universe. If the number density of such sources seen here is typical throughout the cosmos, the total number detectable over the whole sky at this level of sensitivity would be about 300 million (far greater, of course, than the number of sources detected by ROSAT's much bigger, though less sensitive, survey).

Follow-up observations at other wavelengths, including the *Hubble* Deep Field and the SDSS, confirm that some of these objects lie at redshifts >6, corresponding to a time when the universe was less than one billion years old, about 10% of its present age. This is actually a problem because none of the current crop of simulations can account for the quick and comprehensive condensation of mass required in the

---

[3]The Sloan Digital Sky Survey is a project that will produce a detailed digital photometric map of half the northern sky (and corresponding spectroscopy) down to about 23^d magnitude, using an automated wide-field telescope of 2.5-aperture. The complete survey should include about 100 million galaxies and about the same number of quasar candidates. See Kent (1994) for an early description of this program.

[4]The most X-ray luminous normal stars, normal galaxies, BL Lac objects, and AGNs are known to have an X-ray to optical luminosity ratio $\log(L_X/L_{opt})$ of around -1, -0.25, 1.5, and 1.0, respectively (see Stocke et al. 1991). So given the limiting ROSAT survey flux quoted above, these ratios imply that the faintest optical counterparts to X-ray sources in these classes have magnitudes, respectively, of 15, 17, 21, and 20, well within the telescope's capability.

rapidly expanding early universe to produce these objects. Astronomers are still grappling with this major unsolved problem; we will return to it in chapter 12.

2.2 THE SKY BETWEEN 10 keV AND 1 MeV

The sky looks far less crowded higher up the photon energy scale. The next plateau in the hierarchy of celestial surveys occurs at 511 keV, the energy of photons emitted when electrons annihilate with positrons. Other than the sun, which is known to produce positrons via beta decay in its nuclear reaction network, only a couple of 511-keV sources have ever been detected in the sky. One of these, X-ray Nova Muscae (see Figure 10.6), will be featured in chapter 10 after we have learned more about the nature of black-hole binary systems. The other is diffuse emission from the Galaxy's central bulge (see plate 7).

The annihilation radiation reaching us from the center of our Galaxy has been known since the 1970s by virtue of several balloon experiments flown since then.[5] But it was the Oriented Scintillation Spectrometer Experiment (OSSE) aboard the *Compton*-GRO that finally established several important characteristics that define this emission. For example, we now know that this emission is diffuse and steady. There is evidence of an extended source along the galactic plane, but the emission is clearly concentrated toward the center of the Galaxy. For a while, some excitement was generated by the apparent extension of the 511-keV emission toward positive latitudes. But this feature has not been confirmed by INTEGRAL in its replication of the OSSE observations (see figure 13.3). However, INTEGRAL and OSSE do agree on the concentrated emission toward the galactic nucleus.

To this day, the source (or sources) of this radiation has remained a mystery.[6] One can say with certainty that a single point source is ruled out because the bulge emission is resolved. But none of the other obvious source populations have yet been shown to account for the inferred annihilation rate, $\sim 1.5 \times 10^{43}$ positrons s^{-1}, spread out spatially in this fashion.

Positrons can be generated by a number of processes, including radioactive β^+-decay of unstable isotopes produced by stars and supernovae, energetic outflows from compact objects, cosmic ray interactions with molecular clouds in the interstellar medium, and possibly even the annihilation or decay of as yet unidentified dark matter particles. In addition to the total positron injection rate and their spatial distribution, a clue to their origin may be provided by the ratio of positronium to ortho-positronium annihilations, which can reveal the physical characteristics of the medium in which the interactions are taking place.

---

[5]This line was first observed with a NaI scintillator at $\sim$476 keV (Johnson, Harnden, and Haymes 1972), and was subsequently unambiguously identified with a narrow (with full width at half-maximum FWHM <3.2 keV) e^+e^- annihilation line using germanium detectors (Leventhal, MacCallum, and Stang 1978).

[6]See Kurfess (1996) for a summary of OSSE's contribution to this work. For the more recent analysis of the 511-keV annihilation radiation from the galactic center, see Churazov et al. (2005) and Milne (2006).

Electrons and positrons annihilate predominantly after they have formed a bound state, analogous to a hydrogen atom, except that the electron's antiparticle is the positive charge in this couplet, rather than a proton. This can happen either in-flight (rare events, given the particles' high relative velocity) or after the positron has cooled sufficiently (to energies below several hundred eV) for the two antiparticles to combine. However, the electron and positron can orbit each other with their intrinsic half-integer spins either aligned or anti-aligned, which determines the total angular momentum of the final photon state.

The ground state of positronium, with its anti-aligned lepton spins, has zero angular momentum, so two exiting integer-one spin photons can conserve this quantum number. Ortho-positronium, on the other hand, with aligned electron and positron spins, has a total ground-state angular momentum of one, and a minimum of three photons in the final state is required to conserve this dynamical quantity.[7] Compared to previous missions, INTEGRAL's main contribution to this discussion has been its much deeper exposure ($\sim$2 Ms) of the galactic center at 511 keV, permitting a more accurate assessment of ortho-positronium's contribution to the continuum versus the 511-keV positronium annihilation line.

The INTEGRAL results indicate that electron–positron annihilation occurs via the positronium bound state $94 \pm 6\%$ of the time. Future theoretical work may be able to discern which phase (or phases) in the interstellar medium is conducive to the formation of positronium and ortho-positronium in the correct ratio to account for this observed fraction.

In a similar energy range as the OSSE survey of e^+e^- annihilation radiation, though with an entirely different scientific objective, the Burst and Transient Spectrometer Experiment (BATSE) on the *Compton*-GRO produced the γ-ray burst map shown in plate 8. BATSE was an all-sky monitor sensitive in the energy range 20–600 keV, designed primarily to study the spatial distribution of GRBs (see section 11.2). At peak emission, GRBs outshine the entire visible universe before fading away a few seconds later. These transient events remained an enduring mystery for almost four decades after their discovery.

BATSE made significant strides in finally demonstrating that GRBs must be extragalactic. Distributed symmetrically around the GRO satellite, its eight detectors each contained two NaI scintillators: a Large Area Detector (LAD) optimized for sensitivity and directional response, and a Spectroscopy Detector (SD) optimized for energy coverage and energy resolution. During its 9 years of operation, BATSE observed over 2700 bursts, distributed isotropically in the sky, with no preferred structures as might be expected if these events occurred predominantly within the Galaxy, or specific galaxies within the local group, or the Virgo cluster.

This result alone ruled out the main competing model to the extragalactic nature of this phenomenon: galactic neutron stars, possibly magnetized, experiencing a dynamic crustal or magnetospheric disturbance. As we shall see later in this chapter, an essential difference between the galactic and extragalactic models is that whereas the underlying source is involved in the former (spending at most a tiny fraction of

---

[7]A comprehensive description of the slowing down and annihilation of positrons in the interstellar medium is given by Bussard, Ramaty, and Drachman (1979).

its available energy, but probably surviving the burst), it is entirely committed in the latter. The fact that a total energy of $\sim 10^{51}$–10^{52} ergs must be released in the burst if the event is extragalactic means that the source cannot survive the explosion. We will be discussing GRBs as a special category of high-energy objects in chapter 11, so we will defer further discussion of their properties until then.

2.3 SURVEYS UP TO 30 MeV

Moving up the energy scale once more, the next all-sky survey (at 1–30 MeV) was produced by COMPTEL on the *Compton*-GRO, and is shown (again in galactic coordinates) in plate 9. At this energy, γ-ray emission is evident throughout the galactic plane, due in part to collisions between cosmic rays and the gas in the interstellar medium. In addition, the COMPTEL catalog contains 63 objects, 32 of which are steady sources, such as neutron stars and black-hole candidates; the remaining 31 are γ-ray bursts (GRBs) captured serendipitously during COMPTEL's scans.

Two sources of γ-rays that uniquely appear in COMPTEL's energy range are unstable isotopes of titanium (^{44}Ti) and aluminum (^{26}Al), both of which are produced in supernova explosions. The former has a half-life of 60 yr; the latter has a half-life of 700,000 yr. The study of decaying titanium can therefore lead to the discovery of relatively young supernova remnants. Correspondingly, decaying aluminum can point to very old remnants. Both of these γ-ray lines were instrumental in COMPTEL's discovery of a new supernova remnant, known as GRO/RX J0852, which was as bright as the full Moon when the explosion occurred about 700 years ago, though for some reason remained undocumented until COMPTEL's discovery.[8]

Other sources contributing to the emission in plate 9 include pulsars and AGNs. The most powerful AGNs, if fortuitously aligned with respect to our line-of-sight, often emit most of their power in the COMPTEL (and EGRET) energy ranges, so this instrument was ideally suited to this study.

It is worth noting that the COMPTEL catalog took eight years to produce.[9] This time is unusually long and was due to the unique physics of COMPTEL's observable energy band and the mechanics of the instrument itself (see figures 1.8 and 1.9). With this type of telescope, it is more difficult to discern between source radiation and the background, given the various energy-loss interactions suffered by the incoming photon. Modeling the background is always very important, but particularly so in the case of a Compton telescope.

2.4 THE HIGHEST ENERGY MAPS PRODUCED IN EARTH ORBIT

The fourth instrument on the *Compton*-GRO was the Energetic Gamma-Ray Experiment Telescope (EGRET), designed to detect γ-rays in the 20-MeV to 30-GeV

---

[8] See Chen and Gehrels (1999).

[9] The COMPTEL data analysis was performed jointly by the Max Planck Institute for Extraterrestrial Physics, NASA's Goddard Space Flight Center, the Space Research Organization of the Netherlands in Utrecht, Netherlands, the Space Science Department of the Astrophysics Division of ESA/ESTEC in Noordwijk, Netherlands, and the University of New Hampshire.

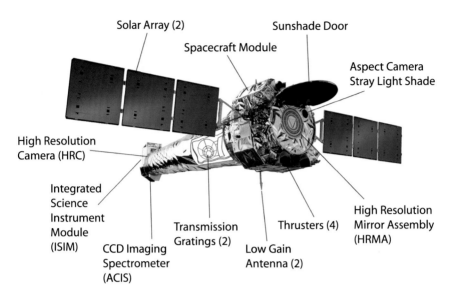

Solar Array (2)   Sunshade Door

Spacecraft Module

Aspect Camera
Stray Light Shade

High Resolution
Camera (HRC)

Integrated
Science
Instrument
Module
(ISIM)

CCD Imaging
Spectrometer
(ACIS)

Transmission
Gratings (2)

Low Gain
Antenna (2)

Thrusters (4)

High Resolution
Mirror Assembly
(HRMA)

Plate 1 Launched toward the middle of 1999, the *Chandra* X-ray Observatory has a resolution ~0.″5, roughly eight times better than that of previous X-ray telescopes, and is able to detect sources more than 20 times fainter. (Image courtesy of NASA)

Plate 2 Artist's concept of XMM-*Newton* in Earth orbit. (Image courtesy of ESA/XMM-*Newton*)

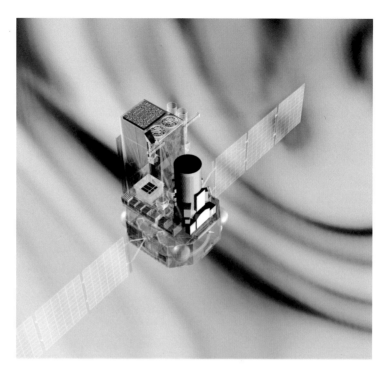

Plate 3 Artist's concept of INTEGRAL, showing its four instruments, which provide simultaneous sensitivity to visible, X-ray, and γ-ray photons. The main goal of the mission is to provide fine spectroscopy (with $E/\Delta E = 500$), while maintaining reasonable spatial resolution ($\sim$12 arcmin FWHM), in the energy range 15 keV to 10 MeV, with concurrent source monitoring in the X-ray (4–35 keV) and optical bands. Note the rectangular coded aperture above the γ-ray detector on the left (see also figure 1.10). Standing 5 m high, the overall satellite weighs in at over 4000 kg. (Image courtesy of ESA/INTEGRAL)

Plate 4 Artist's concept of GLAST in orbit. The Large Area Telescope has a field of view about twice as wide (more than 2.5 steradians) as and sensitivity about 50 times that of EGRET at 100 MeV; the sensitivity will be even greater at higher energies. In a two-year all-sky survey, GLAST's limit for source detection will be 1.6×10^{-9} photons cm^{-2} s^{-1} above 100 MeV. It will also be able to localize sources with an accuracy of 30$''$ to 5$'$. (Image courtesy of NASA/GLAST)

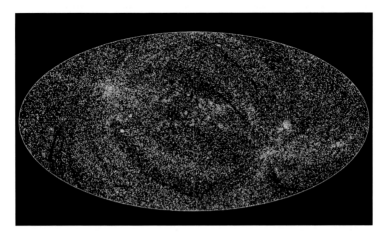

Plate 5 The ROSAT All-Sky Survey, covering the entire celestial sphere in the 0.5- to 2.5-keV range to a typical limiting sensitivity of 5×10^{-13} ergs cm^{-2} s^{-1}. About 80,000 X-ray sources are contained in this data bank, with positional accuracies of about 20″. The sources are shown here in galactic coordinates, with north at the top and longitude increasing toward the left. (Image courtesy of J. Trümper and the ROSAT team, Max-Planck-Institut für Extraterrestrische Physik, Garching)

Plate 6 This *Chandra* Deep Field North image was produced by observing a patch of sky in the constellation of Ursa Major with an area three-fifths the size of the full moon for 23 days. It is the most sensitive X-ray exposure made to date; only one X-ray photon every 4 days reached Earth from each of the faintest sources. Most of the 500 objects seen here are supermassive black holes located in the centers of galaxies. (Image courtesy of G. Garmire, N. Brandt, et al., NASA, and PSU)

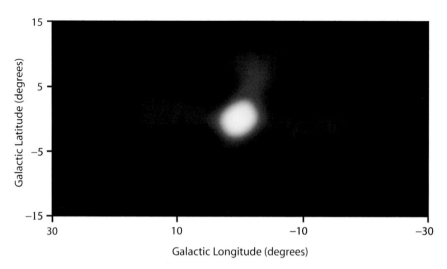

Plate 7 *Compton*-GRO/OSSE map of the distribution of positron annihilation toward the galactic center (at an energy of 511 keV), including an apparent enhancement toward positive latitudes. The brightest feature corresponds to the nucleus of the Galaxy, whereas the horizontal structure lies along the galactic plane. (Image courtesy of D. D. Dixon, W. R. Purcell, UC Riverside, and Northwestern University)

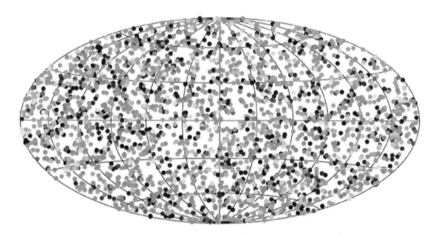

Plate 8 The Burst and Transient Spectrometer Experiment (BATSE) on the *Compton*-GRO was an all-sky monitor sensitive from about 20 to 600 keV. The BATSE Burst Catalog contains eight tables of gamma ray bursts detected by BATSE, covering the period up to and including 26 May 2000. The all-sky gamma-ray burst map produced from these data is shown here in galactic coordinates. (Image courtesy of W. S. Paciesas and the BATSE collaboration)

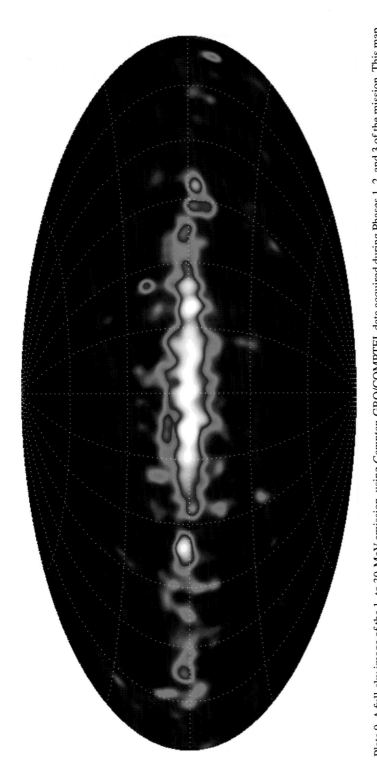

Plate 9 A full-sky image of the 1- to 30-MeV emission, using *Compton*-GRO/COMPTEL data acquired during Phases 1, 2, and 3 of the mission. This map is presented in galactic coordinates, with the galactic center at the origin, north at the top, and galactic longitude increasing toward the left. The galactic plane stands out clearly, indicating the detection of large-scale emission from the Galaxy. (Image courtesy of the COMPTEL collaboration)

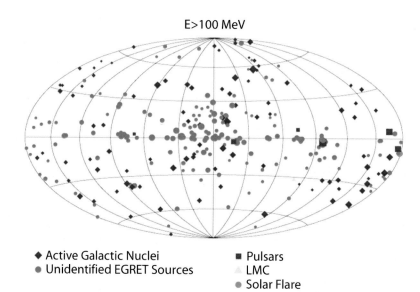

Plate 10 This all-sky map shows the point sources detected above 100 MeV by the *Compton-GRO/EGRET* experiment. The Third EGRET Catalog, from which these data were taken, consists of 271 sources: 5 pulsars, 1 solar flare, 66 high-confidence blazar identifications, 27 possible blazar identifications, 1 likely radio galaxy (Centaurus A), 1 normal galaxy (LMC), and 170 unidentified sources. A sixth EGRET pulsar is shown in the figure for completeness (at galactic coordinates $l = 69°$, $b = 3°$), but is seen only in pulsed data. (Image courtesy of R. C. Hartman and the EGRET collaboration)

Plate 11 The all-sky map produced by *Compton*-GRO/EGRET. The EGRET instrument was sensitive from 20 MeV up to 30 GeV, and was 10 to 20 times larger and more sensitive than previous detectors operating at these high energies. Included in this image are sources of diffuse γ-ray emission, gamma-ray bursts, cosmic rays, as well as pulsars and active galactic nuclei, also called gamma-ray blazars. This all-sky image shows the overall γ-ray emission above 100 MeV. The brightest emission is shown in yellow, while faintest emission is in blue. The plane of the Milky Way galaxy is clearly seen as a strong source of both diffuse and resolved emission. The former is primarily due to cosmic-ray interactions with the interstellar medium. The Vela, Geminga, and Crab pulsars are clearly visible as bright knots of emission toward the right portion of the galactic plane. The gamma-ray blazar 3C 279 is seen as the brightest knot of emission above the plane. (Image courtesy of the *Compton*-GRO/EGRET collaboration)

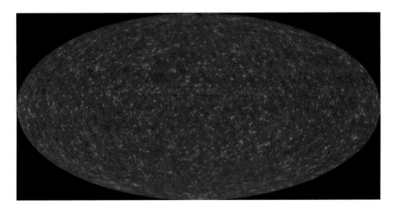

Plate 12 Simulated all-sky map of the background diffuse emission that will be observed with the GLAST mission. The maximum flux for these background sources was set at $5 \times 10^{-8} \, \mathrm{cm}^{-2} \, \mathrm{s}^{-1}$, near or below the EGRET detection threshold. The distribution was cut off at $3 \times 10^{-10} \, \mathrm{cm}^{-2} \, \mathrm{s}^{-1}$ to make the background intensity equal $1.5 \times 10^{-5} \, \mathrm{cm}^{-2} \, \mathrm{s}^{-1} \, \mathrm{sr}^{-1}$. The overall density of background sources is about 4 per square degree. (Image courtesy of S. Digel and the GLAST collaboration)

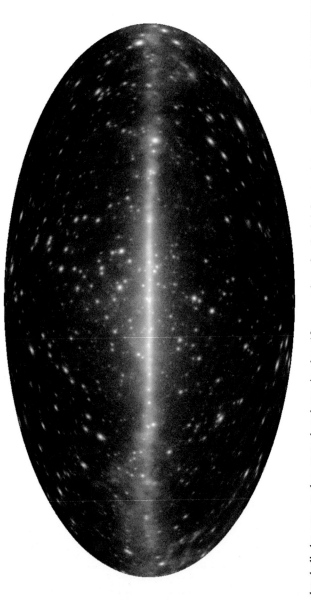

Plate 13 Simulated all-sky map, analogous to that shown in plate 12, except here for the brightest point sources from the Third EGRET Catalog (see plate 10). In many cases, the cataloged fluxes are scaled down by a factor of 2 or so since many of the EGRET sources were detected in flaring states. The colors are chosen such that red represents the flux in the 100- to 400-MeV range, green in the 400- to 1000-MeV range, and blue shows the flux above 1600 MeV. The scaling is logarithmic. (Image courtesy of S. Digel, R. Dubois, J. E. McEnery, and the GLAST collaboration)

range. EGRET holds the current record for the highest energy radiation detected from an orbital platform. Its principal scientific objectives were to perform an all-sky survey in this energy range (portions of which are shown in plates 10 and 11), and to study in detail the various classes of γ-ray emitting sources.

Among its many significant discoveries,[10] EGRET identified blazars (a certain type of AGN whose powerful jet is directed toward us) as prodigious emitters of high-energy radiation. This followed a theoretical prediction of this effect in the late 1980s.[11] These AGNs, it turns out, are quite variable in flux, with flares occurring on timescales of days to hours.

EGRET also detected the high-energy tails of several GRBs, including in one special case GeV photons more than one hour after the burst was seen at lower energies. And among the pulsars shown in plate 10, one is the now well-known Geminga (at galactic coordinates $l = 195.14°$, $b = 4.30°$, on the right-hand-side of this diagram, just above the galactic midplane), confirmed to be a radio-quiet pulsar, the first ever detected in γ-rays.

Plate 11, made possible by EGRET's efficient rejection of background noise, is particularly striking. This is the first sensitive map of the diffuse $\gtrsim$100-MeV emission of the Milky Way, most of which is associated with cosmic ray interactions with the interstellar gas and (via inverse Compton scattering) with low-energy photons. Above and below the plane, we see the isotropic, extragalactic emission. We will have much more to say about the diffuse high-energy background in chapter 13.

But as exquisite as these maps are, the >100-MeV surveys that will be produced by GLAST (see plate 4) in the coming years should be even more spectacular. Plate 12 shows a simulated GLAST all-sky survey for emission >100 MeV, including a diffuse galactic background,[12] the point sources from the Third EGRET Catalog (see plate 10), extended emission models for the Large Magellanic Cloud and M31, and point sources representing some other normal galaxies, several starburst galaxies, a galactic halo of unidentified sources, a sprinkling of unidentified sources along the galactic plane, and a diffuse isotropic background (see section 13.3). The limiting sensitivity for this map corresponds to 85% observing efficiency for a one-year sky survey.

GLAST's version of EGRET's point-source map (see plate 10) is shown in plate 13. The source positions and their brightness for this survey were extracted from the Third EGRET Catalog, though in many cases the registered fluxes were scaled down by a factor 2 or more, since they were measured during flaring states, which are not representative of the time-averaged values.

We have now seen how quickly our view of the sky can change every time a new high-energy window is opened with the deployment of a more sensitive telescope (*Chandra*), or a detector with a bigger collecting area (*Rossi*-XTE), or one that extends our ability to sense photons with energies beyond a GeV (GLAST). Even so,

---

[10]See Hartman et al. (1999).

[11]The original idea behind this was that the narrow range of observed outflow velocities in AGN jets was due to an "inverse Compton braking" effect, in which highly relativistic particles scatter with low-energy ambient radiation and are thereby slowed down, while the radiation itself is boosted into the X-ray and γ-ray domains along the jet. See Melia and Königl (1989).

[12]See Hunter et al. (1997).

our introduction into the various sources one encounters in high-energy astrophysics has been modest at best because the physics we need to describe them stretches us well beyond what is customary in more traditional astronomy.

In the next several chapters, we will concentrate our efforts primarily on establishing the theoretical framework we need to understand what all these new observations are telling us. Then, in chapters 9–13 we shall organize the many sources we have uncovered thus far into distinct subgroups. This delineation will be based in part on their observed characteristics (which is always a useful starting point), but also on our interpretation of the phenomena they produce, which tell us about their underlying nature, e.g., whether they are galactic or extragalactic. Toward the end of this book, we will complete our analysis of how these objects obtain their energy, how they accelerate their particles, and how they emit their radiation.

SUGGESTED READING

One of the most useful functions of groundbreaking instruments, such as ROSAT, is to produce surveys for source identification at other wavelengths and to provide a guide for more focused observations with followup missions. An example of how this works in practice may be found in Lehmann et al. (2001) and Stocke et al. (1991).

An efficient means of acquiring optical counterparts to high-energy sources is the cross-correlation of UV, X-ray, and γ-ray maps with the extensive catalogs being produced, e.g., by the Sloan Digital Sky Survey (SDSS). See Adelman-McCarthy et al. (2007).

The origin of the electron–positron annihilation radiation originating from the Galaxy's central bulge has remained an enduring mystery since its discovery in the 1970s. For a summary of OSSE's contribution to the study of this inexplicable source of high-energy radiation, and for a more recent analysis of its possible origin, see Kurfess (1996), Churazov et al. (2005), and Milne (2006).

One of the more notable discoveries made with the EGRET instrument on the *Compton*-GRO satellite was the powerful γ-ray emission by AGNs whose jet is directed toward us. The possible association between relativistic jets and their high-energy emission was anticipated in Melia and Königl (1989).

In preparation for its two-year all-sky survey, the GLAST team simulated the appearance of the >100-MeV sky expected on the basis of GLAST's unprecedented sensitivity at these energies. See Hunter et al. (1997).

Chapter Three

Relativity

High-energy astrophysics deals with the behavior of matter moving at close to light speed, or trapped within the deep potential well of strongly gravitating sources. In many cases, both of these conditions are present in the emitting regions producing the energetic photons we detect. It is therefore essential to incorporate the effects of special (and, often, general) relativity into our description of the physics underlying the interaction between particles and radiation. In this chapter, we shall develop the language of four-dimensional spacetime, and generalize the simple theory describing systems moving at constant speed to a framework in which the various observers accelerate with respect to one another. We shall see that a transition from special to general relativity is seamless and straightforward when one takes inertial forces into account. And although a discussion of the field equations giving rise to the gravitational field itself is beyond the scope of our streamlined treatment here,[1] we will nonetheless uncover the key elements required for a detailed analysis of all high-energy sources.

3.1 SPECIAL RELATIVITY

The physical laws of *classical* mechanics are invariant under a Galilean transformation, in which the coordinates $\mathbf{x}$ (space) and t (time) of one frame are related to those (primed) of a second frame moving relative to it at constant speed $\mathbf{v}$ according to the expressions

$$\mathbf{x}' = \mathbf{x} - \mathbf{v}t \qquad (3.1)$$

and

$$t' = t. \qquad (3.2)$$

Nonetheless, Galilean relativity has a serious problem in that the Maxwell equations of electrodynamics—themselves forged before the advent of special relativity—are not invariant.[2] To bring consistency to the transformation of all physical laws using the *same* criteria to relate the coordinates of one frame to those of another, it is therefore essential to modify the Galilean relativity theory. At the same time, the new transformation laws must themselves reconcile the observed behavioral

---

[1] Two recommended books on this subject are Weinberg (1972) and Wald (1984).

[2] This can be seen most readily via the application of equations (3.1) and (3.2) to Faraday's law, which changes form from one reference frame to the next. See, for example, Melia (2001a).

disparity of these coordinate relations at low and high velocity. More telling still is the observation that the speed of light c appears to be invariant, for all observers, for all frames, and, apparently, for all time.

It was rather surprising to the early workers in relativity theory that the rest of classical mechanics had to be modified to accommodate the new transformation laws, rather than electrodynamics, which itself remained unchanged when written in the frame-dependent language of a given observer. We shall see that the unifying aspect of the new description is that it permits *all* laws to be written using the invariant vocabulary of four-dimensional spacetime—the underlying framework of the theory of special relativity.

The guidance provided by experiment and observation may be encapsulated within what are now known as the two essential postulates of special relativity: (1) that only *relative* motion is observable, and (2) that the speed of light in vacuum is a *constant*, c, independent of the source and/or observer speed.[3]

It is immediately clear from the second postulate that the Newtonian view of time as a universal fluid advancing independently of change or the circumstance of matter and its environment must be incorrect. Observers in different frames cannot agree on the space and time (*spacetime*) coordinates of an event. To see this, suppose that a pulse of light is emitted at time $t = 0$ when the origins O and O' of the two coordinate systems occupy a common point. Then the observer at O will report that the pulse arrives at another location $\mathbf{x}$ in this space after a time

$$t = r/c \tag{3.3}$$

has elapsed, where

$$r = (x^2 + y^2 + z^2)^{1/2}. \tag{3.4}$$

The elapsed time for the observer at O', on the other hand, is instead

$$t' = r'/c, \tag{3.5}$$

where now

$$r' = \left[(x')^2 + (y')^2 + (z')^2\right]^{1/2}. \tag{3.6}$$

The two origins have separated during the propagation of the pulse, so clearly $r \neq r'$ and $t \neq t'$.

Nonetheless, the fact that the speed of light is constant for all observers means that there must exist certain relations between the coordinates to preserve the invariance of c, even though t and r separately change. The *spacetime interval*

$$(\Delta s)^2 \equiv (c\Delta t)^2 - |\Delta \mathbf{x}|^2 \tag{3.7}$$

is the most useful of these formulations. It is trivially zero along any light path, and every observer measures the same value. In terms of the spherical coordinates we have just described, with $t = t' = 0$ at the origin,

$$(\Delta s)^2 = (ct)^2 - r^2 = 0 = (ct')^2 - (r')^2 = (\Delta s')^2. \tag{3.8}$$

We shall see, however, that the applicability of Δs is far more extensive than this simple mathematical restatement of the second postulate of special relativity.

---

[3] For an early critical review of the experimental basis for the second postulate, see Fox (1965, 1967).

The correct transformation laws of the new relativity theory must relate t and $\mathbf{x}$ to t' and $\mathbf{x}'$ in a such a way that the invariance of $(\Delta s)^2$ is guaranteed. There is no a priori means of knowing what these relations are, of course, except that we do know they must reduce to equations (3.1) and (3.2) when v is much smaller than c. With some guessing, we find that the two sets of coordinates must be related via the so-called Lorentz transformation,

$$x' = x,$$
$$y' = y,$$
$$z' = \gamma \left(z - \frac{v}{c} ct \right),$$
$$ct' = \gamma \left(ct - \frac{v}{c} z \right). \tag{3.9}$$

The Lorentz factor $\gamma \equiv \left[1 - (v/c)^2 \right]^{-1/2}$ has a minimum value of 1, but can increase indefinitely as $v \to c$. (Note that these equations assume the primed frame is moving at constant velocity $\mathbf{v} = v\hat{\mathbf{z}}$ relative to the unprimed coordinate system.) There exit other transformations that also preserve the speed of light—for instance, a rigid rotation. Together, all these possible solutions constitute the homogeneous Lorentz group. However, only the relations in equation (3.9) describe the transformation of coordinates between frames moving relative to each other at constant, uniform velocity.

The first postulate of special relativity requires that there be perfect symmetry between the two observers, so the reciprocal transformation to (3.9) may differ from it only in the sign of $\mathbf{v}$, as seen from the perspective of the second frame:

$$x = x',$$
$$y = y',$$
$$z = \gamma \left(z' + \frac{v}{c} ct' \right),$$
$$ct = \gamma \left(ct' + \frac{v}{c} z' \right). \tag{3.10}$$

To advance from here, and learn more about the internal structure of special relativity theory, we must look to additional clues provided by nature. The transformations embodied by equations (3.9) and (3.10) tell us how to handle quantities that involve dimensions of distance and time. They say nothing about how one observer relates dimensionless numbers to those seen in another frame. A measurement of length and time is essential for the determination of many physical quantities, certainly those that involve velocities and accelerations. Even inertia, for example, is characterized in terms of how much acceleration is induced by a particular force, and because the intervals of length and time change from one observer to the next, inertia's inferred value must therefore depend on the reference frame in which it is measured. But one would think that *dimensionless* quantities, such as the number of events, should be invariant. For example, if a π° decays into two photons in one frame, it must also be seen to decay into two photons in every other frame. Were this not the case, the physics would depend on the observer, and the idea of formulating a set of physical laws that describe nature consistently everywhere would be nonsensical.

One of the most important dimensionless quantities we know—the phase of a wave—characterizes the evolution of many disparate phenomena, from the undulation of the water's surface at the beach, to the wavefunction solution of Schrodinger's equation in quantum mechanics. What all these waves have in common is the connection between position and time, appearing in the phase, that measures where we are on the wave as a fraction of one complete oscillation:

$$\phi \equiv \omega t - \mathbf{k} \cdot \mathbf{x} = \omega' t' - \mathbf{k}' \cdot \mathbf{x}'. \tag{3.11}$$

Constraining the coordinates in this fashion, the phase ϕ clearly represents new physics beyond the constancy of c that gave rise to equation (3.8). Let's follow this lead and see what happens when we introduce the Lorentz transformation equations into this expression:

$$\omega t - kz = \omega' \left(\gamma t - \gamma \frac{vz}{c^2} \right) - k' \left(\gamma z - \gamma v t \right) \tag{3.12}$$

or, collecting terms,

$$\omega t - kz = (\omega' \gamma + \gamma k' v) t - \left(\gamma \frac{\omega' v}{c^2} + \gamma k' \right) z, \tag{3.13}$$

again assuming that the wave is propagating in the z-direction. Given that this must be true for all t and z, it is evident that

$$\omega = \gamma (\omega' + k' v) \tag{3.14}$$

and

$$k = \gamma \left(k' + \frac{v \omega'}{c^2} \right). \tag{3.15}$$

Already an inkling arises that there exists another grouping of physical quantities, aside from the coordinates t and $\mathbf{x}$, that must be transformed together in a translation from one frame to the next. More importantly, these quantities transmute into each other, so that k becomes an ω (equation 3.14), and ω becomes a k (equation 3.15).

When we consider these expressions in the context of the propagation of light, for which the frequency and wavenumber are related as $k = \omega/c$ and $k' = \omega'/c$, it appears that

$$\omega = \gamma \omega' (1 + \beta) \tag{3.16}$$

and

$$\omega' = \gamma \omega (1 - \beta), \tag{3.17}$$

where $\beta \equiv v/c$. More generally, the wavevector $\mathbf{k}$ may not be directed along $\hat{\mathbf{z}}$, in which case only the component parallel to $\mathbf{v}$ is affected. The full expression,

$$\omega' = \gamma \omega (1 - \beta \cos \theta), \tag{3.18}$$

with $\mathbf{k} \cdot \mathbf{v} = \cos \theta |\mathbf{k}| |\mathbf{v}|$, is known as the *Doppler shift formula*.[4]

---

[4]This expression also predicts a *transverse* Doppler shift, which was verified by several early experiments, including one using the Mössbauer effect (Hay, Schiffer, Cranshaw, and Egelstaff 1960).

Something quite remarkable emerges when we compare equations (3.15) and (3.16) with those of (3.9). The quantities $(\omega/c, \mathbf{k})$ have exactly the same (generalized) transformation properties as $(ct, \mathbf{x})$. The implication is that the substructure of special relativity contains 4-groupings of physical variables that couple to each other when one observer compares her measurements with those made in another frame. The four-dimensional groupings

$$x^\mu \equiv (x^0, \mathbf{x}) \quad (\mu = 0, 1, 2, 3) \tag{3.19}$$

and

$$k^\mu \equiv (k^0, \mathbf{k}) \quad (\mu = 0, 1, 2, 3), \tag{3.20}$$

where $x^0 \equiv ct$ and $k^0 \equiv \omega/c$, are examples of four-vectors in the four-dimensional spacetime. That is, x^μ and k^μ *transform* as four-dimensional vectors under a Lorentz boost from one frame to the next. Any quantity that transforms in the same fashion as x^μ is called a four-vector in this space.

Quantities such as x^μ and k^μ involve both a magnitude and a direction in three-space. The constancy of c forces a constraint between the three-space coordinates and t, so it is clear why four components must be involved in any transformation of frames. In cases where a measurable element is independent of direction, however, one would expect the number of transmutable quantities in a grouping to be less than four. For example, the number of marbles in a box is independent of the latter's orientation. In other words, some physical quantities require only denumerability, and for them, a single function of the spacetime coordinates is sufficient. The phase of a wave, ϕ, is another example: even though ϕ depends on all four components of x^μ, it is nonetheless a single function of them. We should thus expect that the four-dimensional spacetime ought to contain bundles comprising just a single function each. Being solitary members of their respective groupings, these physical quantities should therefore be invariant under a transformation from one frame to the next.

One such quantity is the spacetime interval Δs. We've shown that it is invariant for light, but now that we've derived the transformation relations (3.9) and (3.10), it is easy to show that Δs is in fact invariant for any pair of spacetime points, whether or not they are linked with light or slower traveling particles. The difference is that Δs is always zero for photons (or massless matter, which, as we shall see, also moves at speed c), but is greater than zero for all other particles:

$$
\begin{aligned}
(\Delta s)^2 &= (x^0)^2 - (x^1)^2 - (x^2)^2 - (x^3)^2 \\
&= \gamma^2(x'^0 + \beta x'^3)^2 - (x^1)^2 - (x^2)^2 - \gamma^2(x'^3 + \beta x'^0)^2 \\
&= (x'^0)^2 - (x'^1)^2 - (x'^2)^2 - (x'^3)^2 \\
&= (\Delta s')^2.
\end{aligned}
\tag{3.21}
$$

Incidentally, an extension of special relativity theory to situations in which the relative velocity of two frames is changing requires that we consider a Lorentz transformation to be valid only for an infinitesimal interval of time, during which the change in velocity is minimized. Our expectation is that by considering the *differential* elements of physical quantities, the validity of a Lorentz transformation is preserved as long as it is understood that it applies solely to that given instant.

The interval s should then be considered in its infinitesimal limit:

$$(ds)^2 = (c\,dt)^2 - (dx^1)^2 - (dx^2)^2 - (dx^3)^2. \tag{3.22}$$

The fact that ds is a single function of the coordinates and that it is invariant under a Lorentz transformation, means that it is a *scalar* in four-dimensional spacetime. This property has practical value because it allows one observer to compare his intervals of space and time directly with those measured in another frame. The rest frame is particularly useful because $d\mathbf{x} = 0$. The interval ds is then entirely due to the passage of time, $d\tau$ which, for obvious reasons, is called the *proper* time, i.e., the time measured in a frame where the events that delimit it occur at the *same* spatial location:

$$ds = c\,d\tau. \tag{3.23}$$

For an observer in a different frame,

$$\begin{aligned} ds &= dx^0 \left[1 - \left(\frac{d\mathbf{x}}{dx^0} \right)^2 \right]^{1/2} \\ &= c\,dt(1 - \beta^2)^{1/2} \\ &= c\,dt/\gamma, \end{aligned} \tag{3.24}$$

which produces a result known as time dilatation (or stretching) from one frame to the next: $dt = \gamma\,d\tau$.

Correspondingly, there exist certain interactions where four components are insufficient because one must consider the transport of a vector, which itself has three spatial components, along different directions. An example in this category is the stress applied to a surface by an electromagnetic field, for which the individual components of force may vary across the enclosure. For physical quantities such as this, the groupings must involve four-vectors of four-vectors, or a minimum of 16 transmutable components.

In four-dimensional spacetime, these groupings of physical variables that intermix under a transformation are called *tensors*. Another way of expressing our first postulate of special relativity is to say that the laws of nature must be written using tensors in a way that renders their *form* invariant under a Lorentz transformation, even though the components of each tensor may individually change from one frame to the next.

The properties of a tensor depend on its rank k, determined by how it transforms under the translation

$$x \to x'. \tag{3.25}$$

(The four-vector x^0, x^1, x^2, x^3 is sometimes written using the shorthand notation x.) Thus far, we have encountered three members of the tensor class. A *scalar*, i.e., a tensor of rank 0, is a single function of x whose value is *not* changed by the transformation. Thus, the Lorentz interval $ds(x^0, x^1, x^2, x^3)$ is a scalar.

A tensor of rank 1 is a *vector* (or more precisely, a four-vector), a set of four ordered numbers that transform according to the rule

$$V'^{\alpha} = \frac{\partial x'^{\alpha}}{\partial x^{\beta}} V^{\beta}, \tag{3.26}$$

defining a *contravariant* vector, and

$$V'_{\alpha} = \frac{\partial x^{\beta}}{\partial x'^{\alpha}} V_{\beta} \tag{3.27}$$

for a *covariant* vector. In each of these, α and β take on four possible values, viz. $(0, 1, 2, 3)$. In this expression, and throughout this book unless otherwise noted, a repeated index means that the term in which the index appears is to be summed over all its possible values. So, for example, $k^{\alpha} k_{\alpha} \equiv k^0 k_0 + k^1 k_1 + k^2 k_2 + k^3 k_3$.

Not only are the two quantities x^{μ}, k^{μ} therefore tensors of rank 1 but, more importantly, notice that the coefficients in the transformation of equation (3.26) are derived from the chain rule of differentiation for dx'^{α}. Since we know that x^{μ} is a four-vector in this spacetime, and we've determined its transformation properties empirically, we might as well use this knowledge to execute a transformation of all such four-vectors. This is based on an important assumption we are making—that all four-vectors in this spacetime transform in an identical manner to x^{μ}.

The two versions of a four-vector we have introduced here contain the same physics, but they incorporate a sign change in their spatial coordinates that ultimately simplifies the mathematical formalism. The easiest way to recognize the origin of this difference is to compare equations (3.9) and (3.10); a change in sign emerges wherever the velocity appears. In addition, they allow us to easily handle a differentiation with respect to a contravariant coordinate x^{μ}, which evidently produces a covariant four-vector:

$$\frac{\partial}{\partial x'^{\alpha}} = \frac{\partial x^{\beta}}{\partial x'^{\alpha}} \frac{\partial}{\partial x^{\beta}}. \tag{3.28}$$

Given the empirically derived Lorentz transformation laws expressed in (3.9) and (3.10), we can easily derive the coefficients appearing in equations (3.26) and (3.27), and it is more common to see these relations written in matrix form

$$\mathbf{V}' = \mathcal{A} \cdot \mathbf{V}, \tag{3.29}$$

where $\mathbf{V}$ is the column vector

$$\mathbf{V} \equiv \begin{pmatrix} V^0 \\ V^1 \\ V^2 \\ V^3 \end{pmatrix}, \tag{3.30}$$

and $\mathcal{A}$ is the 4×4 matrix of coefficients

$$a^{\alpha}{}_{\beta} \equiv \frac{\partial x'^{\alpha}}{\partial x^{\beta}}. \tag{3.31}$$

When the boost is in the z-direction, as is the case for equations (3.9) and (3.10),

$$A \equiv [a^\alpha{}_\beta] = \begin{pmatrix} \gamma & 0 & 0 & -v\gamma/c \\ 0 & 1 & 0 & 0 \\ 0 & 0 & 1 & 0 \\ -v\gamma/c & 0 & 0 & \gamma \end{pmatrix}. \tag{3.32}$$

For a covariant four-vector, the coefficients are the same as these, though with a sign change that ultimately arises because of the reversed sense of velocity inferred by one observer relative to the other:

$$A^{-1} \equiv [a_\alpha{}^\beta] = \begin{pmatrix} \gamma & 0 & 0 & v\gamma/c \\ 0 & 1 & 0 & 0 \\ 0 & 0 & 1 & 0 \\ v\gamma/c & 0 & 0 & \gamma \end{pmatrix}. \tag{3.33}$$

To avoid confusion, one should keep in mind that the indices in the coefficients $a^\alpha{}_\beta$ and $a_\alpha{}^\beta$ may be either subscripts or superscripts and that their horizontal ordering is not arbitrary. A superscript means that its corresponding coordinate is being differentiated, and the first horizontal position is reserved for the "primed" coordinate. So, for example, $a_\mu{}^\nu = \partial x^\nu / \partial x'^\mu$.

A tensor $T^{\alpha\beta}$ of rank $k = 2$ is a four-vector of four-vectors, and is therefore a grouping of 16 functions of x that transform according to an obvious generalization of the rules for scalars and vectors:

$$T'^{\alpha\beta} = a^\alpha{}_\gamma a^\beta{}_\delta T^{\gamma\delta} \tag{3.34}$$

and, correspondingly,

$$T'_{\alpha\beta} = a_\alpha{}^\gamma a_\beta{}^\delta T_{\gamma\delta}. \tag{3.35}$$

The coefficients $a^\alpha{}_\beta$ and $a_\alpha{}^\beta$ satisfy several simple rules that greatly facilitate algebraic calculations with tensors. For example, the notation in equations (3.32) and (3.33) already suggests that the matrix $a_\alpha{}^\beta$ is the inverse of $a^\alpha{}_\beta$ or, in other words,

$$a^\alpha{}_\beta a_\alpha{}^\gamma = \delta^\gamma_\beta, \tag{3.36}$$

where

$$\delta^\alpha_\beta = \begin{cases} 1 & \text{if } \alpha = \beta \\ 0 & \text{if } \alpha \neq \beta \end{cases} \tag{3.37}$$

is the Kronecker delta. This property may be easily confirmed by using the chain rule of differentiation, which gives

$$\frac{\partial x'^\alpha}{\partial x^\beta} \frac{\partial x^\gamma}{\partial x'^\alpha} = \frac{\partial x^\gamma}{\partial x^\beta} = \delta^\gamma_\beta. \tag{3.38}$$

A possibly even more powerful algebraic relation follows immediately from this property. The *contraction* (i.e., the "dot product") of a covariant four-vector with a contravariant four-vector produces an invariant quantity:

$$\begin{aligned} V'^{\alpha} W'_{\alpha} &= a^{\alpha}{}_{\beta} V^{\beta} a_{\alpha}{}^{\gamma} W_{\gamma} \\ &= \delta^{\gamma}_{\beta} V^{\beta} W_{\gamma} \\ &= V^{\beta} W_{\beta}. \end{aligned} \tag{3.39}$$

(Remember that a repeated index means we must sum over all its values.) In contrast, the product $S_{\alpha\beta} \equiv V_{\alpha} W_{\beta}$ is not a scalar, but rather a tensor of rank 2, since

$$\begin{aligned} S'_{\alpha\beta} &= V'_{\alpha} W'_{\beta} \\ &= a_{\alpha}{}^{\gamma} V_{\gamma} a_{\beta}{}^{\delta} W_{\delta} \\ &= a_{\alpha}{}^{\gamma} a_{\beta}{}^{\delta} S_{\gamma\delta}. \end{aligned} \tag{3.40}$$

It's easy to see how one generalizes from this. Trivially, $T_{\alpha}{}^{\alpha}$ is a scalar, whereas $W_{\alpha} \equiv T_{\alpha}{}^{\beta} V_{\beta}$ is a covariant four-vector, and so forth.

From equation (3.22), we see intuitively that ds is a scalar because it represents a contraction of contravariant and covariant four-vectors. Indeed, we may rewrite it as

$$(ds)^2 = \eta_{\alpha\beta} \, dx^{\alpha} \, dx^{\beta}, \tag{3.41}$$

where $\eta_{\alpha\beta} = \eta_{\beta\alpha}$ is called the *metric tensor*. For now, the introduction of $\eta_{\alpha\beta}$ appears to be one of convenience, simplifying our notation for the sum of terms appearing in equation (3.22). We shall see later, however, when we generalize the theory to include accelerated frames, that this matrix of coefficients is precisely where the inertial terms arising from the noninertial transformation reside.

In a Cartesian coordinate system, $\eta_{\alpha\beta}$ is diagonal:

$$\eta_{\alpha\beta} = \begin{pmatrix} 1 & 0 & 0 & 0 \\ 0 & -1 & 0 & 0 \\ 0 & 0 & -1 & 0 \\ 0 & 0 & 0 & -1 \end{pmatrix}. \tag{3.42}$$

With the definition of ds in equation (3.41) and $\eta_{\alpha\beta}$ in equation (3.42), it is also clear that

$$dx_{\beta} = \eta_{\beta\alpha} \, dx^{\alpha}, \tag{3.43}$$

which demonstrates that a contraction of a four-vector with the metric tensor is the general procedure for converting a contravariant index on any tensor into a covariant one. This also works in reverse, converting a covariant index into a contravariant one. To see this, we note that the coordinates x^{α} are linearly independent, so the condition

$$x^{\alpha} = \eta^{\alpha\beta} (\eta_{\beta\gamma} x^{\gamma}) \tag{3.44}$$

is satisfied as long as

$$\eta^{\alpha\beta} \eta_{\beta\gamma} = \delta^{\alpha}_{\gamma}. \tag{3.45}$$

This means that in special relativity, the coefficients $\eta^{\alpha\beta}$ and $\eta_{\beta\gamma}$ are identical when written in terms of Cartesian coordinates, and so

$$dx^\beta = \eta^{\beta\alpha}\, dx_\alpha. \tag{3.46}$$

It should now be clear how special relativity has advanced our understanding of transformations from one frame to the next. In Galilean relativity, the magnitude of a three-vector must remain unchanged under a rotation or translation, though its individual components may vary; but it contains no mixing of spatial coordinates with time, because the classical view holds that time is a "universal fluid" independent of space or its contents. All observers agree on time's rate of passage, regardless of what changes occur within the observer's frame.

In special relativity, we're confronted from the very beginning—from the very coordinates we must use to execute the transformation—with the empirical fact that time should instead be viewed as a measure of the rate of change, which must then necessarily be tagged to the specific frame of reference. However, nature tells us that even under these circumstances, there still exists an invariant quantity— the contraction $x^\mu x_\mu$—which may be viewed as the four-dimensional analog of the three-dimensional magnitude of $\mathbf{x}$. We learn very quickly that x^μ is but one grouping of variables that intermix under a transformation; the ease with which all other physical quantities fall into their own separate groupings is a measure of the elegance of special relativity. The members of a group may intermix only with others in the same set when one makes a translation between frames.

Not surprisingly, only a description of nature written in the language of four-dimensional spacetime can be made invariant under a Lorentz transformation. Without this invariance, we would have no means of preserving the essence and relevance of physical laws, whose form would then depend on the observer's velocity. But there is no universal frame of reference against which such a velocity could be measured, and so the description of nature would be different for different observers, which is not consistent with what we see. In the next section, we shall see how physical laws must be transcribed between the language of Galilean relativity and that of its successor—special relativity.

3.2 RELATIVISTIC TRANSFORMATION OF PHYSICAL LAWS

The groupings of physical quantities that define the tensors of four-dimensional spacetime are often uncovered by examining the differential equations of classical mechanics. For example, in time-dependent electrodynamics, equations are often written in the Lorenz gauge, defined by the condition

$$\partial\Phi/\partial x^0 + \vec{\nabla}\cdot\mathbf{A} = 0, \tag{3.47}$$

in which the scalar (Φ) and vector ($\mathbf{A}$) potentials satisfy the wave equation (see section 5.4). Equation (3.47) specifies how Φ must change in time in response to a spatial variation of $\mathbf{A}$. According to the first postulate of special relativity, all observers should agree on this connection between the two potentials, which means that the value on the right-hand-side of the equality should be a Lorentz scalar. But

$\partial/\partial x^\alpha$ is a covariant four-vector, and so $A^\alpha \equiv (\Phi, \mathbf{A})$ must be a contravariant four-vector. The grouping A^α is known as the four-vector potential in special relativity.

Developing a description of physical laws using tensors would seem to be quite daunting given that inertia is not an invariant quantity. This presents an obvious difficulty in using forces to quantify the effects on particles due to the influence of others. Not knowing a priori how to calculate the force on a moving object, we have no direct means of translating a physical law known in terms of frame-dependent language into its invariant form mandated by special relativity. But just as we found for the potentials Φ and $\mathbf{A}$, we generally do know how to write down a description of the interaction in at least one frame, usually the frame in which the particle being acted on is at rest. Consequently, the *rest mass*, m, representing the particle's inertia in its own frame, is a quantity upon which all observers can agree, and we shall therefore always mean this value when we refer to the "mass." The variation in the particle's inertia from frame to frame must be represented by something other than m.

We know that as long as the particle's velocity is small, Newton's laws of motion are quite adequate in describing its response to an applied force $\mathbf{F} \equiv (F^1, F^2, F^3)$. For example, an observer in the particle's rest frame would infer the particle's inertia m based on the acceleration she measures when $\mathbf{F}$ is applied: $\mathbf{F} = m\mathbf{a}$. But to turn this law into a form upon which all observers can agree, we must recast it using a combination of four-dimensional tensors. A reasonable guess, which turns out to be the correct one, is to write the four-dimensional version f^α of $\mathbf{F}$ in the form

$$f^\alpha \equiv m \frac{d^2 x^\alpha}{d\tau^2}. \tag{3.48}$$

The rationale behind this is that f^α trivially reduces to $\mathbf{F} = m\mathbf{a}$ in the particle's rest frame and, very importantly, the right-hand-side of equation (3.48) is a four-vector. To see this, note that the rest-frame inertia m and proper time τ are scalar quantities, so the transformation properties of $m\, d^2 x^\alpha/d\tau^2$ are fully specified by those of x^α alone. Had we instead used the frame-dependent time t instead of τ, this would not have worked because f^α as defined would not have been a four-vector.

But what are we to make of f^0? In the particle's rest frame, where $d\tau = dt$, we have

$$\begin{aligned} f^i &= F^i \quad (i = 1, 2, 3) \\ f^0 &= 0, \end{aligned} \tag{3.49}$$

so no new physics emerges yet. However, the fact that the four components of f^α must intermix under a Lorentz transformation means that f^0 is not always going to be zero. Since f^α is a four-vector,

$$f'^\alpha = \alpha^\alpha{}_\beta f^\beta. \tag{3.49}$$

Thus, following the transformation rules summarized in equation (3.9), in which only the time component and the component of the spatial three-vector parallel to the boost direction are modified, we find that

$$f_L^0 = \frac{\gamma}{c} \mathbf{v} \cdot \mathbf{F}$$

$$\mathbf{f}_L = \mathbf{F} + (\gamma - 1)\mathbf{v}\frac{\mathbf{v} \cdot \mathbf{F}}{v^2}, \tag{3.50}$$

where f_L^α is the "four-force" in the frame moving with velocity $\mathbf{v}$ relative to the particle (i.e., in the lab frame). Evidently, f_L^0 is proportional to the power exerted by the force on the *moving* particle.

In classical mechanics, the time rate of change of the linear momentum $\mathbf{p}$ is considered to be an equivalent representation of the applied force $\mathbf{F}$. Following the same arguments as we have just described, it would seem that a reasonable invariant representation of this statement is the following:

$$\frac{dp^\alpha}{d\tau} = f^\alpha, \tag{3.51}$$

where evidently

$$p^\alpha \equiv m\frac{dx^\alpha}{d\tau}. \tag{3.52}$$

This is itself a four-vector, whose components in the particle's rest frame reduce to $p^0 = mc$, $\mathbf{p} = \mathbf{0}$. The fact that $d\tau = dt/\gamma$ means that in any other frame,

$$p^0 = \gamma mc$$

and

$$\mathbf{p} = \gamma m\mathbf{v}, \tag{3.53}$$

but the contraction of p^α with itself always produces an invariant quantity proportional to the particle's rest mass:

$$p^\alpha p_\alpha = \gamma^2 m^2 c^2 - m^2\gamma^2|\mathbf{v}|^2 = m^2 c^2 = \text{constant}. \tag{3.54}$$

We shall see shortly that p^0 is proportional to the particle's energy. Together, the four components $(p^0, \mathbf{p})$ form what is known as the energy-momentum four-vector. The result $\mathbf{p} = \gamma m\mathbf{v}$ is particularly intriguing, because it suggests that the modification introduced to the momentum by special relativity is due entirely to the inertia, which varies with particle speed. The mass of a particle moving at velocity v relative to an observer who determines its inertia is apparently γm, where $\gamma = [1 - (v/c)^2]^{-1/2}$.

Our interpretation of p^0 as the energy component of p^α is based on what we see as $v \to 0$, where

$$\begin{aligned} \gamma mc^2 &= \frac{mc^2}{\sqrt{1 - (v/c)^2}} \\ &\approx mc^2\left[1 + \frac{1}{2}\left(\frac{v}{c}\right)^2 + \frac{3}{8}\left(\frac{v}{c}\right)^4 + \cdots\right] \\ &\approx mc^2 + \frac{1}{2}mv^2. \end{aligned} \tag{3.55}$$

It looks like $p^0 c$ is just the sum of the classical kinetic energy and another form of energy that does not vanish even when $v = 0$. For obvious reasons, we refer to the latter as the particle's *rest energy*, and to the sum

$$E \equiv \gamma mc^2 \qquad (3.56)$$

as its *relativistic energy*. Clearly,

$$p^0 = E/c, \qquad (3.57)$$

and from (3.53) and (3.56), we obtain one of the most celebrated equations in special relativity:

$$E^2 = |\mathbf{p}|^2 c^2 + m^2 c^4. \qquad (3.58)$$

Though we won't go through the exercise here, any other theory or set of physical laws can, with equal validity, be cast in the language of four-dimensional spacetime, just as we have done for classical mechanics. For the next step in this development, we will generalize the new relativity theory to include the principle of equivalence and relative acceleration between frames.

3.3 GENERAL RELATIVITY

Most of us are familiar with scenes of astronauts floating freely about their orbiting spacestation, bounding effortlessly from one wall to the next in seemingly random directions. One of them may release a pad, which remains stationary where she left it, while another squeezes a tube of juice, whose droplets form a straight-line trajectory toward their final destination, splattering against the opposite wall. All of these are signs that the interior of the spacecraft is free of gravity. On the other hand, the spacestation will not break free of Earth's influence.

Our thinking on this is the modern day version of what Einstein called the "happiest thought of my life"—that an observer who is falling freely experiences no gravitational field. Thanks to Galileo's profound insight, we now understand that everything inside the spacecraft is falling at exactly the same rate, since the acceleration these objects experience due to Earth's pull is completely independent of their inertial mass. Be it a pen, a paper clip, or the flight commander, they all exist within a region where no one can tell if anything else is being pulled or if something is pulling them, since every object is accelerated in tandem with everything else.

The profound implication of this phenomenon is that regardless of what actually causes the pull of gravity, its effect is entirely equivalent to an acceleration throughout a given volume of space. Einstein would explain it in this way: imagine that you are standing inside an elevator out in space, very far from any object that can generate a gravitational field. Now you and your belongings really are in a microgravity environment, but suppose that a rocket engine attached to the bottom of the elevator is turned on and it accelerates your capsule. You feel a force "upward" from the floor of the elevator being accelerated in that direction, but inside, you can't tell if it's your capsule that is being accelerated upward, or if you and your belongings are being pulled downward. The two situations would be entirely equivalent, since in both cases you would infer that everything inside the elevator is accelerating downward relative to the floor.

Einstein called this the *principle of equivalence.* But there's a caveat in the story that is often overlooked. What we have just described is based entirely on the behavior of massive particles, because it is for them that the equivalence of inertial and gravitational masses leads to an acceleration independent of mass. What about light, which, as far as we know, is massless? The principle of equivalence, as Einstein formulated it, is all encompassing, meaning that everything within the observer's frame—including massless particles—accelerate at the same rate. This was one of Einstein's inspired guesses, which turned out to be correct, because there is indeed a way to tell if the effects of gravity are equivalent to those produced by an accelerated frame. If you're standing inside an elevator accelerated upward, a beam of light entering horizontally from one side will appear to curve downward, on the same trajectory that any massive particle would follow moving sideways at light speed. But a source of gravity may or may not have the same effect. However, following Eddington's observation during the 1919 solar eclipse—that the Sun's gravity does bend the path of distant starlight—we now know, of course, that the principle of equivalence does apply to massless as well as massive particles. In other words, the bending of light is not really a prediction of general relativity, but it is—through the assumed principle of equivalence—an essential ingredient of it.

Einstein further hypothesized that since all gravitational fields vanish inside a free-falling frame, special relativity ought to apply to all measurements of distances and times within that frame. And in a leap of faith (or inspiration, or both), he argued that the two postulates of the special theory should apply even in cases where we compare the measurements of an observer in this frame with those of a distant observer, for whom the effects of the gravitational field are negligible. This hypothesis, however, is yet to be confirmed experimentally. That is, we do not yet know for sure that the speed of light is completely invariant from frame to frame even when one of them is accelerated by a strong gravitational field, e.g., near the event horizon of a black hole (see below).

But accepting these assumptions for now, we have at our disposal an elegant and all-encompassing way of merging the theory of gravity with special relativity. Gravity is described by its equivalence to an accelerated frame, and its effects are thereby fully incorporated into the laws of physics via the properties of special relativity. This is the essence of the *general* theory of relativity.

The process of merging gravity with special relativity relies on our ability to express the interval ds in terms of an arbitrary coordinate system. Using the notation $X^\alpha \equiv (X^0, X^1, X^2, X^3)$ to represent the generalized coordinates, we can use the chain rule of differentiation to transform equation (3.41) into the following:

$$
\begin{aligned}
(ds)^2 &= \eta_{\alpha\beta}\, dx^\alpha\, dx^\beta \\
&= \eta_{\alpha\beta} \left(\frac{\partial x^\alpha}{\partial X^\gamma}\, dX^\gamma \right) \left(\frac{\partial x^\beta}{\partial X^\delta}\, dX^\delta \right) \\
&= \left(\eta_{\alpha\beta}\, \frac{\partial x^\alpha}{\partial X^\gamma}\, \frac{\partial x^\beta}{\partial X^\delta} \right) dX^\gamma\, dX^\delta.
\end{aligned} \tag{3.59}
$$

In other words,

$$(ds)^2 = g_{\alpha\beta} \, dX^\alpha \, dX^\beta, \tag{3.60}$$

where the metric tensor is now

$$g_{\alpha\beta} \equiv \eta_{\mu\nu} \frac{\partial x^\mu}{\partial X^\alpha} \frac{\partial x^\nu}{\partial X^\beta}, \tag{3.61}$$

in terms of the inertial Cartesian coordinates x^α.

As a simple application of this formalism, let's see what happens when we mix spatial and time coordinates in ds, as would happen when one of the reference frames is accelerated at a rate a relative to the other. Adopting a Cartesian system $X^\alpha \equiv (cT, X, Y, Z)$, we easily infer that the coordinate transformation is given as

$$T = t$$
$$X = x - \tfrac{1}{2}at^2$$
$$Y = y$$
$$Z = z. \tag{3.62}$$

(Notice that we still have a Newtonian mindset here, since $T = t$.) Looking at the definition of $g_{\alpha\beta}$ in equation (3.61), it is clear that

$$g_{00} = 1 - \tfrac{1}{2}(aT/c)^2 - \tfrac{1}{2}(aT/c)^2 = 1 - (aT/c)^2$$
$$g_{11} = -1$$
$$g_{22} = -1$$
$$g_{33} = -1$$
$$g_{01} = g_{10} = -(aT/c). \tag{3.63}$$

All others are zero. Therefore,

$$(ds)^2 = (1 - a^2T^2/c^2)c^2(dT)^2 - (dX)^2 - (dY)^2 - (dZ)^2 - 2aT \, dX \, dT, \tag{3.64}$$

and the metric tensor now has coefficients

$$g_{\alpha\beta} = \begin{pmatrix} (1 - a^2T^2/c^2) & -aT/c & 0 & 0 \\ -aT/c & -1 & 0 & 0 \\ 0 & 0 & -1 & 0 \\ 0 & 0 & 0 & -1 \end{pmatrix}. \tag{3.65}$$

Not only are the coefficients $g_{\alpha\beta}$ functions of the coordinates, but the metric tensor $g_{\alpha\beta}$ is no longer diagonal—though it must always be symmetric (see equation 3.61).

A second situation, in which a transformation is made from Cartesian coordinates x^α into a rotating system $X^\alpha \equiv (cT, r, \theta, Z)$, has some features in common with the Kerr metric describing the spacetime surrounding a spinning object. Here, the transformation rules are

$$t = T$$
$$x = r \cos(\theta + \Omega T)$$
$$y = r \sin(\theta + \Omega T)$$
$$z = Z. \tag{3.66}$$

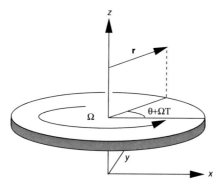

Figure 3.1 Transformation of coordinates from a rotating system (cT, r, θ, Z) to the laboratory Cartesian frame (ct, x, y, z). The accelerated frame is rotating about the z-axis with an angular velocity Ω. The time-dependent angle $\theta + \Omega T$ is measured relative to the x-axis. (Adapted from Melia 2007)

The angular velocity Ω of the accelerated frame is measured relative to the inertial system x^α (see figure 3.1). Leaving the evaluation of the individual coefficients $g_{\alpha\beta}$ as a straightforward exercise for the reader, we note that the corresponding interval here is

$$(ds)^2 = (1 - \Omega^2 r^2/c^2)c^2 (dT)^2 - (dr)^2 - r^2 (d\theta)^2 - (dZ)^2 - 2\Omega r^2 \, d\theta \, dT, \quad (3.67)$$

with

$$g_{\alpha\beta} = \begin{pmatrix} (1 - \Omega^2 r^2/c^2) & 0 & -\Omega r^2 & 0 \\ 0 & -1 & 0 & 0 \\ -\Omega r^2 & 0 & -r^2 & 0 \\ 0 & 0 & 0 & -1 \end{pmatrix}. \quad (3.68)$$

Note, in particular, the appearance of a term proportional to $d\theta \, dT$. If we follow the trajectory of a particle moving radially in the accelerated frame, we still cannot ignore the contribution to ds from an azimuthal motion arising from the angular acceleration. Indeed, we may write the last term in equation (3.67) as $2\Omega r^2 \dot\theta (dT)^2$, allowing us to merge it with the first term, and thereby showing that modifications to this metric arise solely from the nonzero value of $\dot\theta$. This same effect in the Kerr metric will be interpreted as the dragging of spacetime around a spinning compact object.

Accounting for the transformation of intervals between frames in a general relativistic setting follows a very similar procedure as that employed in these examples. The essence of general relativity is the transformation from an inertial frame (usually in free fall) x^α into the accelerated (or laboratory) frame X^α. According to the principal of equivalence, even in the presence of a gravitational field, one can choose a coordinate system—the local inertial frame—within which the laws of nature take the same form as in a Cartesian inertial frame (the CIF) *without* gravitation. The local effects of gravity are then determined by finding the transformation laws for the accelerated coordinate system.

When matter (or energy, since energy and mass are equivalent in special relativity) is present, the gravitational field it produces is described by the metric $g_{\alpha\beta}(x)$, different from the "flat-space" metric $\eta_{\alpha\beta}(x)$. Because this altered $g_{\alpha\beta}(x)$ has an effect on the particle *trajectories*, it is often said that the gravitational field changes the geometry of spacetime.

In an inertial frame x^α, the particle's equation of motion (see equation 3.48) is

$$m\frac{du^\alpha}{d\tau} = f^\alpha, \tag{3.69}$$

where

$$u^\alpha \equiv \frac{dx^\alpha}{d\tau} \tag{3.70}$$

is the four-velocity, and f^α is the four-force. A transformation of this form of Newton's law into another inertial frame obviously leaves it unchanged because that is how special relativity is constructed. But let us now transform it into an accelerated frame, and see how the properties of the $g_{\alpha\beta}(x)$ coefficients bring out the "inertial" forces arising from this acceleration.

According to the transformation law for a four-vector, the force we measure in the noninertial frame is

$$\bar{f}^\delta = \frac{\partial X^\delta}{\partial x^\alpha} f^\alpha$$
$$= m\frac{\partial X^\delta}{\partial x^\alpha}\left(\frac{\partial x^\alpha}{\partial X^\beta}\frac{d^2X^\beta}{d\tau^2} + \frac{\partial^2 x^\alpha}{\partial X^\beta \partial X^\gamma}\frac{dX^\gamma}{d\tau}\frac{dX^\beta}{d\tau}\right)$$
$$= m\frac{d^2X^\delta}{d\tau^2} + m\frac{\partial^2 x^\alpha}{\partial X^\beta \partial X^\gamma}\frac{\partial X^\delta}{\partial x^\alpha}\frac{dX^\gamma}{d\tau}\frac{dX^\beta}{d\tau}. \tag{3.71}$$

So the generalized Newton's law in the accelerated frame of reference is

$$m\frac{d^2X^\alpha}{d\tau^2} = \bar{f}^\alpha - m\Gamma^\alpha{}_{\beta\gamma}\frac{dX^\beta}{d\tau}\frac{dX^\gamma}{d\tau}, \tag{3.72}$$

where the coefficients

$$\Gamma^\alpha{}_{\beta\gamma} = \frac{\partial^2 x^\lambda}{\partial X^\beta \partial X^\gamma}\frac{\partial X^\alpha}{\partial x^\lambda} \tag{3.73}$$

are known as the Christoffel symbols.

The inertial forces appearing in equation (3.72) are easy to evaluate once we know how the free-fall frame coordinates x^α transform into the coordinates X^α in the accelerated frame. For convenience, these Christoffel symbols are often written explicitly in terms of the metric coefficients $g_{\alpha\beta}$:

$$\Gamma^\alpha{}_{\beta\gamma} = \frac{1}{2}g^{\alpha\lambda}\left(\frac{\partial g_{\lambda\beta}}{\partial X^\gamma} + \frac{\partial g_{\lambda\gamma}}{\partial X^\beta} - \frac{\partial g_{\gamma\beta}}{\partial X^\lambda}\right), \tag{3.74}$$

a form that the reader can show is equivalent to (3.73). Note that

$$g^{\alpha\beta}g_{\alpha\gamma} = \delta^\beta_\gamma \tag{3.75}$$

and that $g_{\alpha\beta} = g_{\beta\alpha}$. So $g^{\alpha\beta}$ is the matrix inverse of $g_{\alpha\beta}$. Clearly, $\Gamma^{\alpha}{}_{\beta\gamma} = 0$ in the absence of a gravitational field, since the $g^{\alpha\beta}$ are then all constant coefficients.

To end this section, let us see how this formalism works in practice by applying it to the transformation given in equation (3.62). In that example, the X^{α} frame is accelerated in the positive x-direction, which should therefore represent a gravitational field increasing toward negative values of x. In addition, we'll consider nonrelativistic motion, so that $T \approx \tau$, and $\bar{f}^{1} \approx F^{1}$, the x-component of the Newtonian force. From equation (3.73) we see that

$$\Gamma^{1}{}_{00} = \frac{a}{c^{2}}, \tag{3.76}$$

so that

$$m\frac{d^{2}X^{1}}{d\tau^{2}} \approx m\frac{d^{2}X^{1}}{dT^{2}}$$

$$= \bar{f}^{1} - m\Gamma^{1}{}_{00}\frac{dX^{0}}{d\tau}\frac{dX^{0}}{d\tau}$$

$$= F^{1} - ma. \tag{3.77}$$

This equation displays the correct behavior, in that the "inertial" effect is equivalent to a gravitational force, $-ma$, pointing in the negative x direction.

3.4 STATIC SPACETIMES

We can sample the geometry of spacetime by following the paths taken by particles moving solely under the influence of gravity. These paths, known as *geodesics*, are described by

$$\frac{d^{2}X^{\alpha}}{d\tau^{2}} + \Gamma^{\alpha}{}_{\beta\gamma}\frac{dX^{\beta}}{d\tau}\frac{dX^{\gamma}}{d\tau} = 0, \tag{3.78}$$

which follow from the equation of motion (3.72) in the absence of any other force. Flat spacetime has no curvature because $g_{\alpha\beta} = \eta_{\alpha\beta}$, for which $\Gamma^{\alpha}{}_{\beta\gamma} = 0$. So the particle paths are all straight lines. In the next two sections, we examine the curvature arising from two special metrics, describing the structure of spacetime surrounding a source of gravity.

3.4.1 The Schwarzschild Metric

A source of gravity creates gradients in $g_{\alpha\beta}$, as functions of the coordinates, that translate into nonzero values of $\Gamma^{\alpha}{}_{\beta\gamma}$. Einstein's field equations of general relativity express these gradients in terms of the local source density, in the same way that the Poisson equation in electrodynamics accounts for the gradients in the electric potential due to the presence of local charges. The general relativistic field equations are somewhat more elaborate than Poisson's equation, however, because of two principal features that were not present in Newton's formulation of gravity. We've already touched briefly on the first: since energy and mass are equivalent, one must include the contribution of self-energy to the gradients in $g_{\alpha\beta}$, which makes the differential equations nonlinear. Second, the influence of gravity travels at a finite speed, presumably the speed of light, an effect that was also excluded from the action-at-a-distance approach of the classical theory.

Still, several elegant analytic solutions have been found to Einstein's field equations, the first by Karl Schwarzschild (1873–1916) in 1916, soon after general relativity was founded. The Schwarzschild metric represents the simplest spherically symmetric vacuum description of the spacetime surrounding a mass M:

$$(ds)^2 = B(r)c^2(dT)^2 - B^{-1}(r)(dr)^2 - r^2(d\theta)^2 - r^2 \sin^2\theta(d\phi)^2, \qquad (3.79)$$

where

$$B(r) \equiv \left(1 - \frac{2GM}{c^2 r}\right). \qquad (3.80)$$

The gravitational effects due to the central mass are now accounted for by the inertial terms arising from the metric coefficients

$$g_{\alpha\beta} = \begin{pmatrix} B(r) & 0 & 0 & 0 \\ 0 & -B^{-1}(r) & 0 & 0 \\ 0 & 0 & -r^2 & 0 \\ 0 & 0 & 0 & -r^2\sin^2\theta \end{pmatrix}. \qquad (3.81)$$

This solution and metric describe the relativistic gravitational field everywhere outside a gravitating, nonrotating body, such as a star or a black hole.

Many of the characteristics we infer for high-energy sources are straightforward consequences of the Schwarzschild solution. For example, the passage of time dT at some fixed radius r (and angles θ and ϕ) is related to the proper time $d\tau$ (measured in the inertial, or free-falling, frame) via the interval in equation (3.79):

$$ds = c\,d\tau = \left(1 - \frac{2GM}{c^2 r}\right)^{1/2} c\,dT. \qquad (3.82)$$

Evidently,

$$d\tau = \left(1 - \frac{2GM}{c^2 r}\right)^{1/2} dT, \qquad (3.83)$$

so that clocks run slow in gravitational fields, an effect that produces both a time dilation and a gravitational redshift in radiation emitted near a gravitating body.

This effect is actually not as mysterious as it sounds; its roots may be traced to the dilation of time associated with a Lorentz transformation in special relativity. The distortions to intervals of time and distance are entirely dependent on the relative velocity of two different observers (see equation 3.10). With this as context, one can understand why a relative acceleration causes clocks to run more slowly by thinking about the meaning of time and its dependence on change.

Time elapses when something changes. To measure an interval of time on the clock, the hand must turn across its face. Regardless of how fast the accelerated frame is moving relative to us, its speed will have increased even further during the process of measuring the time interval. Thus, there is no way of avoiding the fact that the other frame's velocity relative to a CIF at the end of the time interval is different compared to its value at the beginning. No matter how small we make the time interval, the starting and ending velocities are always different—that's the nature of acceleration. According to special relativity, there should be an additional time dilation associated with this increase in speed. The effect is greatly magnified when the acceleration is so great that even a tiny interval can bring the magnitude of the falling frame's velocity close to the speed of light. The ensuing time dilation can then appear to freeze the action completely. Thus, as long as we accept the fact that time intervals are altered during a transformation from one frame to another (but always in such a way as to preserve the constancy of the speed of light), we must also accept the conclusion that the acceleration of one frame relative to another itself incurs an additional time dilation. Gravitational fields, therefore, slow down the passage of time as viewed from distant vantage points, and the retardation effect is greater the stronger the field.

In concordance with the time dilation implied by equation (3.83), the frequency of radiation emitted in a gravitational field must also change when viewed by a distant observer, producing what is commonly known as a *gravitational* redshift. Let a source at radius r_e emit radiation with frequency ν_e, as measured in the proper frame (one that is free of any gravitational effects). In the laboratory (or accelerated) frame, we instead measure a frequency

$$\bar{\nu} = \nu_e \left.\frac{d\tau}{dT}\right|_{r_e} = \nu_e \left(1 - \frac{2GM}{c^2 r_e}\right)^{1/2}. \tag{3.84}$$

Correspondingly, a proper-frame observer at r_o will infer a frequency

$$\nu_o = \bar{\nu}\left.\frac{dT}{d\tau}\right|_{r_o}, \tag{3.85}$$

and the overall change in frequency between the points of emission and detection is therefore given by

$$\left(\frac{\nu_o}{\nu_e}\right) = \left(1 - \frac{2GM}{c^2 r_e}\right)^{1/2}\left(1 - \frac{2GM}{c^2 r_o}\right)^{-1/2}. \tag{3.86}$$

Since typically the observer views the high-energy source from a great distance (so that $r_o \gg r_e$),

$$\left(\frac{\nu_\infty}{\nu_e}\right) = \left(1 - \frac{2GM}{c^2 r_e}\right)^{1/2}. \tag{3.87}$$

The third consequence we will investigate has to do with a limitation on r, known as the *static* limit, that also emerges from equation (3.79). The question here is how close can one get to the source of gravity and still be able to maintain a fixed position in space. This condition translates into the statement that for the static observer, only the passage of time should contribute to the interval ds. Clearly, since $(ds)^2 > 0$ or, equivalently, since $(d\tau)^2 > 0$, the observer can remain static ($dr = d\theta = d\phi = 0$) only if $r > 2GM/c^2$. For smaller radii, no force can maintain a constant value of r since a positive value of $(ds)^2$ would then force $dr \neq 0$, and the observer would be pulled inexorably toward the center at $r = 0$. This critical radius,

$$r_S \equiv \frac{2GM}{c^2},$$

(3.88)

is named in honor of Karl Schwarzschild; it specifies the location of the black hole's so-called *event horizon*.

Finally, before we move on to the second analytic solution to Einstein's field equations, we'll take a moment to consider how the Schwarzschild metric affects particle trajectories near the source of gravity. For motion in the $\theta = \pi/2$ plane, the geodesic equations (3.78) yield three separate expressions for r, ϕ, and T. Taking the time to evaluate the Christoffel symbols for the Schwarzschild metric (3.81), the reader can show that the X^3 (or ϕ) equation reduces to

$$r^2 \frac{d\phi}{d\tau} = \text{constant} \equiv \tilde{l},$$

(3.89)

where the integration constant $\tilde{l}$ is the conserved angular momentum per unit mass. The X^0 (or T) equation yields

$$\left(1 - \frac{2GM}{c^2 r}\right) \frac{dT}{d\tau} = \text{constant} \equiv \frac{\tilde{E}}{c^2}.$$

(3.90)

The constant $\tilde{E}$ is the conserved "energy-at-infinity," including the rest mass energy of the particle, per unit mass. The X^1 (or r) equation may be written in terms of the above two constants of the motion, reducing it to the form:

$$c^2 \left(\frac{dr}{d\tau}\right)^2 = \tilde{E}^2 - V_{\text{eff}}(r).$$

(3.91)

We have chosen to write the right-hand side of (3.91) in this fashion to elicit a familiarity with the orbit equations of classical mechanics. The effective potential,

$$V_{\text{eff}}(r) = \left(1 - \frac{2GM}{c^2 r}\right)\left(1 + \frac{\tilde{l}^2}{c^2 r^2}\right)(c^2)^2,$$

(3.92)

is not unlike its classical counterpart,

$$V_N(r) \equiv -\frac{GM}{r} + \frac{\tilde{l}^2}{2r^2},$$

(3.93)

which groups the potential energy together with the specific kinetic energy associated with the angular motion. However, V_{eff} also introduces a new effect whose origin is entirely relativistic in nature.

The potential in a Schwarzschild metric has the additional term $-2GM\tilde{l}^2/r^3$, which can overwhelm even the angular momentum barrier, $c^2\tilde{l}^2/r^2$, at a finite radius. Figure 3.2 shows a plot of $V_{\text{eff}}(r)^{1/2}/c^2$ as a function of $c^2 r/2GM$ for various

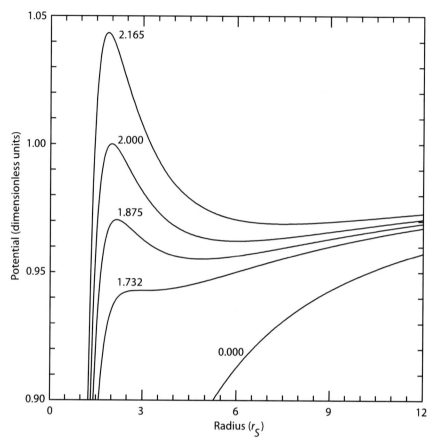

Figure 3.2 A plot of the effective potential, $V_{\text{eff}}(r)^{1/2}/c^2$, versus the radius, $c^2 r/2GM$, for various values of the specific angular momentum, $\tilde{l}c/2GM$, ranging from 0 (i.e., no angular momentum), to 2.165 (representing a high angular momentum barrier). (Adapted from Melia 2007)

values of the specific angular momentum parameter $\tilde{l}c/2GM$, ranging from 0 (i.e., no angular momentum), to 2.165 (representing a high angular momentum barrier).

Given the form of $V_{\text{eff}}(r)$, particles on closed orbits therefore behave quite differently in a Schwarzschild metric compared with their counterparts in Newtonian gravity. First, since the potential is not strictly $1/r$, as in a classical system, no closed orbits are permitted. Instead, orbits that would otherwise have been closed, now have a semi-major axis that *precesses* about the focal point of the ellipse—an effect that has received elegant confirmation with the advancing perihelion of Mercury. Second, particles may be bound, but still possess sufficient energy to overcome the potential hump at small radii (see figure 3.2) and fall into the black hole. There is no analog for these orbits in Newtonian mechanics.

Remarkably, circular orbits are still permitted, because precession for them is not evident. A particle orbit achieves zero eccentricity when the following conditions are met:

$$\frac{\partial V_{\text{eff}}}{\partial r} = 0 \quad \text{and} \quad \frac{dr}{d\tau} = 0, \tag{3.94}$$

which map into two very useful conditions on $\tilde{l}$ and $\tilde{E}$ via equations (3.91) and (3.92):

$$\tilde{l}^2 = \frac{GMr^2}{r - 3GM/c^2} \tag{3.95}$$

and

$$\tilde{E} = \frac{c^2(r - 2GM/c^2)^2}{r(r - 3GM/c^2)}. \tag{3.96}$$

Further analysis shows that circular orbits are not permitted for radii below $3GM/c^2$, since both $\tilde{l}$ and $\tilde{E}$ become excessively large as $r \to 3GM/c^2$. Given that $\tilde{l}$ and $\tilde{E}$ are calculated per unit mass, we interpret this to mean that a circular orbit at this critical radius corresponds to a (massless) photon trajectory.

However, $\tilde{l}^2$ also diverges as $r \to \infty$, meaning that there exists a minimum value of $\tilde{l}^2$ for which a circular orbit may be realized, and this occurs at precisely $r = 6GM/c^2 = 3r_S$. This orbit is called the *last* (or marginally) stable circular orbit. An effective potential with $\tilde{l}$ smaller than this has no minimum, while one with greater angular momentum has a minimum at a larger radius. A particle falling below this radius (due to the dissipative loss of angular momentum, say in an accretion disk) finds itself in progressively less stable orbits and spirals toward the black hole. It no longer has sufficient angular momentum to maintain a circular orbit below $3r_S$.

The last stable circular orbit also provides us with a useful characteristic energy. For a particle on such a trajectory at $r = 3r_S$, $\tilde{E}^2 = \frac{8}{9}c^4$. The amount of energy it loses in reaching this point is therefore $c^2 - \tilde{E} \approx 0.057c^2$ per unit mass. High-energy astrophysicists therefore adopt a value of 6 percent to represent the efficiency with which *disk* accretion converts rest-mass energy into radiative luminosity in black-hole systems possessing large disks.

3.4.2 The Kerr Metric

After Schwarzschild obtained the first (spherically symmetric) solution to Einstein's field equations, it took another half-century before Roy Kerr finally managed to find a description of intervals and time surrounding a *rotating* black hole.[5] So beautiful and surprising was this solution, that the Nobel laureate S. Chandrasekhar described it as the "most shattering experience" of his life, to realize that "an exact solution of Einstein's equations of general relativity" could be found to represent "untold numbers of massive black holes that populate the universe."[6]

The Kerr metric finds greater relevance in nature than Schwarzschild's solution because most objects in the universe spin at least a little—it is virtually impossible to assemble an aggregate of matter without any angular momentum at all. Indeed,

---

[5] Kerr first reported the discovery of this solution in the *Physical Review Letters* in 1963.
[6] See Chandrasekhar (1987).

conservation of angular momentum can produce furiously spinning objects when their progenitors collapse into ultracompact volumes.

Schwarzschild's metric cannot handle rapidly rotating objects (and is only an approximation for slowly spinning ones) because a rotating source of gravity impacts the spacetime around it in unexpected and challenging ways. The interval ds, for example, is no longer given by equation (3.79) for the same reason that a mother watching her child jumping onto a merry-go-round cannot measure the total distance he covers by simply counting his steps. Though she may still be able to monitor the passage of time T by tracking the ticks on her watch, the actual distance traveled by the jumper is now augmented by the merry-go-round's lateral motion, which carries him along for the ride.

In Kerr's metric, the spacetime itself swirls around the spinning object with a speed decreasing with distance from the center. This "frame dragging" forces everything within that spacetime into co-rotation with the source of gravity, even if within that frame these objects are completely stationary. If correct, this interpretation of Kerr's solution implies that even if we could somehow place a particle with zero angular momentum in the vicinity of a spinning black hole, that particle would still appear to be moving laterally from the perspective of a distant observer.

The Kerr solution is completely specified by the central object's mass, M, and its angular momentum, $\mathbf{J}$. It is also possible to find solutions for a mass that contains charge, Q. However, in nature it is difficult to maintain nonzero values of Q over distances associated with the force of gravity, which is much weaker than the electromagnetic force. The spacetime interval for a rotating compact object is given by the expression

$$(ds)^2 = A(r, \theta)c^2(dT)^2 + C(r, \theta)c\, dT\, d\phi - (\Sigma/\Delta)(dr)^2$$
$$- \Sigma(d\theta)^2 - D(r, \theta)\sin^2\theta(d\phi)^2, \tag{3.97}$$

where

$$A(r, \theta) \equiv \left(1 - \frac{2GMr}{c^2\Sigma}\right), \tag{3.98}$$

$$C(r, \theta) \equiv \frac{4aGMr\sin^2\theta}{c^2\Sigma}, \tag{3.99}$$

$$D(r, \theta) \equiv \left(r^2 + a^2 + \frac{2GMra^2\sin^2\theta}{c^2\Sigma}\right), \tag{3.100}$$

with

$$a \equiv \frac{J}{cM}, \tag{3.101}$$

$$\Delta \equiv r^2 - \frac{2GM}{c^2}r + a^2, \tag{3.102}$$

and

$$\Sigma \equiv r^2 + a^2\cos^2\theta. \tag{3.103}$$

The metric coefficients are therefore

$$g_{\alpha\beta} = \begin{pmatrix} A(r) & 0 & 0 & C(r,\theta)/2 \\ 0 & -\Sigma/\Delta & 0 & 0 \\ 0 & 0 & -\Sigma & 0 \\ C(r,\theta)/2 & 0 & 0 & -D(r,\theta)\sin^2\theta \end{pmatrix}. \quad (3.104)$$

We need to point out here that equation (3.97) is written in terms of so-called Boyer-Lindquist coordinates,[7] which reduce to the spherical coordinates of equation (3.79) only when the spin parameter a goes to zero. Specifically, Boyer-Lindquist coordinates are those for which the Cartesian coordinates (x, y, z) may be written

$$x = (r^2 + a^2)^{1/2} \sin\theta \cos\phi,$$
$$y = (r^2 + a^2)^{1/2} \sin\theta \sin\phi,$$
$$z = r\cos\theta. \quad (3.105)$$

One can trivially show that when the spin parameter a is zero, the Kerr solution (equation 3.97) reduces to the spherically symmetric Schwarzschild metric. When a is nonzero, however, two new effects emerge. First, the metric coefficients $g_{\alpha\beta}$ depend on θ, so rotation has broken the spherical symmetry, though we still have azimuthal symmetry. Second, a transformation into a rotating frame inevitably mixes the angle of rotation (here ϕ) with time. So the metric tensor now has a nonzero off-diagonal term, $g_{T\phi}$, analogous to the $g_{T\theta}$ term we considered earlier in (3.68).

Let us first consider what happens along the polar axis, $\theta = 0$, where

$$(ds)^2 = \left[1 - \frac{2GMr}{c^2(r^2+a^2)}\right]c^2(dT)^2 - \frac{r^2+a^2}{\Delta}(dr)^2 - (r^2+a^2)(d\theta)^2. \quad (3.106)$$

Along this direction, the corresponding metric tensor is diagonal because the swirling action of the spinning body has no impact on the spacetime. As we discussed earlier for the Schwarzschild solution, *static* observers are permitted only when $g_{TT} > 0$, i.e., only for speeds smaller than c. The limiting radius for this *timelike* region is therefore given by the condition

$$2GMr < c^2(r^2 + a^2), \quad (3.107)$$

or

$$r > r_+ \equiv \frac{GM}{c^2} + \left(\left[\frac{GM}{c^2}\right]^2 - a^2\right)^{1/2}, \quad (3.108)$$

the negative root corresponding instead to a region inside the limiting radius, which is beyond the derived static limit. The radius r_+ represents the static limit along $\theta = 0$. Note that when $a \to 0$,

$$r_+ \to \frac{2GM}{c^2} = r_S, \quad (3.109)$$

---

[7] See Boyer and Lindquist (1967).

as expected. Interestingly, though, an observer can approach a compact object with $a > 0$ to distances along the polar axis smaller than the Schwarzschild radius, and still be able to maintain causal contact with the rest of the universe. But values of a bigger than GM/c^2 make no sense in the definition of r_+, and so it appears that a "maximally" rotating black hole corresponds to the condition $a = r_S/2$.

In the equatorial plane, we have instead $\theta = \pi/2$, and

$$(ds)^2 = \left(1 - \frac{2GM}{c^2r}\right)c^2(dT)^2 + \frac{4aGM}{c^2r}c\, dT\, d\phi$$

$$- \frac{r^2}{r^2 - 2GMr/c^2 + a^2}(dr)^2 - r^2(d\theta)^2$$

$$- \left[r^2 + a^2 + \frac{2GMa^2}{c^2r}\right](d\phi)^2. \qquad (3.110)$$

So the static region with timelike intervals (i.e., $g_{TT} > 0$) is here Schwarzschild-like:

$$r > r_0 = r_S \equiv \frac{2GM}{c^2}. \qquad (3.111)$$

For any arbitrary angle θ, the static limit (from equation 3.97) is evidently

$$r_0(\theta) = \frac{GM}{c^2} + \left(\left[\frac{GM}{c^2}\right]^2 - a^2\cos^2\theta\right)^{1/2}. \qquad (3.112)$$

However, the second significant difference between the Kerr and Schwarzschild solutions—the mixing of ϕ with T—leads to the surprising conclusion that the static limit in the case that $a \neq 0$ is actually *not* the event horizon. The interval ds can still be greater than zero (i.e., timelike) even when $r < r_0(\theta)$, due to the presence of the off-diagonal term $dT\, d\phi$ in equation (3.97). We encountered something analogous to this before, in the context of the rotating frame solution in (3.67), where there the $dT\, d\theta$ term produced a contribution to ds from the lateral motion of the frame when we made the conversion $dT\, d\theta \rightarrow \dot\theta(dT)^2$.

With a similar conversion here, we see that observers are permitted to exist at a fixed radius r inside $r_0(\theta)$ if the positive contribution to $(ds)^2$ from the term

$$2g_{T\phi}c\, dT\, d\phi = 2g_{T\phi}c(dT)^2\frac{d\phi}{dT} \qquad (3.113)$$

is large enough to offset the negative value of $g_{TT}c^2(dT)^2$. Thus, observers may exist at fixed r inside $r_0(\theta)$ only if they're rotating along with the black hole at a rate $\dot\phi > 0$. This phenomenon, as we have seen, is known as *frame dragging*, and gives rise to another critical radius called the *stationary* limit.

One can see exactly how this arises by rewriting equation (3.97) in the following form:

$$(ds)^2 = c^2 (dT)^2 \left[-\left\{ \left(\frac{r^3 + a^2 r + 2GMa^2/c^2}{r} \right)^{1/2} \frac{d\phi}{c \, dT} \right. \right.$$

$$\left. -\frac{2aGM/c^2}{r^{1/2} \left(r^3 + a^2 r + 2GMa^2/c^2 \right)^{1/2}} \right\}^2$$

$$\left. +\frac{4a^2 (GM/c^2)^2}{r \left(r^3 + a^2 r + 2GMa^2/c^2 \right)} + \left(1 - \frac{2GM}{c^2 r} \right) \right]$$

$$-\frac{r^2}{r^2 - 2GMr/c^2 + a^2} (dr)^2 + r^2 (d\theta)^2. \qquad (3.114)$$

Evidently, $(ds)^2 > 0$ for special values of $\dot{\phi}$, even when $dr = d\theta = 0$. The extremal condition on this effect occurs when the factor inside the curly brackets of this equation reaches a minimum:

$$\left(\frac{d\phi}{dT} \right)_0 = \frac{2aGM/c^2}{r^3 + a^2 r + 2GMa^2/c^2}, \qquad (3.115)$$

which now produces the *stationary* limit, corresponding to the radius at which the coefficient multiplying $(dT)^2$ goes to zero:

$$r = r_+ \equiv \frac{GM}{c^2} + \left(\left[\frac{GM}{c^2} \right]^2 - a^2 \right)^{1/2} \qquad \text{for all } \theta. \qquad (3.116)$$

Since $(d\phi/dT)_0$ is the minimum rotation rate permitted in this region, we interpret it to be the angular velocity forced on a zero angular momentum object by the spin of the black hole.

The radius r_+ is the true event horizon for a Kerr black hole because inside it neither static nor stationary observers are allowed, regardless of the value of $\dot{\phi}$. Stationary observers are permitted within the *ergosphere*, the region bounded by r_+ from below and $r_0(\theta)$ from above, but only if they have the requisite $\dot{\phi}$. Static observers are permitted anywhere outside $r_0(\theta)$. So the static horizon is a flattened sphere, whose semi-minor axis lies parallel to $\mathbf{J}$, whereas the stationary—or the actual event—horizon is a sphere with radius r_+. This horizon matches that of the Schwarzschild metric when the spin parameter, a, is zero, but is otherwise always smaller than r_S. And as we noted before, these relations make sense only so along as $a \leq GM/c^2$.

As far as particle orbits are concerned, there is one additional significant feature introduced by the θ-dependence of the frame dragging rate. Particles on orbits off the equatorial plane experience a restoring force due to the poloidal gradient of gravity, which induces a precession of the particle trajectory around the angular momentum axis of the spinning compact object. In active galactic nuclei, this precession may be the reason why jets are so stable over millions of years.

And we shall see later in this book (see, e.g., section 12.2.1) that the differences between the Kerr and Schwarzschild solutions manifest themselves rather prominently in the high-energy spectra of compact objects. Orbiting matter can radiate to infinity from a region much closer to the central object when the latter is spinning, producing both a higher UV and X-ray flux, *and* photons of higher energy.

SUGGESTED READING

For a more comprehensive introduction to the theories of special and general relativity, the reader is encouraged to read at least the first two chapters of Weinberg (1972).

The importance of early 20th-century electrodynamics in motivating Einstein's theory of special relativity is discussed, e.g., in Melia (2001a).

Einstein's theory of relativity rests on the simple, yet profound, assumption that the speed of light is constant everywhere, including regions of superstrong gravity, e.g., across the event horizon of a black hole. For an early critical review of the experimental basis for this assumption, known as the second postulate of special relativity, see Fox (1965, 1967).

The Mössbauer effect was used in one of the first experiments to verify the transverse shift in frequency predicted by the relativistic Doppler shift formula. See Hay et al. (1960).

The effort that led to the discovery of the Kerr metric for a spinning compact object stands as one of the most remarkable achievements in 20th-century physics. A short history of this work, based on the eyewitness account of Roy Kerr himself, appears in Melia (2009).

Chapter Four

Particle Acceleration

The language of relativity will serve us well as we now seek to understand how particles in high-energy sources accelerate to Lorentz factors required for them to emit X-ray and γ-ray photons. In chapter 13, we will learn firsthand that relatively large structures, even as big as molecular clouds—or even bigger, in the case of galaxy clusters—can contribute significantly to the ambient high-energy intensity. Still, the overwhelming majority of energetic particles seem to be produced in and around dense objects, such as white dwarfs, neutron stars, and black holes. Even in the case of an acceleration site fed directly by an AGN jet (see section 12.3), the ultimate source of energy is the central engine—a supermassive black hole.

Thus, it is quite evident that most high-energy phenomena are associated in one way or another with compact sources, in which mass–energy is squeezed into very small volumes. As we alluded to earlier in this book (see section 1.2), the gravitational acceleration of matter is therefore almost always the primary physical process that initiates the high-energy activity. Of course, particles falling into a potential well possess angular momentum and an understanding of the accretion physics cannot be divorced from a consideration of the disk that inevitably forms under these conditions. We shall devote two entire chapters (7 and 8) to this important topic, so the purpose of the present discussion is not necessarily to assess the relevance of accretion-disk physics to the question of particle acceleration, but rather to compare the optimal efficiency of various fundamental processes in transferring energy from a field—be it gravitational or electromagnetic—to the particles.

4.1 GRAVITY

A particle falling from infinity onto an object of mass M acquires a velocity

$$v(r) = \left(\frac{2GM}{r} \right)^{1/2}. \tag{4.1}$$

Thus, the maximum velocity we can expect from direct gravitational acceleration depends on the compactness M/R of the gravitating source.

White dwarfs are stars whose main pressure support comes from electron degeneracy, which we discuss at greater depth later in this section. Though their mass can be as great as that of the Sun, they are typically about 100 times smaller in radius, with a size that depends on composition.[1] Thus, for a white dwarf, the maximum

---

[1] There are many good reference books covering this topic, including (the original) Chandrasekhar (1939) and Schwarzschild (1958).

velocity attainable by a particle reaching its surface solely under the influence of gravity is $v_{max}(wd) \approx 7.4 \times 10^8$ cm s^{-1}, corresponding to a maximum Lorentz factor $\gamma_{max}(wd) \approx 1.0003$. Clearly, gravity does not produce relativistic particles in white dwarf systems.

We can also estimate the maximum gravitational redshift one might expect for an emission line produced at the white dwarf's surface, with mass $M_{wd} \approx 1\ M_\odot$ and radius $R_{wd} \approx 4.9 \times 10^8$ cm. Defining redshift z as

$$1 + z \equiv \frac{\nu_e}{\nu_o} \qquad (4.2)$$

(see equation 3.87), we infer that

$$z = \left(1 - \frac{2GM}{c^2 r_e}\right)^{-1/2} - 1. \qquad (4.3)$$

When the Schwarzschild radius $r_S \equiv 2GM/c^2$ (see section 3.4.1) is small compared to r_e, we can simplify equation (4.3) with a binomial expansion, keeping only the leading order term

$$z \approx \frac{GM}{c^2 r_e} = \frac{1}{2}\frac{r_S}{r_e}. \qquad (4.4)$$

Thus, $z_{max}(wd) \approx 0.0003$, which is extremely difficult to measure. For example, the Extreme Ultra Violet Explorer (EUVE), a spectroscopic and photometric satellite operating in the 70- to 760-Å band,[2] flew between 1992 and 2001 (see figure 1.13) carrying grazing incidence diffraction gratings to achieve an energy resolution $E/\Delta E$ of about 300. Even with this capability, though, the smallest redshift EUVE could measure was ~ 0.003, ten times bigger than the maximum expected for a white dwarf.

Like white dwarfs, neutron stars are objects whose internal pressure support derives not from the randomly directed momentum of particles moving with kinetic energy $\left(\frac{3}{2}\right)kT$ (where T is the gas temperature), but rather from the Pauli exclusion principle, which prevents identical spin-half particles from occupying the same quantum state. The difference, however, is that this effect is associated with electrons in the former, and neutrons in the latter, which, by virtue of their different masses, result in stellar configurations with vastly different radii.

A discussion of the origin of the Pauli exclusion principle is beyond the scope of this book,[3] but its impact on any aggregate of matter is that the wavefunctions of identical spin-half particles (fermions) will cancel each other out once they occupy the same energy state since they are quantum mechanically indistinguishable. Of course, their intrinsic spins may be aligned in either of two projections when a particular direction is defined (e.g., by a magnetic field), so to be more precise, we should say that there are two fermions per energy state since their different spin alignments permit us to distinguish a pair (but only a pair) of otherwise identical particles.

---

[2]Recall from our discussion in section 1.2 that a white dwarf emits predominantly at UV and "soft" X-ray energies.

[3]The interested reader may find an introductory explanation of the Pauli exclusion principle based on an important, though very surprising, topological property of three-dimensional space in Melia (2003).

Lacking thermal pressure support, a star at the end of its life collapses under gravity and compresses to such a small volume (Earth-sized, in the case of a white dwarf) that electrons fill most of the available energy states below a critical value, designated the Fermi energy; the star is said to be degenerate when this condition is realized. The energy spacing between the levels is greater the more compact the star becomes, and since the overall energy budget is fixed, there is a limit to how much the star can be squeezed because the last few leptons have only a limited supply of energy they can use to occupy the highest levels.

This peculiarity leads to a phenomenon in which the more massive the degenerate star is, the more compact it becomes, since the released gravitational energy is transferred to the constituent particles, which can then occupy progressively higher energy levels. But at the so-called Chandrasekhar limit ($\approx 1.44\ M_\odot$), it actually costs less energy for the electron to fuse with a proton and form a neutron (via inverse beta decay, $e^- + p \rightarrow n + \nu_e$) than to keep climbing the Fermi ladder. So above this mass, the dying star reinitiates its collapse while virtually all of the electrons and protons turn predominantly into a sea of neutrons. This process stops when neutron degeneracy takes over and reconfigures the star into a size where the spin-half neutrons are now the particles filling all of the available energy levels, and the star can shrink no farther.

One may understand heuristically why a neutron star must be much smaller than a white dwarf of the same mass, by noting that (at least classically) a particle's momentum scales as $E^2/2m$, in terms of its energy E and mass m. So for the same energy, the neutron's momentum p_n is roughly proportional to $(m_e/m_n)p_e$. The de Broglie wavelengths of the neutron and electron therefore scale according to $\lambda_n/\lambda_e = p_e/p_n \sim m_n/m_e$. But for a given linear size R of the system, the density of particle states goes as $R/\Delta k$, where $|\Delta k| = \Delta\lambda/\lambda^2$ is the spacing in wavenumber $k = 2\pi/\lambda$. So since $\Delta k_e/\Delta k_n \sim m_n/m_e$ (assuming $\Delta\lambda \sim \lambda$), we gather that the electrons fuse with protons to form a sea of neutrons filling a volume with linear dimension $\sim m_e/m_n$ times smaller than that of the white dwarf.

Thus, whereas a white dwarf ($\sim 4.9 \times 10^8$ cm) is about the size of Earth, a $\sim 1\ M_\odot$ neutron star is typically as big as a large city (~ 10 km); more massive neutron stars can be even smaller. In this case, $v_{max}(ns) \approx c/2$, and $\gamma_{max}(ns) \approx 1.19$. From a relativistic perspective, this is starting to get interesting, and as we saw in section 1.2, the surface of an accreting neutron star glows in X-rays once the liberated gravitational energy is thermalized. But $\gamma_{max}(ns)$ is still far too small for the particles to radiate hard X-rays (~ 5–10 keV) and γ-rays.

The gravitational redshift on the surface of a neutron star with mass $1.44\ M_\odot$ is $z_{max}(ns) \approx 0.2$, an easily measurable quantity. For example, in the low-mass X-ray binary EXO 0748–676 (see section 11.1 and figure 11.1), X-ray burst spectra often contain strong absorption lines identified with iron and oxygen transitions. All are shifted by a fractional amount ~ 0.35, which has been interpreted as a gravitational redshift on the surface of a $\sim 2\ M_\odot$ neutron star.[4] (Remember that degenerate stars get smaller as their mass increases, so z is not just simply proportional to M.)

---

[4]These results are based on XMM-*Newton* observations (see plate 2) carried out during its ~ 500-ksec Commissioning and Calibration phase soon after deployment in 2000. See Cottam et al. (2002).

It is worth noting here that a typical neutron star has a radius $R_{ns} \sim 3r_S$; these objects therefore hover barely above the black-hole precipice. For an actual black hole, the maximum velocity attainable by matter falling inward is c, this being the reason why the reverse process—the escape of matter out through the event horizon—is forbidden by the postulates of special relativity. Thus, $v_{max}(bh) = c$, and $\gamma_{max}(bh) \gg 1$. Correspondingly, the gravitational redshift (from equation 4.3) becomes infinite. This increase in z more than offsets the increase in γ as $r_e \rightarrow r_S$, so very little high-energy emission is expected from regions near the event horizon itself.

4.2 ELECTROMAGNETIC FIELDS

The fact that we see γ-rays with energies $\epsilon_\gamma \sim 30 \, \text{GeV}$ from many AGNs and quasars (see plate 10 and figure 12.15) suggests that the particles producing these photons have a Lorentz factor $\gamma \sim 30 \, \text{GeV}/mc^2$ (equation 3.56), which for electrons is $\sim 6 \times 10^4$. We will learn in chapter 12, particularly through plate 25 and figure 12.17, that the superluminal motion observed in many of these sources provides additional evidence that γ should be much larger than one.

The Lorentz factor may be even higher in pulsars—rapidly spinning, magnetized neutron stars (section 9.1)—reaching values exceeding $\sim 10^7$. It is clear, therefore, that nongravitational acceleration schemes must play a role in energizing the particles in many high-energy sources. The most common of these is electromagnetic acceleration, via at least two methods of energy transfer.

A key ingredient in this discussion is the role played by the magnetic field $\mathbf{B}$. Several different mechanisms may contribute to the acceleration, depending on the field distribution, because charges execute circular motion perpendicular to $\mathbf{B}$ (due to the Lorentz $\mathbf{v} \times \mathbf{B}$ force), and have an additional component parallel to $\mathbf{B}$. When $\mathbf{B}$ is turbulent or random, the principal method of acceleration is the Fermi mechanism (section 4.3). In this process, disordered bundles of magnetic flux act as mirrors to bounce particles back and forth, increasing their energy with every collision. For a well-organized field, however, a more direct acceleration mechanism is based on the idea that a component of $\mathbf{E}$ may be generated parallel to $\mathbf{B}$, where the particle motion is unrestricted. This is the situation we will consider first.

To understand how a magnetic field disturbance energizes the charges, let us for the moment return to some of the basic ideas of magnetohydrodynamics (MHD). It will help us as we go through this physical argument, to have a firm grasp of what a "field" is, and what the empirically derived equations of electrodynamics represent. To begin with, a field is the force experienced by a test particle with unit charge. One of the central ideas of any field theory is that a "source" produces at its location an outflux or influx of some influence that affects another charge. We represent this influence with field "lines," whose density and direction define the force.

It is convenient to define the *divergence* of a vector field (let's say $\mathbf{E}$) to describe how this influence spreads into the neighboring volume. Formally, div $\mathbf{E}$ is the net

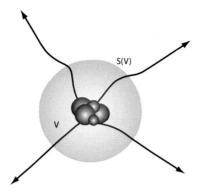

Figure 4.1 The divergence of a vector field is defined as the net outflux per unit volume V through the surface $S(V)$ enclosing that volume.

outflux of $\mathbf{E}$ through the surface area S per unit (enclosed) volume V:

$$\operatorname{div} \mathbf{E} \equiv \lim_{V \to 0} \left\{ \frac{1}{V} \oint_{S(V)} \mathbf{E} \cdot d\mathbf{a} \right\} \qquad (4.5)$$

(see figure 4.1). Two of Maxwell's equations are expressions of this divergence.

Though we won't prove it here formally,[5] equation (4.5) is a three-dimensional derivative of $\mathbf{E}$, and since we know from experiment that $\oint \mathbf{E} \cdot d\mathbf{a}$ is proportional to the total enclosed charge $Q = \int d^3x \, \rho_e(\mathbf{x})$, in terms of the charge density $\rho_e(\mathbf{x})$, we arrive at Gauss's law (also known as the first Maxwell equation):

$$\operatorname{div} \mathbf{E} \equiv \vec{\nabla} \cdot \mathbf{E} = 4\pi \rho_e(\mathbf{x}), \qquad (4.6)$$

written here in Gaussian units. Though we won't need the second Maxwell equation to develop MHD, we can immediately write it down for completeness by analogy with equation (4.6), under the assumption that magnetic monopoles are either nonexistent or very rare:

$$\operatorname{div} \mathbf{B} \equiv \vec{\nabla} \cdot \mathbf{B} = 0. \qquad (4.7)$$

The third Maxwell equation (Faraday's law), expresses the fact that $\mathbf{E}$ is a conservative field. The work done on a unit charge as it moves around a closed loop must be zero as long as the fields are static, in which case $\oint \mathbf{E} \cdot d\mathbf{l} = 0$. However, we also know from Faraday's experiments that when the magnetic flux threading that loop is changing in time, then the *circulation* of $\mathbf{E}$ (this integral over $\mathbf{E}$ around a closed contour) is nonzero, meaning that energy is transferred from the fields to the particle, and vice versa.

As was the case for the surface integral over $\mathbf{E}$, which led to the differential formulation involving a divergence, it is convenient to reduce Faraday's empirically derived macroscopic result for the circulation of $\mathbf{E}$ into its own differential form,

---

[5]A more complete development is presented in Melia (2001a).

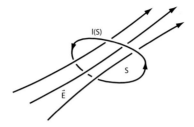

Figure 4.2 The curl of the electric (vector) field is defined as its net circulation around a loop $l(S)$ enclosing a surface area S, as S goes to zero.

via the definition of a *curl*,

$$(\text{curl}\,\mathbf{E})_{\hat{\mathbf{S}}} \equiv \lim_{S \to 0} \left\{ \frac{1}{S} \oint_{l(S)} \mathbf{E} \cdot d\mathbf{l} \right\} \tag{4.8}$$

(see figure 4.2). The notation on the left is used to indicate that the curl must be defined with respect to a given direction $\hat{\mathbf{S}}$, where $\hat{\mathbf{S}}$ is the unit vector normal to the surface S. Therefore, in three-dimensional space, the curl of $\mathbf{E}$ has three components, one for each of the linearly independent coordinates.

Equation (4.8) is also a derivative of $\mathbf{E}$, though in this case the circulation mixes derivative components in different directions. Again, we won't prove the result here formally, but

$$\text{curl}\,\mathbf{E} \equiv \vec{\nabla} \times \mathbf{E}. \tag{4.9}$$

Faraday's contribution was to demonstrate experimentally that $\oint \mathbf{E} \cdot d\mathbf{l} \neq 0$ when the closed path is linked by a changing magnetic flux, in which case

$$\oint \mathbf{E} \cdot d\mathbf{l} = -\frac{1}{c} \frac{d}{dt} \int_{S(l)} \mathbf{B} \cdot d\mathbf{a}. \tag{4.10}$$

And realizing that for a continuous field $\mathbf{B}$,

$$\lim_{S \to 0} \left\{ \frac{1}{S} \int_{S} \mathbf{B} \cdot d\mathbf{a} \right\} = \mathbf{B} \cdot \hat{\mathbf{S}}, \tag{4.11}$$

we arrive at the third Maxwell equation,

$$\text{curl}\,\mathbf{E} = \vec{\nabla} \times \mathbf{E} = -\frac{1}{c} \frac{\partial \mathbf{B}}{\partial t}, \tag{4.12}$$

once we recognize that $d/dt \to \partial/\partial t$ for a magnetic field measured in the loop's rest frame. In other words, the magnetic flux threading $l(S)$ may change only in response to an intrinsic time variation in $\mathbf{B}$, but not via spatial variations in the field because this loop is not moving through it. But the observer would indeed have to include the advective contribution $\mathbf{v} \cdot \vec{\nabla}$ to d/dt if the field used in these equations was that measured in the laboratory frame (which we would have to label differently) rather than that in the loop's coordinate system.

The fourth (and final) Maxwell equation expresses another experimentally known effect, in which the circulation of $\mathbf{B}$ is seen to be proportional to the total current threading the loop:

$$\oint \mathbf{B} \cdot d\mathbf{l} = \frac{4\pi}{c} I. \tag{4.13}$$

We won't repeat the steps that led to equation (4.12); we simply transform equation (4.13) into its differential form by analogy. The result is

$$\text{curl } \mathbf{B} \equiv \vec{\nabla} \times \mathbf{B} = \frac{4\pi}{c} \mathbf{J}, \tag{4.14}$$

where $\mathbf{J}$ is the current density. But as it stands, this equation is not yet complete because it violates charge conservation, for taking the divergence of both sides, we get

$$\vec{\nabla} \cdot (\vec{\nabla} \times \mathbf{B}) = 0 = \frac{4\pi}{c} \vec{\nabla} \cdot \mathbf{J}, \tag{4.15}$$

which can be true only when the system is static. In the general case, conservation of charge leads to the continuity equation,

$$\frac{\partial \rho_e}{\partial t} + \vec{\nabla} \cdot \mathbf{J} = 0, \tag{4.16}$$

which is not consistent with equation (4.15) when $\partial \rho_e / \partial t \neq 0$.

Fixing this problem was one of Maxwell's main contributions to electrodynamics. The solution is simply to add a source term to equation (4.15) to ensure that changes in $\mathbf{J}$ are balanced by corresponding changes in ρ_e, via equation (4.6). The result is known as Ampère's law, which constitutes the fourth Maxwell equation,

$$\text{curl } \mathbf{B} \equiv \vec{\nabla} \times \mathbf{B} = \frac{4\pi}{c} \mathbf{J} + \frac{1}{c} \frac{\partial \mathbf{E}}{\partial t}. \tag{4.17}$$

The equations are now in place for us to develop the essential ideas of magneto-hydrodynamics. In ideal MHD, the conductivity of the medium is very high, which simplifies the behavior of the medium considerably. In particular, if we consider the Lorentz force equation for a charge q under the influence of electric ($\mathbf{E}$) and magnetic ($\mathbf{B}$) fields,

$$\mathbf{F} = q \left(\mathbf{E} + \frac{\mathbf{v}}{c} \times \mathbf{B} \right), \tag{4.18}$$

it is clear that a very high conductivity means the charges move very easily in response to the fields, and it costs very little energy to create a current. The force $\mathbf{F}$ must therefore be close to zero, so

$$\mathbf{E} \approx -\frac{\mathbf{v}}{c} \times \mathbf{B}. \tag{4.19}$$

Strict equality in this equation is attained when the conductivity goes to infinity. Thus, from Faraday's law (4.12), the simple equation satisfied by the magnetic field in these circumstances is

$$\frac{\partial \mathbf{B}}{\partial t} = \vec{\nabla} \times (\mathbf{v} \times \mathbf{B}). \tag{4.20}$$

At the same time, the gas is subject to the mass conservation law,

$$\frac{\partial \rho}{\partial t} + \vec{\nabla} \cdot (\rho \, \mathbf{v}) = 0, \tag{4.21}$$

and the dynamical influence of the electric and magnetic fields. (The mass density ρ is not to be confused with the charge density ρ_e.) That is, starting from the Lorentz force density (which is a continuum generalization of the point particle equation 4.18)

$$\rho \frac{d\mathbf{v}}{dt} = \rho_e \, \mathbf{E} + \frac{1}{c}(\mathbf{J} \times \mathbf{B}), \tag{4.22}$$

we assume that the medium is neutral (i.e., that the charge density ρ_e is zero) and take $\mathbf{J}$ to be given by Ampère's law (equation 4.17), with the result that

$$\rho \frac{\partial \mathbf{v}}{\partial t} + \rho (\mathbf{v} \cdot \vec{\nabla}) \mathbf{v} = -\frac{1}{4\pi} \mathbf{B} \times \left[\vec{\nabla} \times \mathbf{B} - \frac{1}{c} \frac{\partial \mathbf{E}}{\partial t} \right]. \tag{4.23}$$

As we shall see, the acceleration scheme associated with this magnetic field disturbance acts on a small fraction of charges within the plasma (i.e., those particles moving principally in the direction along $\mathbf{B}$). Most of the plasma responds as a fluid with a subrelativistic velocity. Equation (4.23) describes the behavior of this slowly moving gas, and so for a nonrelativistic motion, we reduce this equation to

$$\rho \frac{\partial \mathbf{v}}{\partial t} + \rho (\mathbf{v} \cdot \vec{\nabla}) \mathbf{v} \approx -\frac{1}{4\pi} \mathbf{B} \times \vec{\nabla} \times \mathbf{B}. \tag{4.24}$$

Equations (4.20), (4.21), and (4.24) form a complete set of coupled relations describing the behavior of the fluid and the magnetic field in an interacting environment.

Suppose now that the field $\mathbf{B}$ is jiggled, perhaps due to a disturbance at $z = 0$ as shown in figure 4.3. This field might, for example, be anchored in the disk surrounding a massive, compact object or might be frozen into the surface layers of a highly magnetized neutron star. The solution to equations (4.20), (4.21), and (4.24) that describes this wave propagation with perfect symmetry in the x–y plane is

$$\mathbf{B} = B_0 \hat{\mathbf{z}} + B_A \, \exp(ikz - i\omega t)\hat{\mathbf{x}}, \tag{4.25}$$

$$\mathbf{v} = 0 + v_A \, \exp(ikz - i\omega t)\hat{\mathbf{x}}, \tag{4.26}$$

$$\mathbf{E} = 0 + E_A \, \exp(ikz - i\omega t)\hat{\mathbf{y}}, \tag{4.27}$$

where clearly B_A, v_A, and E_A represent the traveling perturbation, and are labeled in honor of Hannes Alfvén (1908–1995), who carried out much of the early work in this field.[6] It is not difficult to see with direct substitution of (4.25)–(4.27) into the

---

[6]The work of Hannes Alfvén, winner of the 1970 Nobel prize in physics, is today being applied in many branches of science, including the development of particle beam accelerators, controlled thermonuclear fusion, hypersonic flight, rocket propulsion, and the braking of reentering space vehicles. His many contributions in astronomy include an explanation for the Van Allen radiation belt, Earth's magnetosphere, the dynamics of plasma in our Galaxy, and the structure of magnetic fields in intergalactic space.

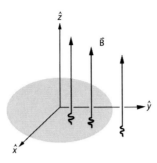

Figure 4.3 The magnetic field at the polar cap of a neutron star may, under some circumstances, be approximated as uniform and parallel to the dipole axis $\hat{\mathbf{z}}$. A crustal disturbance generates oscillations at the base of the magnetosphere, and these travel as waves along $\mathbf{B}$.

coupled equations that

$$\mathbf{B}_A \equiv B_A \exp(ikz - i\omega t)\hat{\mathbf{x}} = -\frac{k}{\omega} B_0 \mathbf{v}. \qquad (4.28)$$

Not surprisingly, this is a plane wave—a so-called Alfvén wave—with phase velocity $v_\alpha = \omega/k$ traveling in the z-direction. But notice (from equation 4.27) that $\mathbf{E}$ is always perpendicular to $\mathbf{B}$ and hence may only accelerate charges perpendicular to the magnetic field. This does not lead to persistent acceleration because of the $\mathbf{v} \times \mathbf{B}$ forces.

Real situations, however, are hardly ideal and perfect plane waves are difficult to sustain. Instead, some x–y structure is expected, and because curl $\mathbf{B}$ is then nonzero, we expect a current and therefore an electric field component along $\hat{\mathbf{z}}$. The MHD equations (4.20, 4.21, and 4.24) also permit the following solution:

$$\mathbf{B} = B_0\hat{\mathbf{z}} + B_A \sin(k_y y) \exp(ikz - i\omega t)\hat{\mathbf{x}}, \qquad (4.29)$$

which we may interpret as one of the frequency components in a Fourier series expansion for a more general structure of the field in the y-direction. Then, from equation (4.17),

$$\frac{\partial \mathbf{E}}{\partial t} \approx c\,\frac{\partial B_x}{\partial z}\,\hat{\mathbf{y}} - c\,\frac{\partial B_x}{\partial y}\,\hat{\mathbf{z}}, \qquad (4.30)$$

which for a harmonic field reduces to

$$\mathbf{E} \approx \frac{ic}{\omega}[ikB_A \sin(k_y y)\,\hat{\mathbf{y}} - k_y B_A \cos(k_y y)\,\hat{\mathbf{z}}] \exp(ikz - i\omega t). \qquad (4.31)$$

The important thing to notice is the appearance of an E_z component, which can accelerate particles along the (local) $\hat{\mathbf{z}}$ direction (see figure 4.4).

The collision frequency ν_c in a typical neutron-star magnetosphere (see chapter 9), where the particle density is $\sim 10^{16}$–10^{26} cm^{-3}, is about 10^7 s^{-1}. Between scatterings, the E_z component accelerates the charges according to

$$\frac{dp_z}{dt} = qE_z, \qquad (4.32)$$

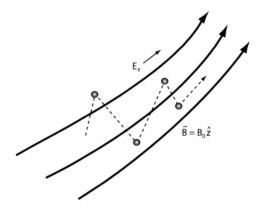

Figure 4.4 The z-component of **E**, pointing along the direction of **B**, accelerates particles along the magnetic field between collisions with other particles. Although these particle–particle interactions dissipate much of the energy transferred from the electromagnetic field, the accelerated charges can nonetheless attain relativistic speeds between collisions.

where $p_z = \gamma m\, v_z$. Thus, for a relativistic electron (with $v_z \approx c$),

$$\frac{d\gamma}{dt} \approx \frac{eE_z}{m_e\, c}. \qquad (4.33)$$

As a rough estimate of how energetic the particles can become, we take these relations and put

$$\gamma_{\max} \sim \frac{eE_z}{m_e c v_c} \approx \frac{ek_y B_A}{m_e \omega v_c}. \qquad (4.34)$$

In typical pulsars, $k_y \sim 2\pi/\lambda_{\text{crust}}$, where the crust scale length λ_{crust} is of order 10 cm, and B_A ranges from 10^8 G to as high as the underlying magnetic field strength $\sim 10^{12}$ G. In addition, $\omega \approx v_\alpha k$, where $k \sim 2\pi/R_{\text{ns}}$ and the stellar radius (as we have said) is $R_{\text{ns}} \approx 10$ km. In principle, this mechanism can therefore accelerate particles to a Lorentz factor well in excess of 10^{10}, since the Alfvén wave *phase* velocity v_α is greater than c. In practice, several damping influences and the creation of lepton pairs set in well before this plateau is reached.

4.3 FERMI ACCELERATION

Gas flowing into a black hole, or onto a white dwarf or a neutron star, is often supersonic (chapters 6 and 7), meaning that the flow exceeds the local sound speed, and sound waves bunch up to form discontinuous jumps across the resultant shock front. We demonstrate in this section that the diffusion of particles across the shock accelerates them and produces a power-law distribution by converting a fraction of the kinetic energy available in the flow. Often, the accelerating particles scatter off the random fluctuations in the magnetic field (a so-called "collisionless" process), rather than with other particles. This mechanism therefore constitutes our second electromagnetic acceleration scheme, complementary to the method we have just been considering in the context of organized magnetospheric fields in pulsars.

It was the particle physicist Enrico Fermi (1901–1954) who first proposed the idea that cosmic rays could be produced via the random scattering of particles between molecular clouds in the interstellar medium.[7] If such clouds move more or less randomly, he argued, the frequency of "head-on" collisions between the cosmic rays and clouds would exceed the rate of "catch-up" encounters, leading to a net energy gain by the particles. This idea was rather compelling, as we shall see, because particles scattering in this manner within a confined environment produce a power-law distribution in energy, mimicking the observations very well.

Let us retrace the steps followed by Fermi, only in one dimension to simplify the mathematics. We consider a large number of massive particles M (the "clouds") moving randomly (though with the same magnitude of velocity V), back and forth along our one-dimensional space, and colliding with a small test particle m, whose velocity is v. For an average separation Δ, the collision time for a head-on collision is

$$\tau_h = \frac{\Delta}{v + V}, \tag{4.35}$$

whereas for a catch-up event it is

$$\tau_c = \frac{\Delta}{v - V}. \tag{4.36}$$

Thus, the probability of a head-on collision is

$$P_h = \frac{1/\tau_h}{1/\tau_h + 1/\tau_c} = \frac{v + V}{2v}, \tag{4.37}$$

and

$$P_c = \frac{v - V}{2v} \tag{4.38}$$

for a catch-up event.

Denoting quantities measured in the center of mass (CM) frame with a prime, we acknowledge that $v_i' = -v_a'$ for an elastic collision, whence

$$v_i' = v + V,$$
$$v_a' = -(v + V),$$
$$v_a = -(v + V) - V, \tag{4.39}$$

where subscripts i and a pertain to the conditions, respectively, before and after the collision between m and M. If we further assume that $M \gg m$ and $V \ll c$, the CM frame is effectively M's frame, and the changes in energy for head-on and catch-up collisions must be, respectively,

$$\Delta E_h = \tfrac{1}{2}m(v + 2V)^2 - \tfrac{1}{2}mv^2, \tag{4.40}$$
$$\Delta E_c = \tfrac{1}{2}m(-v + 2V)^2 - \tfrac{1}{2}mv^2. \tag{4.41}$$

[7] See Fermi (1949).

The net mean change in energy per collision follows from a consideration of equations (4.40) and (4.41), together with the probabilities in equations (4.37) and (4.38). That is,

$$\langle \Delta E \rangle = P_h \Delta E_h + P_c \Delta E_c. \tag{4.42}$$

Some simple algebra leads to the important result that

$$\frac{\langle \Delta E \rangle}{E} = 4 \left(\frac{V}{v} \right)^2, \tag{4.43}$$

clearly a *second-order* effect.

Our initial observation is that yes, the combination of head-on and catch-up collisions does indeed lead to an average increase in the particle's energy over time. Unfortunately, the second-order dependence of $\langle \Delta E \rangle$ on the ratio V/v makes this process extremely slow, since $V \ll v$. Catch-up collisions actually result in an energy loss, partially mitigating the effects of the head-on encounters. But as we shall see later in this section, a shock geometry avoids this difficulty by producing a first-order dependence of $\langle \Delta E \rangle$ on the velocities.

However, before we jump to this new acceleration format, there is still some work to do in establishing the particle spectrum resulting from this stochastic process. Now that we have the change in energy per encounter, we can determine the rate of energy gain with knowledge of the collision rate. Defining ν to be the frequency of collisions, we calculate this quantity as

$$\frac{dE}{dt} = \langle \Delta E \rangle \nu = 4\nu \left(\frac{V}{v} \right)^2 E \equiv \alpha E. \tag{4.44}$$

But this energization must cease once the particle leaves the acceleration zone, and since different particles escape at various times, the energy distribution in the final assembly is not flat.

The energy distribution of particles accelerated by this Fermi statistical process satisfies the diffusion-loss equation, which we now derive. In figure 4.5, let the number of particles within the rectangle be $N(E, x, t) \, dE \, dx$ (for one-dimensional motion along the x-axis). Then, in terms of the spatial (ϕ_s) and energy (ϕ_E) fluxes,

$$\frac{d}{dt}(N \, dE \, dx) = [\phi_s(E, x, t) - \phi_s(E, x + dx, t)] \, dE$$

$$+ [\phi_E(E, x, t) - \phi_E(E + dE, x, t)] \, dx$$

$$+ \frac{\partial N}{\partial t} \, dE \, dx. \tag{4.45}$$

That is,

$$\frac{dN}{dt} = -\frac{\partial \phi_s}{\partial x} - \frac{\partial \phi_E}{\partial E} + \frac{\partial N}{\partial t}. \tag{4.46}$$

The spatial flux ϕ_s may be written in terms of a diffusion coefficient D,

$$\phi_s = -D \frac{\partial N}{\partial x}. \tag{4.47}$$

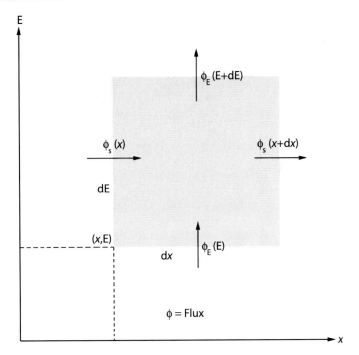

Figure 4.5 Two-dimensional phase–space diagram, showing the diffusion of particles into and out of the square with dimensions (dx, dE), and whose lower left-hand corner lies at coordinates (x, E). The spatial flux is labeled $\phi_s(x)$, whereas the flux in energy is $\phi_E(E)$.

Thus,

$$\frac{dN}{dt} = D\frac{\partial^2 N}{\partial x^2} - \frac{\partial \phi_E}{\partial E} + \frac{\partial N}{\partial t}. \tag{4.48}$$

In three dimensions, this equation generalizes to

$$\frac{dN}{dt} = D\nabla^2 N - \frac{\partial \phi_E}{\partial E} + \frac{\partial N}{\partial t}. \tag{4.49}$$

Let us continue to adopt the simplest set of assumptions and suppose that, rather than diffusing spatially away from the acceleration zone, the particles escape directly when their acceleration terminates, after a characteristic time τ. In that case, $D = 0$, and N is solely a function of the energy E and time t. In addition, we may put

$$\phi_E = \frac{N(E, t)\,\Delta E}{\Delta t}, \tag{4.50}$$

representing the flux ϕ_E in terms of an incremental change in energy ΔE over a small time interval Δt. In the infinitesimal limit, this becomes

$$\phi_E = N(E, t)\frac{dE}{dt}, \tag{4.51}$$

and with equation (4.44), we derive the very useful formulation

$$\phi_E = N(E, t)\alpha E, \tag{4.52}$$

where clearly $\alpha \equiv 4\nu(V/\nu)^2$. With these approximations and assumptions, the diffusion-loss equation (4.48) therefore reduces to the form

$$\frac{dN}{dt} \approx -\frac{\partial}{\partial E}[N(E, t)\alpha E] - \frac{N(E, t)}{\tau}, \tag{4.53}$$

the last term on the right accounting (approximately) for the intrinsic time variation of $N(E, t)$.

Thus, in equilibrium ($dN/dt = 0$), $N(E, t)$ becomes just a function of energy E, and

$$\alpha N(E) + \alpha E \frac{\partial N}{\partial E} \approx -\frac{N}{\tau}, \tag{4.54}$$

or

$$\frac{\partial N(E)}{\partial E} \approx -\left(1 + \frac{1}{\alpha\tau}\right)\frac{N(E)}{E}. \tag{4.55}$$

The solution to this differential equation is trivial,

$$N(E) \approx N_0 E^{-(1+1/\alpha\tau)}, \tag{4.56}$$

and demonstrates the remarkable result that the particle spectrum is a power law in energy as seen, e.g., in cosmic rays (see figure 13.1) and many high-energy sources, with a spectral index that may vary slightly from location to location, depending on the values of α and τ. The simplicity in deriving such an important and universal feature of high-energy plasmas makes this picture of particle acceleration very compelling. What limits its applicability, however, is that the second-order dependence of α on V/ν makes this process very slow.

Let us now turn our attention to an alternative geometry in which the scattering of test particles occurs within a converging flow, a situation that would arise across a shock, as depicted in figure 4.6. The essential difference between this situation and the three-dimensional random scattering environment we have just considered is that here the catch-up collisions (downwind) occur with relative velocities much smaller than the head-on encounters upwind, whereas the average velocity of the scattering centers in the former case is always the same. This small—though essential—difference is sufficient to turn a second-order process into a *first*-order one.

For later convenience, we begin by defining the following quantities: β is the factor by which the average particle energy changes per collision event; $\mathcal{P}$ is the probability that the particle remains in the acceleration zone after one collision; and γ is the ratio of specific heats. Then, after k collisions, the number of particles that are still accelerating is

$$N = N_0 \mathcal{P}^k, \tag{4.57}$$

and their energy is

$$E = E_0 \beta^k. \tag{4.58}$$

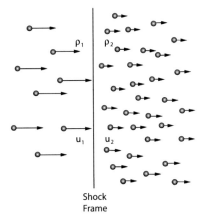

Shock
Frame

Figure 4.6 In the rest frame of the shock, upstream gas approaches the interaction region with velocity u_1 and density ρ_1. After the jump, the gas moves away from the shock at velocity u_2, with corresponding density ρ_2. The shock region is said to contain converging scattering centers because particles bouncing back and forth across the discontinuity experience higher (head-on) collision velocities upstream than (catch-up) velocities downstream.

Thus, eliminating k from these two equations, we get

$$\frac{\ln \mathcal{P}}{\ln \beta} = \frac{\ln(N/N_0)}{\ln(E/E_0)}, \tag{4.59}$$

so that

$$\frac{N}{N_0} = \left(\frac{E}{E_0}\right)^{\ln \mathcal{P}/\ln \beta}. \tag{4.60}$$

Therefore,

$$dN = KE^{\ln \mathcal{P}/\ln \beta - 1}\, dE, \tag{4.61}$$

where K is a normalization constant set by the initial conditions.

We now assume that a steady flow, e.g., into a black hole, sustains a strong shock with $u_1 \gg c_s$, where $c_s^2 \equiv \partial P / \partial \rho$ is the square of the sound speed calculated from the gas pressure P and density ρ. In strong shocks, the sound waves all bunch up to form a discontinuous jump in density, given by

$$\frac{\rho_2}{\rho_1} = \frac{\gamma + 1}{\gamma - 1}, \tag{4.62}$$

where labels "1" and "2" indicate conditions in the pre- and postshock regions, respectively.[8] In a fully ionized gas, $\gamma = \frac{5}{3}$, and therefore $\rho_2/\rho_1 = 4$.

By conservation of mass, the mass flux into the shock must be equal to that leaving from the other side. Thus,

$$\rho_1 u_1 = \rho_2 u_2, \tag{4.63}$$

[8]A good discussion of shock jump conditions may be found in Landau and Lifshitz (1987).

so that $u_1/u_2 = 4$. This is the reason why a particle bouncing back and forth across the discontinuity is squeezed between converging flows and gains energy. A good analogy is a Ping-Pong ball bouncing up and down between the table and a descending paddle. The fact that the paddle is descending results in the ball gaining energy with every encounter.

Now, calculating the average change in energy per cycle across the shock is actually easier here than was the case leading up to equation (4.43). For one thing, the particle executes exactly one head-on and one catch-up collision on every round trip. Thus,

$$\frac{\langle \Delta E \rangle}{E} = \frac{1}{2} \left(\frac{\Delta E_h}{E} + \frac{\Delta E_c}{E} \right), \tag{4.64}$$

and it is trivial to show after some algebra that this reduces to the form

$$\frac{\langle \Delta E \rangle}{E} \approx \frac{2(u_1 - u_2)}{v} \equiv \frac{2\Delta u}{v}, \tag{4.65}$$

which is clearly a first-order effect. Thus,

$$\beta = 1 + \frac{2\Delta u}{v}. \tag{4.66}$$

To get an analogous expression for $\mathcal{P}$, we need to understand how the particles are carried into the acceleration zone and how they escape. If we let N be the number density of accelerating particles, then behind the shock they are swept away with a flux $u_2 N$. The test particles hardly notice the discontinuity, and since they each undergo exactly one scattering upwind and one scattering downwind, N is roughly the same everywhere. (Contrast this with the overall mass density ρ, which undergoes a jump from one side to the other.)

On the other hand, we know from classical kinetic theory that the number of particles crossing a unit surface area per unit time, while traveling at a speed $\lesssim c$, is $\frac{1}{4}Nc$. Thus, in steady state, the flux of test particles that cross back upstream is $\frac{1}{4}Nc - u_2 N$, and the probability that an individual particle remains within the acceleration region after one collision must be

$$\mathcal{P} = \frac{\frac{1}{4}Nc - u_2 N}{\frac{1}{4}Nc} = 1 - \frac{4u_2}{c}. \tag{4.67}$$

But $u_2/c \ll 1$, and therefore

$$\ln \mathcal{P} = \ln \left(1 - \frac{4u_2}{c} \right) \approx -\frac{4u_2}{c} \tag{4.68}$$

and

$$\ln \beta = \ln \left(1 + \frac{2\Delta u}{c} \right) \approx \frac{2\Delta u}{c}. \tag{4.69}$$

So in equation (4.61), the spectral index is

$$\frac{\ln \mathcal{P}}{\ln \beta} = \frac{-4u_2}{2(u_1 - u_2)} = -\frac{2}{3}, \tag{4.70}$$

the principal result we have been seeking.

According to the simple theory we have developed in this section, Fermi shock acceleration is expected to produce a power-law particle distribution

$$dN(E) = KE^{-5/3}\, dE. \tag{4.71}$$

The index is close to, but not quite, the observed "universal" value of ~ -2 (see, e.g., section 12.3). In this analysis, however, we did not take into account the angle of incidence during the scattering process. In other words, the particles don't really bounce straight back and forth across the shock. A more careful treatment, with all the kinematic factors included, would indeed produce a distribution $\propto E^{-2}$.

What we have learned here is that getting a power-law particle spectrum, from either first- or second-order Fermi acceleration, is relatively easy. We will learn later in this book, particularly with GRBs in section 11.2, AGNs in section 12.3, and cosmic rays and supernova shells in section 13.1, that this acceleration scheme has many wide applications.

Now that we have surveyed the three most common acceleration mechanisms encountered in high-energy astrophysics, we understand how relativistic particles may be produced in dynamic environments. We have yet to develop the physics associated with radiative emission, however. This is the subject of the next chapter, where we study the dominant processes believed to be responsible for producing the bulk of high-energy radiation reaching us from outside the solar system.

SUGGESTED READING

The physics of degenerate material is critical to an understanding of the internal structure of white dwarfs and neutron stars. Excellent references on this topic include Chandrasekhar (1939) and Shapiro and Teukolsky (1983).

Degeneracy itself is a manifestation of the Pauli exclusion principle, whose physical explanation is based on an important, though poorly known, topological property of three-dimensional space. Read an account of this surprising aspect of nature in chapter 3 (particularly section 3.3) of Melia (2003).

X-ray astronomy has greatly elucidated the properties of neutron stars, based in part on a study of their spectrum, which sometimes includes gravitationally red-shifted Fe emission lines. For a comprehensive review, see Yakovlev and Pethick (2004).

Magnetic fields are critical to several acceleration schemes, including electromagnetic acceleration in MHD waves and the Fermi process within magnetized shock fronts. The classic treatise on all aspects of magnetic fields in space is the book by Parker (1979).

A giant in the development of magnetohydrodynamics was Hannes Alfvén, winner of the 1970 Nobel prize in physics. He wrote prolifically on this subject, starting in the 1940s. Read an example of his work in Alfvén (1947).

A good theoretical review of (Fermi) diffusive shock acceleration may be found in Jones (1994).

And the properties of shocks, including a discussion of the jump conditions, may be found in Landau and Lifshitz (1987).

Chapter Five

Radiative Processes

As we enter the 21st century, neutrino and gravitational wave detectors hold great promise as major telescope facilities of the future. It is nonetheless still true, however, that nearly all of our knowledge about astronomical objects is obtained through the radiation they produce. The physical characteristics of their emission region are often inferred from features in their spectra, and since we understand radiation processes rather well, the theoretical modeling of these environments often leads to quantitative—not merely qualitative—results.

5.1 THE RADIATION FIELD

A moving charge (figure 5.1) is a source of two types of field: a static (Coulomb-like) field and a radiation field. The relativistically correct expression for $\mathbf{E}$ is rather easy to write down and interpret, though we will not re-create the steps used to derive it, since this development has already been carried out by several authors.[1]

In the full relativistic treatment, one of the most important advances made beyond the purely Coulombic, action-at-a-distance approach employed during the foundational era of electrodynamics was the introduction of a *retarded* time. We want to know the field at vector position $\mathbf{r}$ at time t, due to the influence of a particle following a trajectory $\mathbf{r}_0(t')$, with corresponding velocity $\mathbf{v}(t') = \dot{\mathbf{r}}_0(t')$. But the electromagnetic field travels at finite speed c, so the time of emission, t', cannot be equal to the time of observation, t. Instead, the field we sense at $\mathbf{r}$ at time t was produced at time

$$t' = t - \frac{|\mathbf{r} - \mathbf{r}_0(t')|}{c}, \tag{5.1}$$

where the difference is simply the light-travel time from $\mathbf{r}_0(t')$ to $\mathbf{r}(t)$. The solution[2] to equation (5.1) for t' is known as the retarded time $\tilde{t}$.

The field is always radial and Coulombic (i.e., static and dropping with distance squared away from the source) in the particle's rest frame, but because it is viewed from the laboratory frame, in which the particle may be executing any arbitrary

---

[1] A straightforward derivation appears in Melia (2001a).

[2] But note that equation (5.1) is a quadratic equation (since Pythagoras's theorem must be used to evaluate $|\mathbf{r} - \mathbf{r}_0(t')|$), which means it has two solutions. Physically, this means that any given spacetime point $(ct, \mathbf{r})$ is connected, via lightcones, to regions both in the past and the future. Since we are here interested in a causal connection to activity by the particle at times $t' < t$, the "advanced" solution (corresponding to $t' > t$) will be ignored.

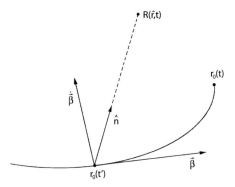

Figure 5.1 A charge moving along the trajectory $\mathbf{r}_0(t')$, with (dimensionless) velocity $\vec{\beta}$ and acceleration $\dot{\vec{\beta}}$, produces a field at the observer's coordinates $(\mathbf{r}, t)$, where $\mathbf{R} \equiv \mathbf{r}(t) - \mathbf{r}_0(t')$, and t' is the retarded time $t - R/c$.

motion, the so-called retarded electric field takes on the following structure:

$$E(\mathbf{r}, t) = q \left[\frac{(\hat{\mathbf{n}} - \vec{\beta})(1 - \beta^2)}{\kappa^3 R^2} \right]_{\tilde{t}} + \frac{q}{c} \left[\frac{\hat{\mathbf{n}} \times \{(\hat{\mathbf{n}} - \vec{\beta}) \times \dot{\vec{\beta}}\}}{\kappa^3 R} \right]_{\tilde{t}}, \qquad (5.2)$$

where

$$\vec{\beta} \equiv \frac{\mathbf{v}}{c}, \qquad (5.3)$$

$$\kappa \equiv 1 - \hat{\mathbf{n}} \cdot \vec{\beta}, \qquad (5.4)$$

and

$$\mathbf{R} \equiv \mathbf{r} - \mathbf{r}_0(t'). \qquad (5.5)$$

As always, the magnetic field can be calculated easily from

$$\mathbf{B}(\mathbf{r}, t) = \hat{\mathbf{n}} \times \mathbf{E}(\mathbf{r}, t), \qquad (5.6)$$

once the electric field is known.

The first term in equation (5.2) is sometimes known as the *velocity* field, falling off as $\sim R^{-2}$. It reduces to Coulomb's law when $v \to 0$. In addition, when $\dot{v} = 0$, this is the only term that survives in $\mathbf{E}$. In that case, the field always points to the particle's current position at time t, even if it was produced at an earlier time and in a different location when $v \neq 0$. The reader can easily convince himself or herself of this, by simply looking at the vector diagram corresponding to the first term in equation (5.2).

The second term is far more important in astrophysics. This is the *acceleration* field, which drops off as R^{-1}. The particle does not have to be moving to produce this field, but it must have an acceleration $\dot{\vec{\beta}} \neq 0$. The fact that it depends on the cross product between $\hat{\mathbf{n}}$ and $\dot{\vec{\beta}}$ means that the radiation field vector will always be perpendicular to the radius vector. This is a well-known result in electrodynamics,

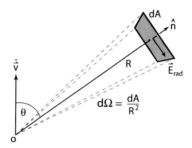

Figure 5.2 Electric field $\mathbf{E}_{\mathrm{rad}}$ produced in the radiation zone by an accelerated charge at the origin. An observer at $\mathbf{R} = R\,\hat{\mathbf{n}}$ measures the Poynting flux carried by this field across an area dA, which subtends a solid angle $d\Omega = dA/R^2$ at the origin.

that radiation fields must always be transverse to the direction of propagation. Note also that $\mathbf{E}$, $\mathbf{B}$, and $\hat{\mathbf{n}}$ form a triad of mutually perpendicular vectors.

Let us for the moment specialize to the case $\beta \ll 1$, and consider only the radiation field, which in this case simplifies to

$$\mathbf{E}_{\mathrm{rad}} = \frac{q}{Rc^2}\,\hat{\mathbf{n}} \times (\hat{\mathbf{n}} \times \dot{\mathbf{v}}). \tag{5.7}$$

The radiation magnetic field associated with $\mathbf{E}_{\mathrm{rad}}$ is

$$\mathbf{B}_{\mathrm{rad}} = \hat{\mathbf{n}} \times \mathbf{E}_{\mathrm{rad}}. \tag{5.8}$$

Thus,

$$|\mathbf{E}_{\mathrm{rad}}| = |\mathbf{B}_{\mathrm{rad}}| = \frac{q\dot{v}}{Rc^2}\sin\theta, \tag{5.9}$$

where θ is the angle between $\hat{\mathbf{n}}$ and the acceleration vector (see figure 5.2).

The electromagnetic flux is given by the Poynting vector

$$\mathbf{S} = \frac{c}{4\pi}\mathbf{E} \times \mathbf{B}, \tag{5.10}$$

which in this case reduces to

$$|\mathbf{S}| = \frac{c}{4\pi}|\mathbf{E}_{\mathrm{rad}}|^2 \equiv \frac{c}{4\pi}E_{\mathrm{rad}}^2, \tag{5.11}$$

pointed in the direction $\hat{\mathbf{n}}$. Often, though, we need to know how the emitted power P is distributed in solid angle Ω. Noting that for a screen area dA the solid angle subtended at the origin is $d\Omega = dA/R^2$, we write

$$\frac{dP}{d\Omega} = R^2 S = \frac{q^2\dot{v}^2}{4\pi c^3}\sin^2\theta. \tag{5.12}$$

The total emitted power is therefore

$$P = \int \frac{dP}{d\Omega}\,d\Omega \tag{5.13}$$

or

$$P = \frac{2q^2\dot{v}^2}{3c^3},$$ (5.14)

the well-known Larmor equation for a single accelerated charge q.

It is interesting to note that the power in this expression depends on the combination $q\dot{v}$, which ultimately has its origin in the field induced when opposite charges are spatially separated to produce a dipole moment. This does not happen with gravity, of course, because (as far as we know) there is only one charge—the mass—so the lowest multipole moment that can radiate in that case is a quadrupole. Here, however, we can begin with a dipole moment

$$\mathbf{d} \equiv qL\hat{z},$$ (5.15)

representing a dipole with charges $-q$ and $+q$ separated by a distance L, pointing in the direction ($\hat{z}$) of positive charge. Differentiating this twice with respect to time, we get

$$\ddot{\mathbf{d}} = q\ddot{L}\hat{z},$$ (5.16)

and we can therefore also write equation (5.14) as

$$P = \frac{2\ddot{\mathbf{d}}^2}{3c^3}.$$ (5.17)

This is the so-called dipole approximation for a single nonrelativistic particle (figure 5.3). The important thing to note here is that the radiation has the characteristic dipole pattern in which $dP/d\Omega \propto \sin^2\theta$. The pattern is symmetric about the acceleration vector, but this is true only for particle speeds small compared to c. Later, when we consider relativistic motion, we shall see that in the instantaneous rest frame of the charge, the pattern still looks like that shown in figure 5.3, but from our perspective in the laboratory, there is an additional effect arising from the boost along the direction $\mathbf{v}$ (see figure 5.7).

For us it is also important to know what the *frequency* distribution of the radiation is, not just how it is distributed in solid angle. We therefore need to determine which frequencies are contributing to (or are associated with) variations in $\mathbf{d}$, which in turn requires a Fourier summation. So let us write the magnitude of $\mathbf{d}$ as an integral in Fourier space,

$$d(t) = \int_{-\infty}^{\infty} e^{-i\omega t} \bar{d}(\omega) \, d\omega,$$ (5.18)

where $\bar{d}$ is the Fourier transform of d. Then

$$\ddot{d}(t) = -\int_{-\infty}^{\infty} \omega^2 e^{-i\omega t} \bar{d}(\omega) \, d\omega,$$ (5.19)

and so from equation (5.9), the magnitude of the radiation field may be written

$$E_{\text{rad}}(t) = -\int_{-\infty}^{\infty} \omega^2 e^{-i\omega t} \bar{d}(\omega) \frac{\sin\theta}{Rc^2},$$ (5.20)

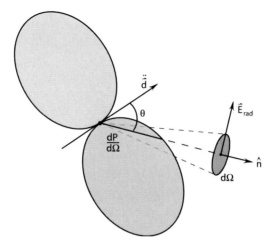

Figure 5.3 The power $dP/d\Omega$ emitted per unit solid angle by a single particle accelerated in the direction **d** exhibits a dipole pattern, in which $dP/d\Omega \propto \sin^2\theta$, in terms of the angle θ relative to **d**. In this diagram, $\mathbf{E}_{\text{rad}}$ is the radiation field, which is always perpendicular to the direction of propagation $\hat{\mathbf{n}}$ of the wave.

and with

$$E_{\text{rad}}(t) = \int_{-\infty}^{\infty} e^{-i\omega t}\, \bar{E}_{\text{rad}}(\omega)\, d\omega, \tag{5.21}$$

we can therefore identify the Fourier transform of E_{rad} as

$$\bar{E}_{\text{rad}}(\omega) = -\omega^2\, \bar{d}(\omega)\, \frac{\sin\theta}{Rc^2}. \tag{5.22}$$

If we now call the total energy W, so that $P = dW/dt$, then

$$\frac{dP}{dA} = |\mathbf{S}| = \frac{c}{4\pi} E_{\text{rad}}^2(t), \tag{5.23}$$

and so

$$\frac{dW}{dA} = \frac{c}{4\pi} \int_{-\infty}^{\infty} E_{\text{rad}}^2(t)\, dt. \tag{5.24}$$

But from Parseval's theorem for Fourier transforms, we know that

$$\int_{-\infty}^{\infty} E_{\text{rad}}^2(t)\, dt = 2\pi \int_{-\infty}^{\infty} |\bar{E}_{\text{rad}}(\omega)|^2\, d\omega, \tag{5.25}$$

and since

$$|\bar{E}_{\text{rad}}(\omega)|^2 = |\bar{E}_{\text{rad}}(-\omega)|^2, \tag{5.26}$$

we can put

$$\frac{dW}{dA} = c \int_{0}^{\infty} |\bar{E}_{\text{rad}}(\omega)|^2\, d\omega. \tag{5.27}$$

Replacing the integrand in this equation with the use of equation (5.22) brings us to the formulation

$$\frac{dW}{dA} = \frac{1}{R^2 c^3} \int_0^\infty \omega^4 |\bar{d}(\omega)|^2 \sin^2 \theta \, d\omega \tag{5.28}$$

or

$$\frac{dW}{d\omega \, d\Omega} = \frac{\omega^4}{c^3} |\bar{d}(\omega)|^2 \sin^2 \theta \, d\omega. \tag{5.29}$$

The angle-integrated distribution is then

$$\frac{dW}{d\omega} = \frac{8\pi \omega^4}{3c^3} |\bar{d}(\omega)|^2. \tag{5.30}$$

We should not forget that, from the beginning, our use of equation (5.7) restricted us to situations where the radiating particle is moving nonrelativistically. As such, equations (5.12) and (5.30) should be applied only to particles moving at speeds $v \ll c$. For example, in subsequent sections we shall use these results to calculate the bremsstrahlung emissivity of a plasma, and it will be understood that those derivations are valid only in the nonrelativistic domain. That does not mean, of course, that relativistic bremsstrahlung is not important. However, the full expression for $\mathbf{E}$ (from equation 5.2) should then be invoked.

5.2 INTENSITY

Before addressing the various radiation mechanisms directly, however, we will first establish some basic definitions and consider how one transforms a description of the photon field from one observer to another.

The specific intensity I_ν at frequency ν is defined by the relation

$$dW = I_\nu \, dA \, dt \, d\Omega \, d\nu, \tag{5.31}$$

where dW is an element of energy in the frequency range $d\nu$ crossing the area dA in time dt (see figure 5.2). Unlike the flux density F_ν, however, which gives the net energy per unit frequency crossing an area per unit time, the intensity represents the energy carried along an individual ray, which may be pointing in any direction relative to the unit vector $\hat{\mathbf{n}}$ normal to dA. It therefore contains more information about the radiation field than F_ν does, since it describes the transport of energy as a function of angle, in addition to the other coordinates also appearing in the flux. In equation (5.31), $d\Omega$ is the solid angle associated with a particular ray, and the units of I_ν are therefore energy per unit area, per unit time, per unit solid angle, per unit frequency.

In terms of I_ν, the flux density is given as

$$F_\nu = \int I_\nu(\theta, \phi) \cos \theta \, d\Omega. \tag{5.32}$$

The multiplicative factor $\cos \theta$ appears here to render F_ν the *net* flux crossing a given area. For example, if I_ν is an isotropic radiation field, each ray carries the same energy, but the net transfer of energy across any surface is zero, since there are as many rays moving in the positive $\hat{\mathbf{n}}$ direction as there are moving in the negative direction. In this case, F_ν would be zero, since $\int \cos\theta \, d\Omega = 0$.

Other useful descriptions of the radiation field follow immediately from equations (5.31) and (5.32) with an appropriate manipulation of these quantities. For example, one sometimes simply needs to know the total energy flux or intensity, which are defined as

$$F = \int F_\nu \, d\nu \tag{5.33}$$

and

$$I = \int I_\nu \, d\nu, \tag{5.34}$$

respectively.

In high-energy astrophysics, we often encounter situations where we must know how to transform the radiation field from one frame to another, either involving high relative velocities between the two, or a relative gravitational acceleration, or both. This is most conveniently done in terms of the specific intensity I_ν.

We recall from our discussion in section 3.1 that all observers must agree on the value of a dimensionless quantity. Thus, even though the intensity itself is not invariant, we can nonetheless still use it to define a photon number that is the same in every frame. Let N be the number of photons in the frequency interval $(\nu, \nu + d\nu)$, passing through an element of area dA perpendicular to the direction $\hat{\mathbf{z}}$ and at rest in the laboratory, into a solid angle $d\Omega$ along $\hat{\mathbf{r}}$ making an angle θ relative to this axis, during a time interval dt. Mathematically, this is written

$$N = \left[\frac{I_\nu(\mu)}{h\nu}\right] d\Omega \, d\nu \, dA \, \cos\theta \, dt, \tag{5.35}$$

where $dA \cos\theta$ is the projected area "seen" by the photons, and $\mu \equiv \cos\theta$.

Now suppose that a second observer is moving with velocity $\mathbf{v} = v\hat{\mathbf{z}}$ relative to the laboratory. According to him, the area dA moves to the left (i.e., to a smaller value of z) during the time that the photons (moving to the right) are counted; the travel time incurred by this motion must be taken into account when he adds up all the photons that have crossed dA.

Thus, all told, the moving observer counts a photon number

$$N' = \left[\frac{I'_{\nu'}(\mu')}{h\nu'}\right] d\Omega' \, d\nu' \left[dA' \, \cos\theta' \, dt' + \frac{v}{c} dA' \, dt'\right]. \tag{5.36}$$

The second term inside the brackets represents the additional volume swept out by the area dA' as it moves to the left. With our choice of axes, and the fact that $N = N'$, we therefore have

$$\left[\frac{I_\nu(\mu)}{\nu}\right] d\Omega \, d\nu \, \cos\theta \, dt = \left[\frac{I'_{\nu'}(\mu')}{\nu'}\right] d\Omega' \, d\nu' \left[\cos\theta' + \frac{v}{c}\right] dt' \tag{5.37}$$

or, rearranging,

$$I_\nu(\mu)\,\mu\,d\Omega\,dt\,\frac{d\nu}{\nu} = I'_{\nu'}(\mu')\,(\mu'+\beta)\,d\Omega'\,dt'\,\frac{d\nu'}{\nu'}, \tag{5.38}$$

using the standard notation we introduced in chapter 3.

Later in this chapter, we will consider the transformation of angles under a Lorentz boost, a phenomenon known as aberration. Borrowing the results from equations (5.88) and (5.90), which lead to the angle transformation formula in equation (5.91), we infer that

$$\mu = \frac{\mu'+\beta}{1+\beta\mu'} \tag{5.39}$$

and

$$d\Omega = \sin\theta\,d\theta\,d\phi = \frac{d\Omega'}{\gamma^2(1+\beta\mu')^2}. \tag{5.40}$$

With dA at rest in the laboratory, we have $dt' = \gamma\,dt$. Thus, altogether

$$I_\nu(\mu) = \gamma^3(1+\beta\mu')^3\,I'_{\nu'}(\mu'). \tag{5.41}$$

A more elegant form of this expression may be obtained from the Doppler shift formula (equation 3.18), which allows us to express the multiplicative factor on the right-hand side as a ratio ν/ν'. Evidently,

$$\frac{I_\nu(\mu)}{\nu^3} = \frac{I'_{\nu'}(\mu')}{\nu'^3}, \tag{5.42}$$

so the Lorentz invariant is not I_ν, but rather I_ν/ν^3. This elegant and powerful Lorentz scalar finds many uses in high-energy astrophysics, including, as we shall see in section 12.3.3, a direct application to the radiative flux produced in relativistic jets.

5.3 THERMAL BREMSSTRAHLUNG

There is a fundamental reason why an isolated charged particle cannot spontaneously emit (or absorb) radiation. According to equation (3.58), all of a photon's energy appears as momentum, whereas a fraction of a massive particle's energy is locked up in its inertia and cannot contribute directly to **p**. Therefore, as long as both energy and momentum are independently conserved, it is impossible to extract energy and momentum from the charged particle in just the right balance to render the photon dynamically independent. However, this transfer is possible when another source of energy and momentum lies nearby. This reservoir could be the virtual photons in a pulsar's magnetosphere, or in the Coulomb field of an ion. Under the guise of electron–proton scattering, the latter is one of the most common manifestations of bremsstrahlung radiation in astronomy. As we shall see, when the reservoir is a magnetic field, this type of bremsstrahlung is more commonly known as synchrotron radiation. Some of the older texts sometimes referred to this process as "magnetic bremsstrahlung," as a more specific characterization of what was actually taking place.

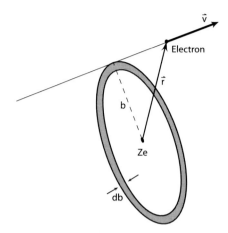

Figure 5.4 Bremsstrahlung geometry for a single electron accelerated in the Coulomb field of an ion Ze. The impact parameter b indicates the distance of closest approach of the two particles, estimated on the basis that the ion, being much heavier than the electron, essentially remains stationary during the encounter. The annulus of width db is used in the integral determination of the overall emissivity due to an ensemble of electrons flowing past the ion with a range of impact parameters.

We shall return to synchrotron emission in section 5.3. For the time being, we will restrict our attention to the radiation emitted by a single electron moving at velocity $\mathbf{v}$ under the influence of a static electric field produced by an ion with charge Ze (see figure 5.4). (We assume here that the ion is so much more massive than the electron that its motion may be ignored during the interaction.) In that case, $\ddot{\mathbf{d}} = -e\mathbf{v}$, so from equation (5.18), we know that the Fourier transform of d is

$$\bar{d}(\omega) = -\frac{1}{2\pi\omega^2} \int_{-\infty}^{\infty} e^{i\omega t'} \ddot{d}(t') \, dt' \tag{5.43}$$

or

$$\bar{d}(\omega) = \frac{e}{2\pi\omega^2} \int_{-\infty}^{\infty} e^{i\omega t'} \dot{v}(t') \, dt'. \tag{5.44}$$

If we assume further that most of the electron's change in velocity Δv occurs over a collision time $\tau \ll 1/\omega$, then $\exp(i\omega t')$ is always close to one during the main contribution[3] to the integral in equation (5.44), so

$$\bar{d}(\omega) \approx \frac{e}{2\pi\omega^2} \Delta v. \tag{5.45}$$

Thus, the energy spectrum of the radiation (see equation 5.30) emitted by the accelerated electron is

$$\frac{dW}{d\omega} = \frac{2e^2}{3\pi c^3} |\Delta v|^2. \tag{5.46}$$

[3]Note that for times $\gg 1/\omega$, the integrand in equation (5.44) oscillates so rapidly that any additional contribution to the integral is small anyway.

But for small path deviations, the acceleration in the direction parallel to the segment b (the distance of closest approach) is roughly $(b/r)(Ze^2/r^2 m_e)$. This is the only component of acceleration that maintains the same direction throughout the electron's motion. The other component—parallel to $\mathbf{v}$—reverses sign as the electron crosses the midpoint in its path, and therefore cancels out when the effect is integrated over the whole trajectory. Therefore,

$$\Delta v \approx \frac{Ze^2}{m_e} \int_{-\infty}^{\infty} \frac{b\,dt}{(b^2 + v^2 t^2)^{3/2}}, \tag{5.47}$$

with solution

$$\Delta v \approx \frac{2Ze^2}{m_e b v}. \tag{5.48}$$

And this leads immediately to the single-particle bremsstrahlung spectrum for an impact parameter b,

$$\frac{dW(b)}{d\omega} = \frac{8Z^2 e^6}{3\pi c^3 m_e^2 v^2 b^2}. \tag{5.49}$$

This expression is sufficient for us to calculate the overall emissivity and spectrum of any population of particles, as long as we know their distribution in b. As a concrete example, suppose we have a monoenergetic distribution of electrons scattering off a population of ions with a much smaller characteristic velocity (i.e., $v_i \ll v_e \approx v$).

The flux of electrons (number per unit area, per unit time) passing by a given ion is $n_e v$, where n_e is the electron number density (and we denote the ion number density as n_i). Therefore, the number of electrons passing through an annulus of width db at b (see figure 5.4) is $n_e v\, 2\pi b\, db$. The frequency-dependent emissivity (energy per unit volume, per unit time, per unit frequency) from this system is thus

$$\left(\frac{dW}{d\omega}\right)\frac{1}{\Delta V\, \Delta t} = n_e n_i Z^2\, 2\pi v \int_{b_{min}}^{\infty} \frac{dW(b)}{d\omega} b\, db. \tag{5.50}$$

Given that the integrand is proportional to $1/b$ (with a substitution from equation 5.49), the result is conveniently expressed in terms of the *Gaunt* factor

$$g_{ff}(v, \omega) = \frac{\sqrt{3}}{\pi} \ln\left(\frac{b_{max}}{b_{min}}\right), \tag{5.51}$$

the upper limit for the integral in equation (5.50) having been replaced by a more manageable maximum value b_{max}. The corresponding minimum value of the impact parameter is labeled b_{min}.

The Coulomb force due to the ion drops off as $1/r^2$, but is never zero, so it is not always clear where the upper limit to the integral in equation (5.50) should be taken. However, based on our earlier discussion concerning the collision time τ (which we argued should be $\ll 1/\omega$), we believe that a reasonable estimate for the ion's range of influence ought to be $b_{max} \sim v\tau$, or $b_{max} \sim v/\omega$ if we ignore factors of order one. It is equally difficult to pin down a precise value for b_{min}, but here we may invoke a quantum mechanical argument based on the uncertainty principle. With $\Delta x\, \Delta p \gtrsim \hbar$ and $p_e = m_e v$, where Δx and Δp are, respectively, the uncertainty in position and momentum, and $\hbar$ is Planck's constant divided by 2π, we see that $b_{min} \sim \Delta x \sim \hbar/m_e v$.

With the Gaunt factor thus known, the left-hand side of equation (5.50) becomes

$$\frac{d^3 W}{d\omega \, dV \, dt} = \frac{16\pi e^6}{3\sqrt{3}\, c^3 m_e^2 v} n_e n_i Z^2 \, g_{ff}(v, \omega), \tag{5.52}$$

in which we have also taken the infinitesimal limit for the volume V and time t.

It is not difficult to generalize this useful expression to produce a formulation with broader applicability in astrophysics. When the particles belong to a thermal population, their velocity profile may be characterized in terms of the probability $\Pi(v)$ of finding a particle within the velocity range $d^3\mathbf{v} = 4\pi v^2 \, dv$ at $\mathbf{v}$. Invoking a Maxwell-Boltzmann distribution, we write

$$d\Pi(v) \propto \exp\left(-\frac{m_e v^2}{2kT}\right) 4\pi v^2 \, dv. \tag{5.53}$$

Therefore, putting $\omega = 2\pi v$ in equation (5.52), we obtain the thermal-velocity averaged emissivity

$$\epsilon_{ff}(v) \equiv \frac{d^3 W}{dv \, dV \, dt} = \frac{32\pi^2 e^6 n_e n_i Z^2}{3\sqrt{3}c^3 m_e^2} \, \text{Int}(v, T), \tag{5.54}$$

where

$$\text{Int}(v, T) \equiv \left[\int_{v_{min}}^{\infty} g_{ff}(v, v)\, v \, \exp\left(-\frac{m_e v^2}{2kT}\right) dv\right]$$
$$\div \left[\int_0^{\infty} v^2 \exp\left(-\frac{m_e v^2}{2kT}\right) dv\right]. \tag{5.55}$$

Note that the first integral in equation (5.55) has a lower cutoff because the accelerated electron must have sufficient energy to produce a photon with frequency v before it can actually radiate. Thus, $v_{min} = (2hv/m_e)^{1/2}$.

The evaluation of these elementary integrals produces the final result, written in CGS units (ergs cm^{-3} s^{-1} Hz^{-1}),

$$\epsilon_{ff}(v) = 6.8 \times 10^{-38} n_e n_i Z^2 T^{-1/2} \exp\left(-\frac{hv}{kT}\right) \bar{g}_{ff}, \tag{5.56}$$

where $\bar{g}_{ff}$ is the velocity-averaged Gaunt factor. For the vast majority of astrophysical thermal plasmas, $5 \gtrsim \bar{g}_{ff} \gtrsim 1$, and one may therefore set $\bar{g}_{ff} \sim 1$ for practical applications of equation (5.56).

As we analyze the emitted spectrum in this expression, several important characteristics emerge. Among them is the fact that the spectrum is flat for photon energies hv much smaller than the thermal energy kT. As we shall see later in this chapter, such a profile is unique among the various emission mechanisms, providing a distinct observational signature. In addition, the spectrum turns over at $hv \sim kT$ and drops off exponentially at even higher energies (see figure 5.5). Not only does this trait help to make the spectral shape more distinctive but, more importantly, it also provides us with the means of measuring the temperature of the emitting

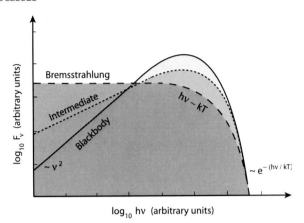

Figure 5.5 The optically thin bremsstrahlung spectrum (dashed curve) is flat, and has a characteristic rollover at photon energy $h\nu \sim kT$. The other two curves show the systematic progression from optically thin emission to a blackbody spectrum (solid curve), resulting from a gradual increase in optical depth. The key process responsible for this is inverse Compton scattering, which shifts the low-frequency radiation toward higher energy as the photons are brought into equilibrium with the ambient plasma. The peak in the blackbody curve lies at $h\nu \sim 2.82kT$.

plasma, for once we have identified the photon energy $h\nu_{\text{knee}}$ corresponding to this bremsstrahlung *knee*, we know that it must be equal to kT.

By now, it must be obvious that we have not yet said anything about the transport of radiation through the medium once the radiation is emitted. Our derivation leading to equation (5.56) assumed all along that the photons would leave the system unaffected once they were produced. For this reason, the spectrum we have just considered is called the optically thin bremsstrahlung spectrum, since no property of the medium is imprinted on it other than the temperature T of the plasma.

But this is not always the situation we encounter in high-energy astrophysics. Often, the medium is not thin enough for the radiation to escape without a further interaction with the plasma that produced it, and the emerging spectrum is modified in ways that reveal the underlying physical conditions. This evolution is demonstrated schematically in figure 5.5, which shows the transition from optically thin (bremsstrahlung) to optically thick (blackbody) emission.

To make this discussion more quantitative, we note that the radiating medium is optically thin when the photon's mean free path $\langle l \rangle$ for scattering or reabsorption is large compared with the size R of the system, i.e., when $R/\langle l \rangle \lesssim 1$. And since the photon mean free path may be written $\langle l \rangle \sim 1/n_e\sigma$, where σ is the cross section (equal to the Thomson cross section σ_T in the case of pure scattering), the requirement for the medium to be optically thin is that the quantity $\tau \equiv n_e\sigma R$ (known as the *optical depth*) should be less than one.

On the other hand, when $\tau \gg 1$, the typical photon in the emitting medium undergoes numerous scatterings (and/or absorptions) before leaving the system. The particles with which the photons scatter have energies $\sim kT$, and therefore the low-frequency photons (to the left in figure 5.5) tend to be upscattered to energies

$\sim kT$, as we shall see in our discussion of inverse Compton scattering in section 5.6. So the spectrum starts off flat when $\tau \ll 1$, and then develops a prominent bump near kT at the expense of lower-energy photons whose number depletes more and more as τ increases. Eventually, the optical depth is so large that the radiation and plasma reach a local thermodynamic equilibrium, and the radiation emerges with the intensity of a blackbody at temperature T,

$$B_\nu(T) = \frac{2h\nu^3/c^2}{\exp(h\nu/kT) - 1}. \tag{5.57}$$

This has a $\sim \nu^2$ dependence at low energies, peaks near $2.82kT$, and then falls off exponentially like the bremsstrahlung spectrum.

5.4 SINGLE-PARTICLE SYNCHROTRON EMISSIVITY

The energy and momentum transferred to the radiating electron by an ion's Coulomb field may also be provided by a magnetic field $\mathbf{B}$, which accelerates a charge q according to the Lorentz force in equation (4.18). A nonrelativistic electron's motion in the presence of $\mathbf{B}$ is a superposition of a translational path with (constant) velocity $v_\parallel = v \cos \alpha$ (see figure 5.6), and a circular (accelerated) component with velocity $v_\perp = v \sin \alpha$. This acceleration acts to change only the direction of $\mathbf{v}$, not its magnitude. Classically, however, the electron's orbit decays because it loses energy to the radiation field it produces. (In a quantum mechanical description, the electron radiates as its orbit degenerates through the various Landau energy levels.)

From Newton's law of motion, we know that

$$m_e \frac{v_\perp^2}{r_{\text{gyr}}} = \left| -\frac{ev_\perp B}{c} \right| = \left| -\frac{ev \sin \alpha \, B}{c} \right|, \tag{5.58}$$

where r_{gyr} is the gyration radius, which may be written

$$r_{\text{gyr}} = \frac{v \sin \alpha}{\omega_{\text{gyr}}}, \tag{5.59}$$

in terms of the gyration frequency

$$\omega_{\text{gyr}} \equiv \frac{eB}{m_e c} \approx 1.8 \times 10^7 \left(\frac{B}{1\,\text{G}} \right). \tag{5.60}$$

Thus, according to Larmor's equation (5.14), the power emitted by a single electron, accelerated at a rate $a = \omega_{\text{gyr}} v \sin \alpha$, is

$$P = \frac{2e^2}{3c^3} \omega_{\text{gyr}}^2 v^2 \sin^2 \alpha. \tag{5.61}$$

We see right away that the power emerges as monochromatic radiation with angular frequency ω_{gyr}. This type of emission is known as *cyclotron* radiation, a name derived from the cyclotron, a type of particle accelerator first used in the 1930s to produce highly energetic particles for scattering experiments. The charges within this device

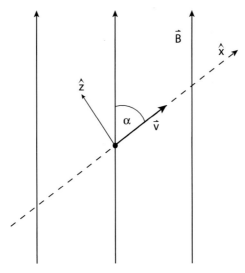

Figure 5.6 Geometry of a single particle gyrating with velocity **v**, making a pitch angle α relative to the magnetic field **B**.

were kept in circular orbits by the **v** × **B** forces generated from the magnetic field imposed on them inside the instrument.

The observer in the laboratory frame will see the same frequency ω_{gyr} as that measured in the particle's rest frame because the Doppler shift (equation 3.18) has only a very minor effect on ω when the particle speed v is subrelativistic. As a consequence, the monochromatic spectrum associated with cyclotron radiation is essentially independent of viewing angle, and its angular distribution in the laboratory is essentially that viewed in the particle's rest frame—the dipole pattern shown in figure 5.3—since the boosting effects in the direction of **v** are minimal. This situation will change dramatically, however, when $v \to c$.

A more interesting development arises when the motion is relativistic, for then the electron follows a path distorted by time-dilation and length-contraction effects. The fact that the emitted power is angle dependent in this case means that a single frequency can no longer adequately describe the entire spectrum. But for us to properly handle the particle dynamics in terms of physical quantities transformed from one frame to the next, we first need to understand what happens to the electromagnetic field under a relativistic boost, so we will need to consider the equations of section 4.2 written in the language of four-dimensional spacetime.

The charge-conservation equation (4.16) may be written as an invariant physical law with the identification of a four-vector for the current,

$$J^\alpha \equiv (c\rho_e, \mathbf{J}), \tag{5.62}$$

with which the continuity equation then becomes

$$\frac{\partial J^\alpha}{\partial x^\alpha} \equiv \partial_\alpha J^\alpha = 0. \tag{5.63}$$

As the contraction of two four-vectors, the left-hand side of this equation is clearly a four-scalar, and since it reduces to the correct form (equation 4.16) in any given observer's frame, this expression is the appropriate Lorentz-invariant representation for charge conservation.

In dealing with Maxwell's equations, it is often convenient to write the fields in terms of potentials. For example, Gauss's law for **B** (equation 4.7) shows immediately that there must exist a vector potential **A** such that

$$\mathbf{B} = \vec{\nabla} \times \mathbf{A}. \tag{5.64}$$

Faraday's law (equation 4.12) then gives

$$\text{curl } \mathbf{E} = -\frac{1}{c} \frac{\partial}{\partial t} \text{curl } \mathbf{A} \tag{5.65}$$

or

$$\text{curl} \left(\mathbf{E} + \frac{1}{c} \frac{\partial \mathbf{A}}{\partial t} \right) = 0. \tag{5.66}$$

Since the quantity in parentheses must be the gradient of a scalar function, it is evident that

$$\mathbf{E} = -\frac{1}{c} \frac{\partial \mathbf{A}}{\partial t} - \vec{\nabla} \Phi, \tag{5.67}$$

where Φ is the scalar potential. In electrostatics (i.e., time-independent situations), the first term on the right-hand side of this equation is zero, and Φ reduces to the more familiar potential from which the conservative field **E** may then be derived without reference to **B** or **A**.

It is rather easy to show that the contraction $\partial_\alpha A^\alpha$ is a Lorentz scalar, where

$$A^\alpha \equiv (\Phi, \mathbf{A}). \tag{5.68}$$

Thus, like J^α, the quantity A^α must also be a four-vector. We may now return to equation (5.67), and consider each component i individually,

$$
\begin{aligned}
E^i &= -\frac{1}{c} \frac{\partial A^i}{\partial t} - \frac{\partial \Phi}{\partial (-x_i)} \\
&= -\frac{\partial A^i}{\partial x^0} - \frac{\partial A^0}{\partial (-x_i)} \\
&= -\frac{\partial A^i}{\partial x_0} + \frac{\partial A^0}{\partial x_i} \\
&= -\left(\partial^0 A^i - \partial^i A^0 \right).
\end{aligned}
\tag{5.69}
$$

With a similar derivation, we find that

$$
\begin{aligned}
B_i &= -\varepsilon_{ijk} \frac{\partial A^j}{\partial (-x_k)} \\
&= \varepsilon_{ijk} \partial^k A^j,
\end{aligned}
\tag{5.70}
$$

where ε_{ijk} is defined by

$$\varepsilon_{ijk} = \begin{cases} +1 & \text{if } ijk \text{ forms an even permutation of 123} \\ -1 & \text{if } ijk \text{ forms an odd permutation of 123} \\ 0 & \text{otherwise.} \end{cases} \qquad (5.71)$$

Together, these two equations suggest that the components of **E** and **B** are elements of a second-rank, antisymmetric *field-strength tensor:*

$$F^{\alpha\beta} = \partial^\alpha A^\beta - \partial^\beta A^\alpha. \qquad (5.72)$$

Written out explicitly, we have

$$F^{\alpha\beta} = \begin{pmatrix} 0 & -E^x & -E^y & -E^z \\ E^x & 0 & -B^z & B^y \\ E^y & B^z & 0 & -B^x \\ E^z & -B^y & B^x & 0 \end{pmatrix}. \qquad (5.73)$$

This field-tensor knits the two 3-vector fields **E** and **B** into a single entity, resolvable on the four dimensions of spacetime in ways that convert electric into magnetic fields, and vice versa, merely by viewing the *electromagnetic field* from relatively moving frames. Its primary usefulness will be to permit us to cast the Maxwell equations into an explicitly covariant form.

Gauss's law (equation 4.6) becomes

$$\partial_i F^{i0} = \frac{4\pi}{c} J^0, \qquad (5.74)$$

which generalizes to

$$\partial_\alpha F^{\alpha\beta} = \frac{4\pi}{c} J^\beta. \qquad (5.75)$$

Though not immediately obvious, the various new components of this equation actually correspond to those of equation (4.17), as one may verify by using the definition of J^α and the field tensor in equation (5.73). The decomposition of equation (5.75) into its different components may be made with equal validity in any frame, consistent with the relativity principle. This is the four-dimensional composition of the inhomogeneous Maxwell equations in special relativity.

Using the language of four-dimensional spacetime, the homogeneous Maxwell equations may be rewritten with comparable ease and validity. For example, Gauss's law for **B** (equation 4.7) becomes

$$\partial^1 F^{32} + \partial^2 F^{13} + \partial^3 F^{21} = 0. \qquad (5.76)$$

And, again, a generalization to the form

$$\partial^\alpha F^{\beta\gamma} + \partial^\beta F^{\gamma\alpha} + \partial^\gamma F^{\alpha\beta} = 0 \qquad (5.77)$$

produces the rest of the components that correspond to Faraday's equation (4.12).

The successful formulation of electrodynamics using the electromagnetic field tensor $F^{\alpha\beta}$, which blends together the components of **E** and **B**, shows that the concept of a pure electric or a pure magnetic field is not consistent with Lorentz invariance. It is precisely for this reason that we refer to it as the *electromagnetic field*—one quantity, unified in the special theory of relativity. This point is affirmed compellingly when we examine how the electromagnetic field behaves under a Lorentz transformation.

For example, in figure 5.6, the boost from the electron's rest frame to the laboratory frame is along the $\hat{x}$-direction. Thus, the transformation matrix of coefficients $\mathcal{A}$ in equation (3.32) here becomes

$$\mathcal{A} \equiv [a^{\alpha}{}_{\beta}] = \begin{pmatrix} \gamma & -v\gamma/c & 0 & 0 \\ -v\gamma/c & \gamma & 0 & 0 \\ 0 & 0 & 1 & 0 \\ 0 & 0 & 0 & 1 \end{pmatrix}. \tag{5.78}$$

Therefore, using our prescription for transforming contravariant tensors (see equation 3.34), in which a prime denotes quantities in the rest frame, we find that

$$\begin{aligned} E^{x'} &= E^x \\ E^{y'} &= \gamma(E^y - vB^z/c) \\ E^{z'} &= \gamma(E^z + vB^y/c). \end{aligned} \tag{5.79}$$

Thus, since $\mathbf{E} = 0$,

$$\begin{aligned} E^{x'} &= 0 \\ E^{y'} &= -\frac{\gamma v}{c} B \sin \alpha \\ E^{z'} &= 0. \end{aligned} \tag{5.80}$$

And so in the electron's rest frame, an application of the Lorentz force equation (4.18) results in the acceleration

$$\dot{\mathbf{v}}' = \frac{e\gamma vB \sin \alpha}{m_e c} \tag{5.81}$$

or

$$\dot{\mathbf{v}}' = \gamma v\omega_{\text{gyr}} \sin \alpha, \tag{5.82}$$

which is exactly γ times the classical value.

According to the Larmor equation (5.14), the power radiated by the electron in its own rest frame is therefore

$$P' = \frac{2e^2}{3c^3} \gamma^2 \omega_{\text{gyr}}^2 v^2 \sin^2 \alpha. \tag{5.83}$$

Using the definition of the Thomson cross section,

$$\sigma_T \equiv \frac{8\pi r_0^2}{3}, \tag{5.84}$$

in terms of the classical electron radius $r_0 \equiv e^2/m_e c^2$, we can also write this as

$$P' = \tfrac{4}{3}\sigma_T c\beta^2 \gamma^2 u_B, \tag{5.85}$$

where $u_B \equiv B^2/8\pi$ is the magnetic energy density, and we have averaged over all pitch angles, with $\int 2\pi \sin^3 \alpha \, d\alpha = 8\pi/3$.

The fact that energy ($= cp^0$) and time ($= x^0/c$) transform in exactly the same way under a relativistic transformation means that the ratio E/t must be a Lorentz invariant. Thus, $P = P'$, and we conclude that the total synchrotron power from an accelerated electron in the laboratory must be

$$P_{\text{sync}} = \tfrac{4}{3}\sigma_T c\beta^2\gamma^2 u_B. \tag{5.86}$$

Now that we know how much power may be produced by a single charge in the presence of a magnetic field, we must follow the same procedure we devised for bremsstrahlung emission, and consider both the frequency dependence of this radiation and the spectrum produced by an ensemble of charges. Of course, we need to do this to be able to recognize the pattern of synchrotron emission in specific high-energy sources. It is here that we recognize the importance of *beaming*, which arises as a result of aberration—effects due to the relativistic transformation of angles.

The radiation pattern in the electron's rest frame is that shown in figure 5.3. Though the velocity in this frame is, by definition, zero, the charge experiences an acceleration (parallel to $\dot{\mathbf{d}}$) evaluated in equation (5.82), for which our derivation of $dP/d\Omega$ in equation (5.12) then carries over to this frame, with the proper identification of the power P' and solid angle Ω'. But the angular distribution of the radiation cannot be symmetric about $\mathbf{d}$ in the laboratory frame, because in general $\mathbf{v}$ and $\dot{\mathbf{v}}$ are not parallel.

For the sake of specificity, let us assume that $\mathbf{v}$ and $\dot{\mathbf{v}}$ are perpendicular to each other, as would occur for motion in a uniform magnetic field (or for any circular motion for that matter). Then, $dP'/d\Omega' \propto \sin^2\theta'$, where θ' is the rest-frame angle corresponding to θ shown in figure 5.7. The quantity $dP/d\Omega$ is not the same as $dP'/d\Omega'$ specifically because $\theta' \neq \theta$, the effect known as aberration. The most straightforward way to find the relationship between these two angles is to use the fact that c is invariant.

Suppose two frames are moving relative to each other with velocity $\mathbf{v}$ along the z-axis. Then, according to equation (3.10), the component $u_z = dz/dt$ of a velocity u measured in the laboratory frame is

$$\frac{dz}{dt} = \frac{dz' + v\,dt'}{dt' + v\,dz'/c^2} \tag{5.87}$$

or

$$u_z = \frac{u_z' + v}{1 + vu_z'/c^2}. \tag{5.88}$$

Similarly, in the (x- or) y-direction,

$$u_y = \frac{dy}{dt} = \frac{dy'}{\gamma\,(dt' + v\,dz'/c^2)}, \tag{5.89}$$

which also simplifies to the form

$$u_y = \frac{u_y'}{\gamma\,(1 + vu_z'/c^2)}. \tag{5.90}$$

Since we wish to determine boosting effects relative to $\mathbf{v}$, let us define the angle $\psi \equiv \pi/2 - \theta$. Then, using this angle convention with $u = c$, we infer from

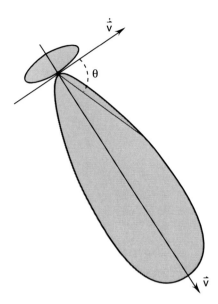

Figure 5.7 When the radiating particle moves relativistically, its dipole pattern (which would otherwise be symmetric about the acceleration vector as indicated in figure 5.3) becomes distended in the direction of motion due to the effects of Doppler blueshifting forward and redshifting backward.

equation (5.90) that

$$\sin \psi = \frac{\sin \psi'}{\gamma (1 + \beta \cos \psi')}. \tag{5.91}$$

A ray leaving the electron in a direction $\theta' = \pi/4$ to $\ddot{\mathbf{d}}$ has a $dP'/d\Omega'$ equal to half its maximum possible value (which occurs at $\theta' = \pi/2$). However, as seen in the laboratory frame, this ray points in a direction much closer to $\mathbf{v}$. According to equation (5.91),

$$\sin \psi \approx \psi \approx \frac{1}{\gamma}. \tag{5.92}$$

Thus, whereas the power is radiated nearly isotropically in the particle's rest frame, most of it is *beamed* into a narrow cone with half-opening angle $\sim 1/\gamma$ as seen in the laboratory (see also figure 5.8).

Because of this beaming, an observer in the laboratory does not see a uniform, sinusoidal variation of the electric field as the particle gyrates. Instead, the emission is pulsed every time the cone sweeps around the line of sight. In Fourier space, this pulsed emission is represented as a superposition of many frequencies, unlike the purely monochromatic spectrum produced by a nonrelativistic charge. Calculating the distribution in frequency of the emitted radiation is not difficult, but it requires an understanding of a subtle, though important, point. This has to do with how we

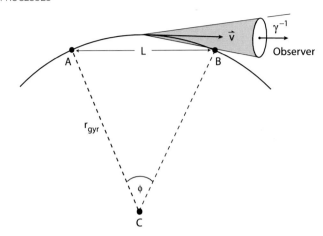

Figure 5.8 The effect highlighted in figure 5.7 manifests itself in several important ways, including the range of frequencies spanned by the emitted radiation. When its dipole pattern is heavily boosted in the forward direction, the radiation from a gyrating particle is visible predominantly within a narrow cone whose half-opening angle is ~$1/\gamma$. Instead of seeing a sinusoidally modulated wave with a single frequency corresponding to the gyration rate, an observer now sees a spike in emission for only a portion of the orbit which, in Fourier space, is represented by a range of frequencies as indicated in figure 5.9.

measure the time of emission in the laboratory frame versus the actual interval over which the particle is radiating.

With reference to figure 5.8, let us assume that γ is so large that $L \approx r_{\mathrm{gyr}}\phi$. We infer that the particle is beaming its radiation toward us only for as long as our line of sight lies within the cone of emission, i.e., for a time t^{0}_{AB}. But because of light travel-time effects, t^{0}_{AB} is actually shorter than $t_{AB} \equiv L/v$. During the time t_{AB} that the particle has moved from A to B, the radiation emitted at A has traveled a distance cL/v. So we interpret the distance over which emission has taken place to be $cL/v - L$. Thus, the arrival (or observed) time is $(cL/v - L)/c$, or

$$t^{0}_{AB} = \left(1 - \frac{v}{c}\right) t_{AB}. \tag{5.93}$$

But

$$\frac{L}{v} \approx \phi \frac{r_{\mathrm{gyr}}}{v} = \frac{2}{\gamma} \frac{1}{\omega^{\mathrm{rel}}_{\mathrm{gyr}}}, \tag{5.94}$$

where $\omega^{\mathrm{rel}}_{\mathrm{gyr}}$ is the relativistic angular gyration frequency, which differs from ω_{gyr} in equation (5.60) due to time-dilation and length-contraction effects.

From equations (3.50), (3.51), and (4.18), we know that the components of the electron's equation of motion in terms of its charge $q = -e$ are

$$\frac{d}{dt}(\gamma m_e \mathbf{v}) = \frac{q}{c}\mathbf{v} \times \mathbf{B} \tag{5.95}$$

and

$$\frac{d}{dt}(\gamma m_e c^2) = q\mathbf{v} \cdot \mathbf{E} = 0. \tag{5.96}$$

The second of these forces the condition $\gamma = $ constant, so that in the first equation,

$$m_e \gamma \frac{d\mathbf{v}}{dt} = \frac{q}{c}\mathbf{v} \times \mathbf{B}. \tag{5.97}$$

That is,

$$\frac{d\mathbf{v}_\parallel}{dt} = 0 \tag{5.98}$$

and

$$\frac{d\mathbf{v}_\perp}{dt} = \frac{q}{\gamma m_e c}\mathbf{v}_\perp \times \mathbf{B}, \tag{5.99}$$

where $\mathbf{v}_\parallel$ is parallel to $\mathbf{B}$, and $\mathbf{v}_\perp$ is the perpendicular component of velocity. It is evident that the electron's relativistic angular frequency of gyration must be

$$\omega_{\mathrm{gyr}}^{\mathrm{rel}} \equiv \left|\frac{qB}{\gamma m_e c}\right|, \tag{5.100}$$

which is exactly $\omega_{\mathrm{gyr}}/\gamma$ (see equation 5.60).

Returning now to equations (5.93) and (5.94), we see that

$$t_{AB}^0 \approx (1-\beta)\frac{2}{\omega_{\mathrm{gyr}}}, \tag{5.101}$$

and since $(1-\beta) \approx 1/2\gamma^2$ for relativistic motion,

$$t_{AB}^0 \approx \frac{1}{\gamma^2 \omega_{\mathrm{gyr}}}. \tag{5.102}$$

The point here is that, whereas in the nonrelativistic case the relation $t_{AB}^0 = 2\pi/\omega_{\mathrm{gyr}}$ results in a single frequency of emission $\nu = \nu_{\mathrm{gyr}} = \omega_{\mathrm{gyr}}/2\pi$, the relativistic arrival time is much shorter than an orbital period so that the dominant frequency of the radiation (which goes as $1/t_{AB}^0$) is much *higher* than ν_{gyr}. And because the observed field is no longer sinusoidal when beaming is important, the spectrum is associated not with a single frequency, but with a range of frequencies spread across $1/t_{AB}^0$. In addition, because $\omega_{\mathrm{gyr}}^{\mathrm{rel}} = \omega_{\mathrm{gyr}}/\gamma$, the spacing between adjacent frequencies is much smaller than it would be in the nonrelativistic case. The net result of all these modifications to the classical cyclotron emissivity is a summation of many discrete emission lines that blend together near $\gamma^2 \nu_{\mathrm{gyr}}$ to form a continuum when $\gamma \to \infty$, as shown for the single particle spectrum in figure 5.9.

Our discussion of synchrotron emission thus far has been heuristic for the purpose of identifying the essential principles involved in this important physical process. Of course, it is possible to go beyond this simple treatment and determine the shape of the curve in figure 5.9 precisely. To do this, we need to calculate the variation of $\mathbf{E}$ as the particle moves around its orbit, and then extract its Fourier transform according to the procedure outlined in section 5.1. This eventually leads

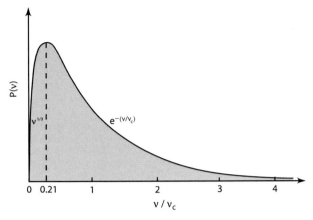

Figure 5.9 Single-particle spectrum in the relativistic domain, showing the spread in frequencies resulting from a departure from pure sinusoidal emission in the dipole limit. The peak emissivity occurs at $\nu \approx 0.29\nu_c$, where ν_c is roughly the inverse-Compton-scattered frequency of a virtual photon with $\nu = \nu_{\text{gyr}}$.

to a precise formulation of the single-particle spectrum in equation (5.30).[4] Taking all the Fourier components into account results in the following expression for the total power emitted per unit frequency by a single electron:

$$P(\nu)\, d\nu = \frac{\sqrt{3} e^3 B \sin\alpha}{m_e c^2} \frac{\nu}{\nu_c}\, d\nu \int_{\nu/\nu_c}^{\infty} K_{5/3}(\eta)\, d\eta, \qquad (5.103)$$

where $K_{5/3}$ is the Bessel function of order $\frac{5}{3}$ (representing the "summation" over all frequency components), and

$$\nu_c \equiv \tfrac{3}{2}\gamma^2 \nu_{\text{gyr}} \sin\alpha \qquad (5.104)$$

is the *critical frequency*, corresponding roughly to $1/t_{AB}^0$.

The maximum of $P(\nu)$ may be found by differentiating equation (5.103) with respect to ν (remembering that the limits on the integral themselves depend on ν). The turning point occurs at

$$\nu_{\text{max}} \approx 0.29\nu_c \approx 0.45\gamma^2 \nu_{\text{gyr}} \sin\alpha. \qquad (5.105)$$

For frequencies $\nu \gg \nu_c$, the Bessel function asymptotes to an exponential function, so in this limit, $P(\nu) \to \exp(-\nu/\nu_c)$. At low frequencies, $P(\nu) \to \nu^{1/3}$. These simplified functional forms are easily identified in figure 5.9.

We note, however, that the single-particle spectrum still depends on the pitch angle α, through its appearance in the critical frequency ν_c. Because of this, and the fact that most synchrotron emission from astrophysical sources is produced by an ensemble of particles, the use of equation (5.103) is rather limited. One often has to make some assumption concerning the electron distribution, e.g., whether they are thermal

---

[4]This derivation was carried out in detail by Pacholczyk (1970) and we cannot improve on his beautiful presentation here. Instead, the reader is referred to his now classic book on the many important aspects of radio astronomy.

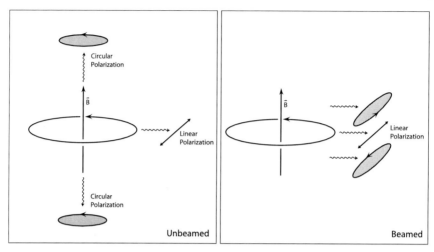

Figure 5.10 Forward beaming in relativistic motion blends together the circular and linear polarized components, which would otherwise be visible separately at different angles in the case of pure dipole radiation. In the synchrotron limit, the right and left circularly polarized waves cancel, producing a linearly polarized radiation field.

or nonthermal, and their distribution in α. We shall consider two special cases in sections 5.4 and 5.5. But, first, there is another important observational characteristic of synchrotron emission that we ought to consider here—the polarization.

Bremsstrahlung emission is a purely stochastic process. Electrons and ions accelerate each other in a completely random fashion and the electric and magnetic fields produced by the charges blend together in a merged, outwardly traveling electromagnetic wave. The best one can hope to measure from such a source is the average field intensity, which provides us with a spectrum and an overall power emitted by the plasma. In synchrotron emission, on the other hand, the magnetic field may contain a global component with some degree of uniformity—not all sources are completely turbulent, as we learned in chapter 4. This is crucial insofar as the electric field produced by the charges is concerned, because the electron's motion is restricted by the direction of **B**.

Synchrotron radiation is indeed polarized, but in a very simple way, as it turns out. To understand why, let us consider the two cases depicted in figure 5.10, which show cyclotron radiation from an accelerated electron on the left and synchrotron radiation on the right. The particle gyrates with a constant angular frequency (ω_{gyr} on the left and ω_{gyr}^{rel} on the right). In cyclotron emission, an observer in the plane of the electron's orbit sees pure linear polarization because the charge appears to swing back and forth in a straight line. From above or below, the electron appears to be executing pure circular motion, so the polarization is purely circular with positive helicity (as viewed from above) or negative helicity (seen from below).

The relativistic limit is considerably different (right-hand panel). Since the radiation is strongly beamed in the forward moving direction, all of the radiation components, linearly polarized and circularly polarized with positive and negative helicity, are now merged together in the forward cone. A simple inspection of figure 5.10 easily convinces us that the two circularly polarized components cancel each other out, leaving behind only a contribution to the linear polarization. The net result is a *linearly* polarized wave. For a source with a uniform magnetic field, the overall degree of polarization can be as high as ∼70%, once the canceled component is taken into account. Of course, in reality, we very rarely encounter objects in nature with such a well-structured, global magnetic field. More typically, the field will have both a turbulent component and a contribution from a large-scale uniform component. This reduces the overall degree of polarization in synchrotron sources to values around a few percent. The detection of polarized emission (usually in the radio portion of the spectrum) is a powerful diagnostic indicating a synchrotron origin for the radiation.

5.5 THERMAL SYNCHROTRON

The total emissivity (power per unit frequency per unit volume) from an ensemble of particles with number density $N(E) \, dE$ between energies E and $E + dE$ is

$$P_{\text{tot}}(\nu) = \int_{E_1}^{E_2} P(\nu) N(E) \, dE. \tag{5.106}$$

The single-particle spectrum $P(\nu)$ (equation 5.103) is itself a function of energy through the Lorentz factor γ in ν_c.

A relativistic Maxwellian distribution of charges is described by the function

$$N(E) \, dE = N_0 E^2 \exp(-E/kT) \, dE, \tag{5.107}$$

expressed in terms of E rather than velocity v to keep it correct for arbitrary values of γ. Writing $E = \gamma m_e c^2$, and assuming an isotropic distribution, we can easily integrate equation (5.106), obtaining the expression

$$P_{\text{tot}}(\nu) \approx \frac{\sqrt{3} e^3 n_e B}{8\pi m_e c^2} \left(\frac{\nu}{\nu_T} \right) I \left(\frac{\nu}{\nu_T} \right), \tag{5.108}$$

where n_e is the electron number density, and

$$\nu_T \equiv \frac{3e B (kT)^2}{4\pi m_e^3 c^5}. \tag{5.109}$$

For thermal synchrotron, the frequency integral over the Fourier components in $P(\nu)$ is subsumed into the function $I(\nu/\nu_T)$, described in greater detail below. Note that all reference to γ has been replaced by T because the energies in this particle distribution are apportioned by thermal equipartition, characterized only by temperature.

Much of the calculational effort in determining $P_{\text{tot}}(\nu)$ is expended in the evaluation of $I(\nu/\nu_T)$, shown in figure 5.11. Formally,

$$I(x) \equiv \frac{1}{x} \int_0^\infty u^2 e^{-u} F \left(\frac{x}{u^2} \right) \, du, \tag{5.110}$$

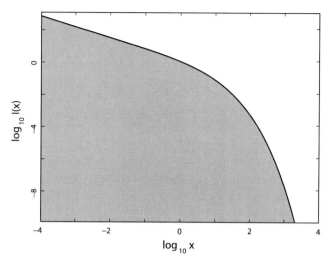

Figure 5.11 The dimensionless function $I(x)$ appearing in the expression for the total synchrotron emissivity by a thermal plasma.

where

$$F(y) \equiv y \int_y^\infty K_{5/3}(u)\, du. \tag{5.111}$$

Thankfully, however, $I(x)$ may be approximated as a power law over most of its range, though with different indices above and below $x \sim 1$.

Historically, synchrotron-emitting sources have relied on nonthermal particles to produce their radiation, rather than the more commonly encountered thermal populations. The reason for this is that the synchrotron power increases with particle energy or, equivalently, with temperature. But thermal processes are not very efficient, since a large fraction of the available energy is transferred to low-energy particles, and unless the plasma temperature is very high ($\gg m_e c^2 / k$), thermal synchrotron sources are simply too faint to be seen easily. But this situation has changed in recent years, as the instrument sensitivity has continued to improve. Later in this book, in section 12.2, we shall study the supermassive black hole at the center of our Galaxy, an excellent example of an object whose spectrum includes a thermal synchrotron component, and one that has emerged as an important high-energy source only in recent years.

Still most synchrotron sources, including the relativistic jets in AGNs, and the shells in supernova remnants, are nonthermal radiators. We shall consider the physics of nonthermal synchrotron emission in the next section.

5.6 NONTHERMAL SYNCHROTRON

Many high-energy sources accelerate particles to relativistic velocities, well above the mean thermal energy present in their environment. We already saw in chapter 4 that several mechanisms can transfer power efficiently to a subset of the charges, producing a power-law distribution on top of the underlying population. Thus, when we talk of "nonthermal synchrotron" radiation, we usually mean that the emitting particles belong to a distribution

$$N(E)\,dE = KE^{-x}\,dE,\tag{5.112}$$

where K is a normalization constant, and x is an index that varies over the range $\sim$2–2.5.

As before, we may find the total emissivity from such a distribution by evaluating the integral in equation (5.106). However, for instructional purposes, we can use the fact that the single-particle spectrum $P(\nu)$ is highly peaked near the critical frequency ν_c (see figure 5.9) to simplify this procedure considerably. We argue that to a good approximation most of the energy is radiated at $\nu_c(E)$, where

$$\nu_c \approx \gamma^2 \nu_{\text{gyr}} = \left(\frac{E}{m_e c^2}\right)^2 \nu_{\text{gyr}},\tag{5.113}$$

with

$$\nu_{\text{gyr}} \equiv \frac{eB}{2\pi m_e c}\tag{5.114}$$

(cf. equation 5.60). But the integral in equation (5.106) is over energy, so let us invert the expression in (5.113) to

$$E = \gamma m_e c^2 \approx \left(\frac{\nu}{\nu_{\text{gyr}}}\right)^{1/2} m_e c^2,\tag{5.115}$$

so that

$$dE \approx \frac{m_e c^2}{2(\nu_{\text{gyr}}\nu)^{1/2}}\,d\nu.\tag{5.116}$$

Since we're treating the single-particle emission as if it all comes out at ν_c, we will also use the synchrotron power in equation (5.86) instead of the frequency-dependent spectrum $P(\nu)$. In that case,

$$P = -\frac{dE}{dt} = \frac{4}{3}\sigma_T c\beta^2\gamma^2 u_B,\tag{5.117}$$

or (with $\beta \approx 1$)

$$-\frac{dE}{dt} \approx \frac{4}{3}\sigma_T c\left(\frac{E}{m_e c^2}\right)^2 \frac{B^2}{8\pi}.\tag{5.118}$$

And therefore the evaluation of $P_{\text{tot}}(\nu)$ in equation (5.106) yields

$$P_{\text{tot}}(\nu) = \int \left(-\frac{dE}{dt}\right) N(E)\,dE$$
$$\propto B^{(x+1)/2}\nu^{-(x-1)/2}.\tag{5.119}$$

With a full evaluation of $P_{\text{tot}}(\nu)$, using the frequency-dependent single-particle emissivity, we get the same functional form as equation (5.119), and an accurate determination of the proportionality constant. The complete expression is

$$P_{\text{tot}}(\nu) = 1.7 \times 10^{-21} a(x) K B^{(x+1)/2}$$

$$\times \left(\frac{6.26 \times 10^{18} \text{ Hz}}{\nu} \right)^{(x-1)/2} \text{ergs cm}^{-3} \text{s}^{-1} \text{Hz}^{-1}. \tag{5.120}$$

The function $a(x)$ depends only weakly on x, with values $(x, a) = (1, 0.283), (1.5, 0.147), (2, 0.103), (2.5, 0.0852), (3, 0.0742), (4, 0.0725)$, and $(5, 0.0922)$.

The most important property of nonthermal synchrotron emission is that a power-law particle distribution with index x produces a power-law radiation spectrum with index $(x - 1)/2$. But unlike thermal synchrotron, we cannot assume that this kind of spectrum extends with equal validity over all frequencies. In a thermal distribution, the particle energies are partitioned correctly subject to the limitation provided by the total energy available in the system. Instead, a quick inspection of equation (5.112) shows that, if unchecked, the power-law distribution diverges at low energies. It may not be obvious why this is a serious problem until one compares the spectrum in equation (5.120) with the Planck (blackbody) function in figure 5.5. As one moves to lower and lower frequencies, $P_{\text{tot}}(\nu)$ eventually exceeds the so-called blackbody limit ($\propto \nu^2$ in the Rayleigh-Jeans portion of the spectrum), which is not permitted on physical grounds.

By definition, a blackbody is a system that completely absorbs all of the radiation incident upon it. In 1859, Gustav Kirchhoff (1824–1887) used general considerations of thermodynamic equilibrium to propose a fundamental law of thermodynamics—that the emissivity and absorptivity of such a medium are equal.[5] An important corollary follows from this when one includes the conservation of energy, for then the emissivity of any plasma cannot exceed the limit imposed by the maximal rate at which that plasma can absorb energy in the first place—which therefore constitutes a *blackbody limit*.

A quick solution to the problem of divergence at low frequencies would be to argue that the nonthermal synchrotron spectrum should be a power law (equation 5.120) everywhere except below some critical (or "break") frequency ν_b, at which the synchrotron intensity first crosses the Planck function (equation 5.57). In that case, the spectrum would go as $\nu^2 T$ below ν_b. But, in fact, these synchrotron-emitting particles are not thermal, so the use of T is not warranted, and the restricted emissivity is not quite $\propto \nu^2$.

But we can still understand how a *self-absorbed* spectrum is produced using the blackbody analogy, though with a more accurate assessment of which particle energy corresponds to frequency ν. For this purpose, let us introduce a frequency-dependent "temperature" $T(\nu)$, at which the radiation with frequency ν and the particles are in equilibrium. In that case,

$$\gamma m_e c^2 \approx \tfrac{3}{2} k T(\nu), \tag{5.121}$$

where

$$\gamma \approx \left(\frac{\nu}{\nu_{\text{gyr}}} \right)^{1/2}. \tag{5.122}$$

[5] See Lifshitz (1980).

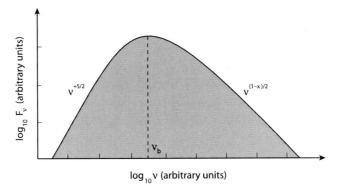

Figure 5.12 The optically thin nonthermal synchrotron spectrum increases indefinitely toward lower frequencies. Eventually, the emission rate exceeds that due to blackbody radiation, which is maximally set by the balance between emission and reabsorption by the radiating particles. The nonthermal synchrotron spectrum is therefore self-absorbed below some break frequency ν_b, whose value depends on the properties of the medium.

That is,

$$T(\nu) \sim \frac{m_e c^2}{3k} \left(\frac{\nu}{\nu_{\text{gyr}}} \right)^{1/2}. \tag{5.123}$$

The Rayleigh-Jeans portion of the spectrum below ν_b therefore goes as

$$P_{\text{tot}}(\nu) \sim \nu^{5/2}. \tag{5.124}$$

In our discussion of bremsstrahlung emission, we pointed out several key observational characteristics of the spectrum that not only help us identify it unambiguously, but also provide us with empirical evidence of the physical state of the system, such as its temperature. The nonthermal synchrotron spectrum produced by a self-absorbed source (figure 5.12) is equally unique and informative. The $\sim \nu^{5/2}$ behavior at low frequencies is a clear indication that the radiation is nonthermal synchrotron, and the break frequency ν_b is a measure of the optical depth through the emitter. At higher frequencies, the spectral slope provides us with the power-law index x of the emitting particles. Finally, equation (5.120) may be used to estimate the magnetic field B. All in all, this is quite a substantial amount of information to gather from just one spectrum!

5.7 COMPTON SCATTERING

The final interaction between matter and radiation that we will consider in this book is Compton scattering.[6] In reality, all three of the processes we are studying here are quite similar, since in every case we are dealing with the collision between

---

[6]Sometimes the term *inverse* Compton scattering is used to signify that the charge is more energetic than the photon, so the radiation is upscattered by the collision.

a charge (usually an electron) and a photon. What changes from one mechanism to the next is the nature of the prescattered radiation. In bremsstrahlung emission, the electron plows through the virtual (Coulomb) photon field of the ion Ze, and in magnetic bremsstrahlung (or synchrotron), the electron scatters with the virtual photons comprising the magnetic field. In Compton scattering, on the other hand, the collision occurs between a charge and a *real* (i.e., free) photon, dynamically distinct from the charge that produced it.

We begin, as for synchrotron emission, with a simplified classical treatment. An electron subject to the electromagnetic field of an incoming photon accelerates with a rate

$$a = \frac{e}{m_e} E. \tag{5.125}$$

According to the Larmor equation (5.14), the electron should then radiate a power

$$P = \frac{2}{3} \frac{e^4}{c^3 m_e^2} E^2. \tag{5.126}$$

Physically, the photon field represents an energy flux incident on the electron that scatters a fraction of it consistent with the overall power in equation (5.126).

Letting u_{rad} represent the radiation energy density, we write the incident flux as

$$f = c u_{\text{rad}} \tag{5.127}$$

or

$$f = \frac{cE^2}{4\pi}, \tag{5.128}$$

once we recognize that $u_{\text{rad}} = 2u_E$ and $u_E = E^2/8\pi$. Thus,

$$P = \frac{8\pi}{3} \frac{e^4}{c^4 m_e^2} f. \tag{5.129}$$

The appearance of $e^4/c^4 m_e^2$ in this expression is rather interesting historically because the classical electron radius r_0 is defined by its rest mass energy, according to $e^2/r_0 = m_e c^2$. Evidently,

$$P = \frac{8\pi}{3} r_0^2 f, \tag{5.130}$$

which represents an energy flux (f) times an area ($\propto r_0^2$).

Indeed, the cross section for scattering—in this case the Thomson cross section—is defined to be

$$\sigma_T \equiv \frac{P}{f}, \tag{5.131}$$

the total scattered power divided by the incoming flux. Thus,

$$\sigma_T = \frac{8\pi}{3} r_0^2 \approx 6.65 \times 10^{-25} \text{ cm}^2. \tag{5.132}$$

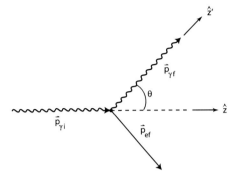

Figure 5.13 When a photon is very energetic, the electron's recoil momentum must be taken into account during their collision. This diagram shows the three momenta that must be considered during a Compton event.

Note that in classical Thomson scattering, the emitted power is independent of frequency—a great simplification that often lends itself to very useful "back-of-the-envelope" estimates.

When the incoming photon is very energetic, however, we cannot ignore the recoil of the electron (figure 5.13). In the general case of electron–photon scattering, we must make use of the four-momentum (equation 3.52), which will permit us to simultaneously consider the conservation of energy and momentum. With reference to figure 5.13, we define the following four-momenta:

$$p_{\gamma i} = \left(\frac{\epsilon}{c}, \frac{\epsilon}{c}\hat{\mathbf{z}}\right)$$

$$p_{\gamma f} = \left(\frac{\epsilon'}{c}, \frac{\epsilon'}{c}\hat{\mathbf{z}}'\right)$$

$$p_{ei} = (m_e c, \mathbf{0})$$

$$p_{ef} = (\gamma m_e c, \mathbf{p}), \qquad (5.133)$$

where ϵ is the photon energy, and a prime denotes quantities in the scattered state.

Conservation of energy and momentum requires that

$$p_{ef} = p_{ei} + p_{\gamma i} - p_{\gamma f}, \qquad (5.134)$$

which in component form gives

$$\gamma m_e c = m_e c + \frac{\epsilon}{c} - \frac{\epsilon'}{c} \qquad (5.135)$$

and

$$|\mathbf{p}|^2 = \left|\frac{\epsilon}{c}\hat{\mathbf{z}} - \frac{\epsilon'}{c}\hat{\mathbf{z}}'\right|^2$$

$$= \left(\frac{\epsilon}{c}\right)^2 + \left(\frac{\epsilon'}{c}\right)^2 - 2\frac{\epsilon\epsilon'}{c^2}\cos\theta. \qquad (5.136)$$

But from equation (3.58) we know that

$$c^2(\gamma m_e c)^2 = c^2|\mathbf{p}|^2 + m_e^2 c^4, \tag{5.137}$$

so equations (5.135) and (5.136) may be merged into the form

$$\left(\frac{\epsilon}{c} - \frac{\epsilon'}{c}\right)^2 c^2 + 2m_e c^2 \left(\frac{\epsilon}{c} - \frac{\epsilon'}{c}\right) c = \epsilon^2 + \epsilon'^2 - 2\epsilon\epsilon' \cos\theta. \tag{5.138}$$

Some further reduction leads to the final result,

$$\epsilon' = \frac{\epsilon}{1 + (\epsilon/m_e c^2)(1 - \cos\theta)}. \tag{5.139}$$

This shift in photon energy, absent in the classical Thomson case, is entirely due to the recoil of the electron. Notice that when the incoming photon has an energy $\epsilon \ll m_e c^2$, this expression reduces to the classical limit in which $\epsilon' = \epsilon$. However, when $\epsilon \gg m_e c^2$, the scattered photon ends up with roughly the electron's rest mass energy.

When we invoke relativity in this manner, we must also contend with the effects of time dilation, which reduce the time the photon spends in the electron's range of influence. The cross section is correspondingly reduced, as was first demonstrated almost a century ago.[7] We won't reproduce the derivation here, but it is useful for us to know the frequency-dependent total cross section in two limits,

$$\sigma \approx \sigma_T \left(1 - \frac{2h\nu}{m_e c^2} + \cdots\right) \quad \frac{h\nu}{m_e c^2} \ll 1, \tag{5.140}$$

$$\sigma \approx \frac{3}{8}\sigma_T \left(\frac{m_e c^2}{h\nu}\right)\left(\ln\frac{2h\nu}{m_e c^2} + \frac{1}{2}\right) \quad \frac{h\nu}{m_e c^2} \gg 1. \tag{5.141}$$

Before we go on to consider the power emitted by an energetic electron moving through a photon field, let us first consider what happens during a typical encounter between just one electron and one photon. Physical interactions are best considered in the rest frame of the massive particle (or in the center-of-momentum frame, if both particles are massless). For a photon trajectory making an angle θ relative to that of the electron, and an energy ϵ_0 in the laboratory, the rest-frame photon energy is given by the Doppler shift equation (3.18),

$$\epsilon_0' = \gamma\epsilon_0(1 - \beta\cos\theta). \tag{5.142}$$

Thus (with $\theta \sim \pi/2$),

$$\epsilon_0' \approx \gamma\epsilon_0. \tag{5.143}$$

In many cases, $\epsilon_0' \ll m_e c^2$, and the scattering in the electron's rest frame occurs in the Thomson limit. That is, the scattered photon in such cases has an energy $\epsilon_{sc}' \approx \epsilon_0'$, moving in a direction $\theta' \sim \pi/2$. The transformation back into the laboratory frame

---

[7]The energy-dependent differential cross section for this process was one of the first results from quantum electrodynamics. See Klein and Nishina (1929).

thus produces a scattered photon with energy

$$\epsilon_{sc} \approx \gamma \epsilon'_{sc} \approx \gamma^2 \epsilon_0. \tag{5.144}$$

The Compton power, driven by this strong γ^2 dependence, can therefore have an enormous impact on the emission of X-rays and γ-rays. For example, in the interstellar medium GeV cosmic ray electrons often interact with photons ($\epsilon_0 \sim 10^{-3}$ eV) in the Cosmic Microwave Background.[8] In this case, $\gamma \gtrsim 10^6/511 \approx 2 \times 10^3$. These scattering events convert microwaves into X-rays, producing photons with energy $\epsilon_{sc} \sim 4$ keV.

This is about as far as we can go analyzing scattering events between energetic particles and photons without introducing additional information regarding their distributions. It should already be evident that, like synchrotron emission, Compton scattering is an effective process only for highly relativistic encounters. Thus, we are led again to consider nonthermal (mostly power-law) particle populations.

An energetic electron moving through a "bath" of photons encounters a radiative energy flux $\gamma^2 c u_{rad}$ in its own rest frame. The overall factor of γ^2 arises from a combination of the energy transformation in equation (5.142), and a length contraction (which leads to a volume reduction by a factor γ or, equivalently, a number density enhancement by the same factor).

Thus, the radiated power in the rest frame is

$$P' \approx \sigma_T \gamma^2 c u_{rad}. \tag{5.145}$$

Converting back into the laboratory frame, we remember that power is an invariant (see discussion following equation 5.85), so

$$P \approx \sigma_T \gamma^2 c u_{rad}. \tag{5.146}$$

Of course, in using equation (5.143) for the energy boost, we ignored the angle dependence $(1 - \beta \cos \theta)$, and similarly for the boost back into the laboratory frame. Including the full Doppler shift formula (equation 5.142), we derive the (slightly modified) single-particle total Compton power,

$$P_{Comp} = \tfrac{4}{3} \sigma_T c \beta^2 \gamma^2 u_{rad}, \tag{5.147}$$

equivalent to P_{sync} in equation (5.86).

The similarity between P_{sync} and P_{Comp} is not a coincidence. At the very beginning of this section, we pointed out that all three of the emission processes we are considering in this book are based on the same physical interaction—the collision between a charge and a photon. Thus, the single-particle total power should depend only on the density of photons with which the charge can interact, $u_B = B^2/8\pi$ in the case of synchrotron, and u_{rad} in the case of a real (or free) ambient radiation field.

Later, when we apply this theory to specific classes of high-energy objects, such as the jets in AGNs (see section 12.3), we will appreciate this similarity because of the power it affords us in probing the physical conditions at the source. For example, if we can measure both a radio and a γ-ray spectrum in the same object, then the ratio P_{sync}/P_{Comp} permits us either to infer the magnetic field intensity if we know

---

[8]We shall return to this topic in section 13.1.

the local u_{rad} or, vice versa, to probe the ambient photon field intensity if we have a measure of **B** (say from Faraday rotation measurements).

Compton sources, like their synchrotron counterparts, are detectable only if an ensemble of many particles are involved in the emission process. Thus, in the final analysis, we must also know something about the underlying particle distribution to fully understand how the high-energy spectrum is produced. In Compton scattering, thermal distributions are very inefficient radiators, since even at $T \sim 10^9$ K, a typical particle has a velocity $v = (3kT/m_e)^{1/2} \sim (2/3)c$, with a Lorentz factor $\gamma \approx 1.4$. There are exceptions, however, and in chapter 12 we will learn that the supermassive black hole at the galactic center is one of the sources producing both a P_{sync} and a P_{Comp}. More to the point, its X-ray flux appears to be due to the inverse Compton scattering of radio and infrared photons by a very hot plasma ($T \gg 10^9$ K) with a significant thermal population of leptons.

Even so, the majority of Compton sources are nonthermal emitters. To describe them, we will employ a slightly different version of the distribution in equation (5.112), using γ as the independent variable rather than energy. Letting

$$N(\gamma)\,d\gamma = K\gamma^{-x}\,d\gamma \tag{5.148}$$

represent the number of particles per unit volume between γ and $\gamma + d\gamma$, we calculate the total Compton power (energy per unit volume, per unit time) as

$$P_{tot} = \int_{\gamma_1}^{\gamma_2} P_{Comp}(\gamma)N(\gamma)\,d\gamma. \tag{5.149}$$

For example, with $\gamma_1 = 1$ and $\gamma_2 \equiv \gamma_{max}$, we get

$$P_{tot} = \frac{4}{3}\sigma_T c u_{rad} \frac{K}{3-x}\left(\gamma_{max}^{3-x} - 1\right) \tag{5.150}$$

or

$$P_{tot} \approx \frac{4}{3}\sigma_T c u_{rad} \frac{K}{3-x}\gamma_{max}^{3-x}, \tag{5.151}$$

when $\gamma_{max} \gg 1$.

This expression will find several applications in chapters 9–13, where we will learn that many high-energy sources harbor an inverse Compton scattering component in their spectrum. As with bremsstrahlung and synchrotron, spectral measurements of a Compton source can provide valuable clues regarding the physical state of the system. These include the spectral index x and a high-energy cutoff γ_{max}. We will see that such a turnover in the spectrum is an indication of significant cooling processes, delimiting the efficiency of particle acceleration. Or it might be evidence that the source has only been active for a short time, so that its particles have not yet been accelerated to higher energies.

With this, we have reached the end of our discussion concerning radiative processes in high-energy sources. In the past few chapters, we have concentrated primarily on the microphysics of high-energy emission. Our attention will now shift to the macroscopic aspects of dynamical phenomena in these objects, beginning in the next chapter with the physics of accretion.

SUGGESTED READING

The retarded electric field, containing both the Coulomb (or velocity) field and the radiation (or acceleration) field can be found using either classical physics or special relativity. The two approaches produce the same result because Maxwell's equations are correct in either framework. To see this, compare the two derivations in sections 4.6 and 7.2 of Melia (2001a).

The reader is encouraged to study the properties of intensity and the physics of radiation transport at greater depth. Two highly recommended texts on this topic are Rybicki and Lightman (1985) and Mihalas and Mihalas (1999).

An elegant and comprehensive discussion of both thermal and nonthermal synchrotron emission may be found in Pacholczyk (1970).

Chapter Six

Accretion of Plasma

As we move away from the physics of individual particles and their interactions and begin to consider the behavior of gases and plasmas under the influence of strong fields, one of the first questions we must address is whether or not a fluid description is appropriate. A hydrodynamics approach is valid if the mutual interaction time between the constituent particles is shorter than the time over which the external field changes the flow dynamics in the bulk. These timescales depend on the ionization state of the medium and on the depth of the potential well. In this chapter, we develop the theory of accretion onto compact objects under the assumption that a fluid approximation is tenable. In cases where the timescales are not consistent with this approach, it is usually adequate to treat the particles individually, since in those situations they respond primarily to the external influences rather than to their mutual interactions.

6.1 HYDRODYNAMICS

The principal equations of hydrodynamics are formal representations of the conservation of mass, momentum, and energy. We begin by considering the mathematical implications of the statement that the total mass within a system is conserved. Let $\rho(t)$ be the mass density inside a cube positioned with one corner at the coordinates $x(t)$, $y(t)$, $z(t)$. If the cube's dimensions are $\delta x(t)$, $\delta y(t)$, and $\delta z(t)$, its volume is given by the expression

$$\delta V(t) = \delta x(t)\,\delta y(t)\,\delta z(t). \tag{6.1}$$

Conservation of mass inside the cube requires that

$$\frac{dm}{dt} = 0, \tag{6.2}$$

where $m = \rho\,\delta V$. Thus,

$$\frac{d}{dt}(\rho\,\delta V) = \delta V \frac{d\rho}{dt} + \rho \frac{d(\delta V)}{dt} = 0. \tag{6.3}$$

But

$$
\begin{aligned}
d(\delta V) = & \left[-v_x\,dt\,\delta y\,\delta z + \left(v_x + \frac{\partial v_x}{\partial x}\delta x \right) dt\,\delta y\,\delta z \right] \\
& + \left[-v_y\,dt\,\delta x\,\delta z + \left(v_y + \frac{\partial v_y}{\partial y}\delta y \right) dt\,\delta x\,\delta z \right] \\
& + \left[-v_z\,dt\,\delta x\,\delta y + \left(v_z + \frac{\partial v_z}{\partial z}\delta z \right) dt\,\delta x\,\delta y \right].
\end{aligned} \tag{6.4}
$$

With appropriate cancellations, this expression simplifies considerably and reduces to the form

$$d(\delta V) = \left(\frac{\partial v_x}{\partial x} + \frac{\partial v_y}{\partial y} + \frac{\partial v_z}{\partial z} \right) \delta V \, dt. \tag{6.5}$$

That is, the relative rate of change in the volume element equals the divergence of the velocity, so

$$\frac{1}{\delta V} \frac{d(\delta V)}{dt} = \vec{\nabla} \cdot \mathbf{v}. \tag{6.6}$$

Thus, from equations (6.3) and (6.6), we get

$$\frac{d\rho}{dt} + \rho(\vec{\nabla} \cdot \mathbf{v}) = 0. \tag{6.7}$$

However, as it stands, this equation is not always very practical because the total time derivative includes contributions from both the actual time dependence of ρ (through $\partial\rho/\partial t$) and variations resulting from the motion through a nonuniform medium (via the advective derivative, $\mathbf{v} \cdot \vec{\nabla}$).

We therefore write

$$\frac{\partial\rho}{\partial t} + \mathbf{v} \cdot \vec{\nabla}\rho + \rho(\vec{\nabla} \cdot \mathbf{v}) = 0, \tag{6.8}$$

which simplifies to

$$\frac{\partial\rho}{\partial t} + \vec{\nabla} \cdot (\rho\mathbf{v}) = 0. \tag{6.9}$$

This is the first equation of hydrodynamics. Recalling our discussion leading up to equations (4.5) and (4.6), we immediately recognize the physical meaning of the second term in this expression. The quantity $\rho\mathbf{v}$ is the mass flux (mass per unit area per unit time) anywhere in the system. Its divergence is the net influx or outflux of mass per unit volume. Equation (6.9) is telling us that ρ can change in time (with $\partial\rho/\partial t \neq 0$), but only if there is a corresponding net flux of mass entering or leaving the point at which the partial time derivative is being calculated. Otherwise, $\partial\rho/\partial t = 0$, and nothing changes. Mass is being conserved because any change in density must be coupled to a net influx or outflux of mass to compensate for that variation.

The fluid's momentum is also conserved, an empirically motivated statement embodied in Newton's second law of motion,

$$\frac{d\mathbf{p}}{dt} = \mathbf{F}, \tag{6.10}$$

where $\mathbf{p}$ is the three-momentum and $\mathbf{F}$ is the net force. We employ the same reasoning as above, and put

$$\mathbf{p} = m\mathbf{v} = \rho \, \delta V \, \mathbf{v}, \tag{6.11}$$

where δV and the other symbols all have their usual meanings. Thus,

$$
\frac{d\mathbf{p}}{dt} = \frac{d}{dt}(\rho\,\delta V\,\mathbf{v})
$$

$$
= \delta V \left[\frac{d}{dt}(\rho\mathbf{v}) + \rho\mathbf{v}\,\frac{1}{\delta V}\frac{d(\delta V)}{dt} \right]. \tag{6.12}
$$

In a fluid, the total force $\mathbf{F}$ has at least two contributions—from pressure gradients within the medium itself and from an external field (such as gravity), which produces what we normally call a "body" force. Let us define the stress tensor element σ_{ij} to be the i-component of momentum density flux across the surface whose unit normal points in the j-direction. The net force due to pressure is then the difference in σ_{ij} from one face of our cube to the next. That is,

$$
\mathbf{F} = \hat{\mathbf{x}} \left[\sigma_{xx}\,\delta y\,\delta z - \left(\sigma_{xx} + \frac{\partial\sigma_{xx}}{\partial x}\delta x\right)\delta y\,\delta z + \sigma_{xy}\,\delta x\,\delta z - \left(\sigma_{xy} + \frac{\partial\sigma_{xy}}{\partial y}\delta y\right)\delta x\,\delta z \right.
$$

$$
\left. + \sigma_{xz}\,\delta x\,\delta y - \left(\sigma_{xz} + \frac{\partial\sigma_{xz}}{\partial z}\delta z\right)\delta x\,\delta y \right]
$$

$$
+ \hat{\mathbf{y}} \left[\sigma_{yx}\,\delta y\,\delta z - \left(\sigma_{yx} + \frac{\partial\sigma_{yx}}{\partial x}\delta x\right)\delta y\,\delta z + \sigma_{yy}\,\delta x\,\delta z - \left(\sigma_{yy} + \frac{\partial\sigma_{yy}}{\partial y}\delta y\right)\delta x\,\delta z \right.
$$

$$
\left. + \sigma_{yz}\,\delta x\,\delta y - \left(\sigma_{yz} + \frac{\partial\sigma_{yz}}{\partial z}\delta z\right)\delta x\,\delta y \right]
$$

$$
+ \hat{\mathbf{z}} \left[\sigma_{zx}\,\delta y\,\delta z - \left(\sigma_{zx} + \frac{\partial\sigma_{zx}}{\partial x}\delta x\right)\delta y\,\delta z + \sigma_{zy}\,\delta x\,\delta z - \left(\sigma_{zy} + \frac{\partial\sigma_{zy}}{\partial y}\delta y\right)\delta x\,\delta z \right.
$$

$$
\left. + \sigma_{zz}\,\delta x\,\delta y - \left(\sigma_{zz} + \frac{\partial\sigma_{zz}}{\partial z}\delta z\right)\delta x\,\delta y \right] + \mathbf{F}_b. \tag{6.13}
$$

Many of these terms cancel in pairs, and the expression simplifies to

$$
\mathbf{F} = -\vec{\nabla}\cdot\vec{\sigma}\,\delta V + \mathbf{F}_b, \tag{6.14}
$$

where $\vec{\sigma}$ is the tensor of coefficients $\{\sigma_{ij}\}$. Thus, the conservation of momentum equation becomes

$$
\frac{d}{dt}(\rho\mathbf{v}) + \rho\mathbf{v}\left(\vec{\nabla}\cdot\mathbf{v}\right) = -\vec{\nabla}\cdot\vec{\sigma} + \frac{1}{\delta V}\mathbf{F}_b \tag{6.15}
$$

or

$$
\frac{\partial}{\partial t}(\rho\mathbf{v}) + \vec{\nabla}\cdot(\rho\mathbf{v}\mathbf{v}) = -\vec{\nabla}\cdot\vec{\sigma} + \frac{1}{\delta V}\mathbf{F}_b, \tag{6.16}
$$

once we separate the total time derivative into its two constituent parts.

Notice that equations (6.9) and (6.16) have the same structure. The temporal variation in momentum density ($\rho\mathbf{v}$) is balanced by the divergence of the momentum flux density ($\rho\mathbf{v}\mathbf{v}$), which here functions as a tensor. (Since momentum is a vector, we can transport any of its independent components in any of the independent coordinate directions.) The appearance of the "product" $\mathbf{v}\mathbf{v}$ may seem strange, but remember that there is a vector dot product which reduces the second term on the left-hand side to a three-dimensional vector.

The right-hand side of these conservation equations is always the source of the quantity being conserved on the left-hand side. However, there is an important difference between the mass and momentum conservation equations. Whereas mass has no source, momentum may be added or subtracted to the system by the action of a force. Thus, the source on the right-hand side of equation (6.9) is zero, but it can be nonzero in the momentum equation if there are pressure gradients in the system, or if an external agent is acting on the fluid. Incidentally, we call gravity a body force, $\mathbf{F}_b$, because it acts at every point in the fluid regardless of what other parts of the medium are doing. It is the opposite of a pressure gradient force, which depends entirely on how the pressure varies from point to point within the fluid.

Equation (6.16) is the second equation of hydrodynamics. The third equation expresses the conservation of energy, which we derive from Newton's second law of motion (equation 6.10) by taking the dot product of each side with the velocity $\mathbf{v}$:

$$\mathbf{v} \cdot \frac{d\mathbf{p}}{dt} = \mathbf{v} \cdot \mathbf{F}. \tag{6.17}$$

Thus, combining this equation with (6.12),

$$\mathbf{v} \cdot \frac{d}{dt}(\rho\mathbf{v}) + \rho v^2(\vec{\nabla} \cdot \mathbf{v}) = -\mathbf{v} \cdot (\vec{\nabla} \cdot \vec{\sigma}) + \mathbf{v} \cdot \left(\frac{\mathbf{F}_b}{\delta V}\right). \tag{6.18}$$

This expression is still not quite useful because it blends terms that look like a power $(\mathbf{v} \cdot \mathbf{F}_b)$, with terms that still retain the appearance of a momentum density $(\rho\mathbf{v})$. So let us manipulate the first term to cast it into the form of an energy. We use the Einstein convention throughout, where a repeated index means summation (in three-space) from 1 to 3:

$$\mathbf{v} \cdot \frac{d}{dt}(\rho\mathbf{v}) = v_i \frac{d}{dt}(\rho v_i)$$
$$= \frac{d}{dt}\left(\frac{1}{2}\rho v_i v_i\right) - \frac{1}{2}v_i v_i \frac{d\rho}{dt} + v_i v_i \frac{d\rho}{dt}$$
$$= \frac{d}{dt}\left(\frac{1}{2}\rho v^2\right) - \frac{1}{2}\rho v^2\left(\vec{\nabla} \cdot \mathbf{v}\right), \tag{6.19}$$

the last step following from equation (6.9). With this transformation, the energy equation is now

$$\frac{d}{dt}\left(\frac{1}{2}\rho v^2\right) + \frac{1}{2}\rho v^2\left(\vec{\nabla} \cdot \mathbf{v}\right) = -\mathbf{v} \cdot (\vec{\nabla} \cdot \vec{\sigma}) + \mathbf{v} \cdot \left(\frac{\mathbf{F}_b}{\delta V}\right). \tag{6.20}$$

But the fluid contains internal energy, as well as kinetic, and to broaden the conservation of energy, we must resort to the first law of thermodynamics,

$$\frac{du}{dt} = T\frac{ds}{dt} + \frac{P}{\rho^2}\frac{d\rho}{dt}, \tag{6.21}$$

where u is the internal energy per unit mass, s is the *entropy* per unit mass, and P is the thermal pressure. That is,

$$\frac{d}{dt}(\rho u) = u\frac{d\rho}{dt} + \rho\frac{du}{dt}$$

$$= \rho T\frac{ds}{dt} + \left(u + \frac{P}{\rho}\right)\frac{d\rho}{dt}, \tag{6.22}$$

or

$$\frac{d}{dt}(\rho u) + \rho u(\vec{\nabla}\cdot\mathbf{v}) = \rho T\frac{ds}{dt} - P(\vec{\nabla}\cdot\mathbf{v}). \tag{6.23}$$

Defining the *enthalpy* per unit mass

$$w \equiv u + \frac{P}{\rho}, \tag{6.24}$$

we may also write this equation as

$$\frac{d}{dt}(\rho u) = \rho T\frac{ds}{dt} + w\frac{d\rho}{dt}, \tag{6.25}$$

or

$$\frac{d}{dt}(\rho u) = \rho T\frac{ds}{dt} - w\rho\left(\vec{\nabla}\cdot\mathbf{v}\right). \tag{6.26}$$

The fact that we now have both equations (6.20) and (6.23) highlights an important difference between the conservation of energy and momentum. Whereas there is only one momentum, there are various forms of energy. Ideally, we should have just a single energy equation, so we will proceed to combine these two expressions into a single relation. We will also separate the various contributions to the stress tensor $\vec{\vec{\sigma}}$, and write it as a sum of two parts—the isotropic pressure P, and the rest:

$$\vec{\vec{\sigma}} = P\vec{\mathbf{I}} + \vec{\vec{\sigma}}_0, \tag{6.27}$$

where $\vec{\mathbf{I}}$ is the unit matrix. Thus,

$$\left(\vec{\nabla}\cdot\vec{\vec{\sigma}}\right)_i = \frac{\partial\sigma_{ij}}{\partial x_j} = \frac{\partial}{\partial x_j}(P\delta_{ij} + \sigma_{0,ij})$$

$$= \frac{\partial P}{\partial x_i} + \frac{\partial}{\partial x_j}\sigma_{0,ij}. \tag{6.28}$$

Finally, we shall also write the gravitational force density as

$$\frac{\mathbf{F}_b}{\delta V} = -\rho\vec{\nabla}\Phi_g, \tag{6.29}$$

where Φ_g is the gravitational potential.

With these refinements, our final form of the momentum equation becomes

$$\frac{\partial}{\partial t}(\rho\mathbf{v}) + \vec{\nabla}\cdot(\rho\mathbf{v}\mathbf{v}) = -\vec{\nabla}P - \vec{\nabla}\cdot\vec{\vec{\sigma}}_0 - \rho\vec{\nabla}\Phi_g. \tag{6.30}$$

Further, adding equations (6.20) and (6.23), we get

$$\frac{d}{dt}\left[\rho\left(\frac{1}{2}v^2+u\right)\right]+\rho\left(\frac{1}{2}v^2+u\right)\left(\vec{\nabla}\cdot\mathbf{v}\right)$$
$$=-\vec{\nabla}\cdot(P\mathbf{v})+\rho T\frac{ds}{dt}-\mathbf{v}\cdot\left(\vec{\nabla}\cdot\vec{\sigma}_0\right)-\rho\mathbf{v}\cdot\vec{\nabla}\Phi_g. \qquad (6.31)$$

Our derivation of the energy equation will be complete once we have separated out the contributions to the full time derivative as before, and use the fact that

$$-\rho\mathbf{v}\cdot\vec{\nabla}\Phi_g=-\vec{\nabla}\cdot\left(\rho\mathbf{v}\Phi_g\right)-\frac{\partial}{\partial t}\left(\rho\Phi_g\right)+\rho\frac{\partial\Phi_g}{\partial t}. \qquad (6.32)$$

The result is

$$\frac{\partial}{\partial t}\left[\rho\left(\frac{1}{2}v^2+u+\Phi_g\right)\right]+\vec{\nabla}\cdot\left[\rho\mathbf{v}\left(\frac{1}{2}v^2+u+\frac{P}{\rho}+\Phi_g\right)\right]$$
$$=\rho T\frac{ds}{dt}-\mathbf{v}\cdot(\vec{\nabla}\cdot\vec{\sigma}_0). \qquad (6.33)$$

We have dropped the term proportional to $\partial\Phi_g/\partial t$ because the use of a time-dependent gravitational potential is very rare.

In the next section we will begin to apply the hydrodynamic equations (6.9), (6.30), and (6.33) to the problem of accretion onto a compact object. In most situations, the stress $\vec{\sigma}_0$ is zero, or insignificant compared to P, and we may ignore it. We will also learn that several different mechanisms contribute to the heat flux ($\rho T ds/dt$), including radiative diffusion and heat conduction, so we will need to customize this term accordingly, case by case.

6.2 BONDI-HOYLE ACCRETION

Matter rarely falls in radially toward a central source of gravity. Compact objects tend to be surrounded by an accretion disk. Nonetheless, spherical accretion is perhaps the simplest way to think of matter accreting into a deep potential well, and often we can learn much about a high-energy source by considering situations in which the plasma's azimuthal component of velocity may be ignored. For example, a supermassive black hole accreting from the interstellar medium captures matter at $r\sim 1$ lyr or more, depending on its mass. At these distances, the impact of angular momentum on the gas dynamics is greatly suppressed compared to what happens much closer in, where the gas settles into Keplerian orbits. The Bondi-Hoyle theory of accretion, which assumes that radial motion is dominant, is useful both historically—because it was the first attempt at quantifying the properties of accretion—and as a relatively simple physical description yielding quick, reliable answers.

We will develop Bondi-Hoyle accretion for several reasons, one of which is to establish the accretion rate as a function of the thermodynamic variables at infinity. The hydrodynamic equations we need are the mass continuity equation (6.9), the slightly modified version of the momentum equation (6.16),

$$\frac{\partial}{\partial t}(\rho\mathbf{v}) + \vec{\nabla}\cdot(\rho\mathbf{v}\mathbf{v}) = -\vec{\nabla}P + \mathbf{f}_b, \tag{6.34}$$

and the modified version of the energy equation (6.33),

$$\frac{\partial}{\partial t}\left[\rho\left(\frac{1}{2}v^2 + u\right)\right] + \vec{\nabla}\cdot\left[\rho\mathbf{v}\left(\frac{1}{2}v^2 + u + \frac{P}{\rho}\right)\right]$$
$$= -\vec{\nabla}\cdot\mathbf{F}_{\mathrm{rad}} - \vec{\nabla}\cdot\mathbf{q} + \mathbf{v}\cdot\mathbf{f}_b. \tag{6.35}$$

In these expressions, $P = nkT$ is the gas pressure in terms of the particle number density n, and $\mathbf{f}_b = -\rho\vec{\nabla}\Phi_g$ is the body force per unit volume. Also, the heat transfer rate $\rho T\,ds/dt$ has been supplanted by the gradient in radiative flux $(-\vec{\nabla}\cdot\mathbf{F}_{\mathrm{rad}})$ and the gradient in conductive flux $(-\vec{\nabla}\cdot\mathbf{q})$. And since we are keeping the gravitational potential as a "source" of energy on the right-hand side, the conserved quantity in equation (6.35) is the specific energy $v^2/2 + u$.

In terms of the intensity of radiation $I_v(\hat{\mathbf{n}}, \mathbf{r})$ (energy per unit area, per unit time, per unit frequency, per unit steradian) at position $\mathbf{r}$ in the direction $\hat{\mathbf{n}}$, we write

$$\mathbf{F}_{\mathrm{rad}} = \int dv \int d\Omega\,\hat{\mathbf{n}}\,I_v(\hat{\mathbf{n}}, \mathbf{r}). \tag{6.36}$$

The term $-\vec{\nabla}\cdot\mathbf{F}_{\mathrm{rad}}$ gives the rate at which radiant energy is being lost by emission, or gained by absorption, per unit volume of gas.

In the special case where the gas is optically thin, so that radiation can escape freely,

$$-\vec{\nabla}\cdot\mathbf{F}_{\mathrm{rad}} = -4\pi \int j_v\,dv, \tag{6.37}$$

where j_v is the emissivity (energy per unit volume, per unit time, per unit frequency, per unit steradian).

In the other extreme, where the gas is optically thick and the particles and radiation are in thermodynamic equilibrium, $\mathbf{F}_{\mathrm{rad}}$ is described well by the blackbody law. Consider the following somewhat heuristic argument. In local thermodynamic equilibrium, the radiation pressure is

$$P_{\mathrm{rad}} = \tfrac{1}{3}aT^4. \tag{6.38}$$

Thus, the radiative force per unit volume is

$$\mathbf{f}_{\mathrm{rad}} = -\vec{\nabla}P_{\mathrm{rad}} = -\tfrac{4}{3}aT^3\vec{\nabla}T. \tag{6.39}$$

Now, since a photon's momentum is $p_\gamma = \epsilon_\gamma/c$ (in terms of its energy ϵ_γ), the radiative momentum flux through the medium is $\mathbf{F}_{\mathrm{rad}}/c$. Thus, if all of this momentum flux is absorbed within a distance $\langle l\rangle$, we must have

$$\frac{1}{\langle l\rangle}\frac{\mathbf{F}_{\mathrm{rad}}}{c} = \mathbf{f}_{\mathrm{rad}} = -\frac{4}{3}aT^3\vec{\nabla}T. \tag{6.40}$$

But

$$\langle l\rangle = \frac{1}{n\sigma} \equiv \frac{1}{\kappa\rho}, \tag{6.41}$$

where σ is the cross section and κ is known as the *opacity*. Therefore,

$$\mathbf{F}_{rad} = -\frac{4ac}{3\kappa\rho}T^3\vec{\nabla}T. \tag{6.42}$$

Equations (6.36) and (6.42) bracket the range of possible radiative fluxes one might encounter in spherical accretion problems. Correspondingly, the quantity **q** measures the rate at which random motions carry thermal energy through the medium. Fortunately, we can often ignore this term.

Let us now assume that all the relevant quantities are independent of angles and t, the latter implying the flow is steady. Then the mass continuity equation becomes

$$\frac{1}{r^2}\frac{d}{dr}\left(r^2\rho v\right) = 0, \tag{6.43}$$

where $\mathbf{v} = (v_r, 0, 0) \equiv (-v, 0, 0)$. Thus, $r^2\rho v = $ constant, which has the units of mass per unit time. That is,

$$\dot{M} = 4\pi r^2\rho(-v), \tag{6.44}$$

where $\dot{M}$ is the (constant) accretion rate.

In the momentum equation (6.34), $\mathbf{f}_b = (f_r, 0, 0)$, where

$$f_r \equiv -\frac{GM\rho}{r^2}. \tag{6.45}$$

Therefore,

$$v\frac{dv}{dr} + \frac{1}{\rho}\frac{dP}{dr} + \frac{GM}{r^2} = 0. \tag{6.46}$$

Insofar as the conservation of energy is concerned, it is often possible to get a fairly accurate solution by replacing the full differential equation (6.35) with the *polytropic* relation

$$P = D\rho^\Gamma, \tag{6.47}$$

since this allows us to treat both adiabatic ($\Gamma = \frac{5}{3}$) and isothermal ($\Gamma = 1$) flows, and just about everything in between these limits. Here, Γ is the ratio of specific heats, and D is a normalization constant. For example, the former is valid when the heating timescale is long compared to the free-fall time, whereas the latter is appropriate when local sources of heat maintain a constant temperature faster than the dynamics can change it.

The problem is now to solve equation (6.46) with (6.47), using equation (6.44) to determine $\dot{M}$. Put

$$\frac{dP}{dr} = \frac{dP}{d\rho}\frac{d\rho}{dr} = c_s^2\frac{d\rho}{dr}, \tag{6.48}$$

where

$$c_s \equiv \left(\frac{dP}{d\rho}\right)^{1/2} \tag{6.49}$$

is the sound speed. Then

$$v\frac{dv}{dr} + \frac{c_s^2}{\rho}\frac{d\rho}{dr} = -\frac{GM}{r^2}. \tag{6.50}$$

But from the mass continuity equation, we know that

$$\frac{1}{\rho}\frac{d\rho}{dr} = -\frac{1}{vr^2}\frac{d}{dr}(vr^2), \tag{6.51}$$

so

$$v\frac{dv}{dr} - \frac{c_s^2}{vr^2}\frac{d}{dr}(vr^2) + \frac{GM}{r^2} = 0. \tag{6.52}$$

A simple rearrangement produces the final result,

$$\frac{1}{2}\left(1 - \frac{c_s^2}{v^2}\right)\frac{d}{dr}(v^2) = -\frac{GM}{r^2}\left(1 - \frac{2c_s^2 r}{GM}\right). \tag{6.53}$$

This now-famous equation was first derived to describe the behavior of the solar wind.[1] At large radii, the right-hand side is greater than zero, since $c_s \to c_s(\infty)$. Thus, since $v \to 0$ as $r \to \infty$, $dv^2/dr < 0$, which is consistent with $c_s^2/v^2 > 1$. That is, the flow is subsonic far from the accretor.

At the other extreme, as $r \to 0$, the multiplicative factor in parentheses on the right-hand side of equation (6.53) becomes less negative and eventually goes to zero. (The temperature would have to be extraordinarily large to make c_s increase rapidly enough to compensate for r decreasing.) Thus, since dv^2/dr is expected to always be negative, $c_s^2/v^2 < 1$ near the compact object, and therefore $v^2 > c_s^2$, i.e., the flow is supersonic near the origin.

The profile we have just described, a flow that begins subsonically at large radii and becomes supersonic near the compact object, is one of several possible solutions to the Parker wind equation. It is *transonic*, one of two such flows that pass through a *sonic* point, r_s, where $v(r_s) = c_s(r_s)$. The other possibility is actually the wind solution, for which equation (6.53) was first devised. All six types of solution realized with appropriate choices of boundary conditions are shown schematically in figure 6.1. Other than the two transonic flows, the rest are unphysical, either because they never attain escape velocity (types 3 and 6), or they never accrete (types 4 and 5).

The sonic (also called "critical") point is where the flow becomes transonic with $v \to c_s$. At this radius, we clearly need

$$1 - \frac{2c_s^2(r_s)r_s}{GM} = 0, \tag{6.54}$$

for otherwise equation (6.53) becomes singular. This equation immediately identifies the location where the sonic condition must be manifested,

$$r_s = \frac{GM}{2c_s^2(r_s)}. \tag{6.55}$$

---

[1]This equation is now also known as the Parker wind equation, in honor of Eugene Parker, who first discovered it in the early 1960s. See Parker (1960).

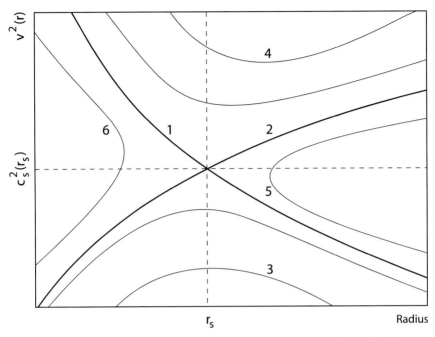

Figure 6.1 The Parker wind equation admits six classes of solution, as indicated in this figure. Classes 1 and 2 are transonic fluids that represent, respectively, inflows and outflows. The other four classes are unphysical, either because the plasma never escapes the system (classes 3 and 6), or because it never accretes onto the central object (class 5), or because it corresponds to an unrealizable thermodynamic state, in which the temperature (and therefore the sound speed) is either too low (class 4) or too high (class 3). (Adapted from Parker 1960)

On the other hand, if the flow is not transonic, then $dv^2/dr = 0$ at r_s, which results in solutions of type 3 or 4. Interestingly,

$$r_s = \frac{1}{4}\left(\frac{c}{c_s}\right)^2 r_S, \tag{6.56}$$

where r_S is the Schwarzschild radius defined in equation (3.88). For example, an object accreting from the interstellar medium in which the composition is mostly hydrogen at a temperature $T = 10^7$ K, will produce a sonic transition at $r_s = 2.7 \times 10^5 r_S$.

With this new insight on transonic flows, we may now return to equation (6.46) to begin the process of calculating the accretion rate $\dot{M}$ in terms of the physical conditions at infinity. Integrating this equation directly, we have

$$\frac{1}{2}v^2 + \int \frac{dP}{\rho} - \frac{GM}{r} = \text{constant}. \tag{6.57}$$

But for a polytrope

$$dP = D\Gamma\rho^{\Gamma-1}\,d\rho, \tag{6.58}$$

so

$$\frac{1}{2}v^2 + \frac{D\Gamma}{\Gamma - 1}\rho^{\Gamma - 1} - \frac{GM}{r} = \text{constant} \tag{6.59}$$

(as long as $\Gamma \neq 1$). From the equation of state, we know that

$$D\Gamma\rho^{\Gamma - 1} = \Gamma P/\rho = c_s^2, \tag{6.60}$$

and, therefore,

$$\frac{1}{2}v^2 + \frac{c_s^2}{\Gamma - 1} - \frac{GM}{r} = \text{constant}. \tag{6.61}$$

For our type 1 solution (see figure 6.1), $v \to 0$ at infinity, so

$$0 + \frac{c_s^2(\infty)}{\Gamma - 1} - 0 = \text{constant}, \tag{6.62}$$

and equation (6.61) becomes

$$\frac{1}{2}v^2 + \frac{c_s^2}{\Gamma - 1} - \frac{GM}{r} = \frac{c_s^2(\infty)}{\Gamma - 1}. \tag{6.63}$$

Thus, at the sonic point, where $v^2 = c_s^2$ and $r_s = GM/2c_s^2$, we have

$$c_s^2(r_s)\left(\frac{1}{2} + \frac{1}{\Gamma - 1} - 2\right) = \frac{c_s^2(\infty)}{\Gamma - 1}. \tag{6.64}$$

This means that the sound speed at r_s is completely determined by the sound speed at infinity and the thermodynamic state of the flow through the ratio of specific heats:

$$c_s(r_s) = c_s(\infty)\left(\frac{2}{5 - 3\Gamma}\right)^{1/2}. \tag{6.65}$$

The accretion rate $\dot{M}$ follows immediately from an application of equation (6.44) at $r = r_s$, where

$$\dot{M} = 4\pi r_s^2 \rho(r_s)c_s(r_s). \tag{6.66}$$

With $D\Gamma\rho^{\Gamma - 1} = c_s^2$, we infer that

$$\rho(r_s) = \rho(\infty)\left[\frac{c_s(r_s)}{c_s(\infty)}\right]^{2/(\Gamma - 1)}, \tag{6.67}$$

and therefore

$$\dot{M} = \pi G^2 M^2 \frac{\rho(\infty)}{c_s^3(\infty)}\left[\frac{2}{5 - 3\Gamma}\right]^{(5 - 3\Gamma)/2(\Gamma - 1)}. \tag{6.68}$$

This is what we set out to find—a determination of $\dot{M}$ in terms of the physical conditions at infinity.

As an illustration of what this means observationally, consider the case of an isolated neutron star accreting from the interstellar medium. There, $\rho(\infty) \approx 10^{-24}\,\text{g cm}^{-3}$ and $c_s(\infty) \approx 10\,\text{km s}^{-1}$, corresponding to a temperature

$T(\infty) \approx 10^4$ K. A neutron star in this medium accretes at a rate

$$\dot{M} \approx 1.4 \times 10^{11} \left(\frac{M}{M_\odot}\right)^2 \left[\frac{\rho(\infty)}{10^{-24} \text{ g cm}^{-3}}\right] \left[\frac{c_s(\infty)}{10 \text{ km s}^{-1}}\right]^{-3} \text{ g s}^{-1}. \tag{6.69}$$

Its accretion luminosity is therefore

$$L_{\text{acc}} = \frac{GM\dot{M}}{R_{\text{ns}}} \approx 2 \times 10^{31} \text{ ergs s}^{-1}, \tag{6.70}$$

which is detectable all the way out to about 1 kpc with current instrumentation (see section 1.5).

With today's X-ray telescopes, it is also possible to resolve the *capture* region surrounding some compact objects, where gas in the interstellar medium becomes bound to the accretor and funnels inward toward the origin. Bondi-Hoyle accretion theory can help us determine the so-called capture (or accretion) radius r_{acc}, in addition to the accretion rate $\dot{M}$.

From equation (6.63), the plasma in the interstellar medium first becomes bound to the central object when $c_s \to c_s(\infty)$ and

$$\frac{1}{2}v^2 - \frac{GM}{r} \approx 0. \tag{6.71}$$

But at large radii, $v \to c_s(\infty)$, so

$$\frac{c_s^2(\infty)}{2} \approx \frac{GM}{r_{\text{acc}}}, \tag{6.72}$$

or

$$r_{\text{acc}} \approx \frac{2GM}{c_s^2(\infty)}. \tag{6.73}$$

The physical interpretation of the accretion radius is that for $r > r_{\text{acc}}$, the thermal energy per unit mass $(c_s^2[\infty])$ is larger than the gravitational binding energy per unit mass (GM/r), and the gas does not accrete. A quick estimate of $\dot{M}$ may therefore be made with the supposition that gas flowing past the accretor encounters a gravitational cross section $\sim \pi r_{\text{acc}}^2$, which then results in an accretion rate

$$\dot{M} \sim \pi r_{\text{acc}}^2 c_s(\infty)\rho(\infty), \tag{6.74}$$

where $c_s(\infty)\rho(\infty)$ is the mass flux (mass per unit area, per unit time). A substitution for r_{acc} from equation (6.73) then produces a solution for $\dot{M}$ impressively close to equation (6.68), though, of course, lacking the factor arising from the internal gas dynamics,

$$\dot{M} \sim \pi G^2 M^2 \frac{\rho(\infty)}{c_s^3(\infty)}. \tag{6.75}$$

In section 12.2, we will apply this rather simple theory to the class of nearby supermassive black holes, and learn that quantities such as $\dot{M}$ and r_{acc} can be estimated reliably via the Bondi-Hoyle prescription for direct comparison with the observations, even without the more esoteric and detailed numerical simulations one can now carry out with modern computers. In cases where $v \gg c_s(\infty)$ far from the central object, the capture radius is determine not by the thermal speed in the interstellar medium, but rather by the flow velocity past the accretor. So in those situations, equation (6.75) is transformed accordingly to

$$\dot{M} \sim \pi G^2 M^2 \frac{\rho(\infty)}{v^3(\infty)}. \tag{6.76}$$

For example, the supermassive black hole at the center of our Galaxy has a mass $M \approx 3.7 \times 10^6 \, M_\odot$, and is surrounded by a medium with an average particle density $n \sim 10^3 \, \text{cm}^{-3}$, and a flow velocity $v(\infty) \sim 1000 \, \text{km s}^{-1}$. Its accretion radius is therefore $r_{\text{acc}} \sim 10^{17} \, \text{cm}$, which at the $\sim$8-kpc distance to the galactic center subtends an angle of $\sim 1''$. This is easily resolvable with X-ray telescopes such as *Chandra*, whose resolving power is $\sim 0.''5$. The accretion flow in supermassive black holes is therefore no longer a mere intellectual curiosity, for we are now in a position to test accretion theories directly with the observations.

6.3 ROCHE LOBE GEOMETRY IN BINARIES AND ACCRETION FROM A COMPANION STAR

Many compact objects become visible to us primarily because of their accretion of plasma from the surrounding medium. This is certainly the case for supermassive black holes in the nuclei of galaxies, and isolated neutron stars and white dwarfs. But in astrophysics, we encounter another common situation in which high-energy sources grow by absorbing matter—when they are members of tight binaries, especially with a companion star whose size approaches or exceeds the tidal radius, at which the gravitational field of the accretor may strip the outer layers of its hapless partner.

Roche lobe theory concerns itself with a determination of test particle orbits within the full gravitational potential of two massive bodies orbiting each other (see figure 6.2). With a the binary separation (between masses M_1 and M_2) and P_{orb} the binary period, Kepler's law tells us that

$$4\pi^2 a^3 = G(M_1 + M_2) P_{\text{orb}}^2. \tag{6.77}$$

We consider a description of the motion in terms of coordinates fixed to the rotating (binary) frame. Thus, in addition to the gravitational force from the two stars, the test particle also experiences a fictitious force from the acceleration of the frame itself. The three accelerations we must include in a full description of the particle dynamics are, therefore,

$$\mathbf{a}_1 = -\frac{GM_1}{|\mathbf{r} - \mathbf{r}_1|^3} (\mathbf{r} - \mathbf{r}_1), \tag{6.78}$$

$$\mathbf{a}_2 = -\frac{GM_2}{|\mathbf{r} - \mathbf{r}_2|^3} (\mathbf{r} - \mathbf{r}_2), \tag{6.79}$$

and the centrifugal acceleration

$$\mathbf{a}_f = -\vec{\omega} \times (\vec{\omega} \times \mathbf{r}). \tag{6.80}$$

Radii are measured relative to an origin placed at the center of mass of the system.

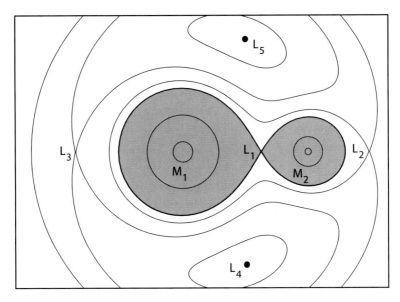

Figure 6.2 Sections in the orbital plane of the Roche equipotentials $\Phi_R(\mathbf{r}) = $ constant, for a binary system with a mass ratio $M_2/M_1 = 0.2$. Beginning with the inner circular contours surrounding M_1 and M_2, the equipotentials move outward with increasing Φ_R. Thus, the saddle point L_1 (at the inner Lagrangian point) forms a preferred pass for matter flowing from one Roche lobe to the other (the two portions of the figure-eight surface). The "Trojan asteroid" points L_4 and L_5 are local maxima of Φ_R.

It is often useful to express these quantities in terms of the potential. Equations (6.78) and (6.79) are easy, for writing

$$\mathbf{a}_i = -\vec{\nabla}\Phi_i, \tag{6.81}$$

we infer that

$$\Phi_i = -\frac{GM_i}{|\mathbf{r} - \mathbf{r}_i|}. \tag{6.82}$$

For equation (6.80), we note that

$$-\vec{\omega} \times (\vec{\omega} \times \mathbf{r}) = -\tfrac{1}{2}\vec{\nabla}(\vec{\omega} \times \mathbf{r})^2, \tag{6.83}$$

and so in total (summing over all all three effects) we have

$$\Phi_R(\mathbf{r}) = -\frac{GM_1}{|\mathbf{r} - \mathbf{r}_1|} - \frac{GM_2}{|\mathbf{r} - \mathbf{r}_2|} - \frac{1}{2}\vec{\nabla}(\vec{\omega} \times \mathbf{r})^2, \tag{6.84}$$

known as the *Roche* potential. We point out, however, that terms that depend on the particle's velocity, such as the Coriolis force, have been ignored since they tend to be small compared to the other three accelerations included in equation (6.84).

In trying to understand the behavior of our test particle subject to the influence of Φ_R, it is best to examine the equipotential surfaces shown in figure 6.2. These surfaces are defined by the condition that tangential gradients in Φ_R are zero (i.e.,

$\vec{\nabla}\Phi_R \cdot \hat{\mathbf{t}} = 0$, where $\hat{\mathbf{t}}$ is the unit vector tangential to the surface), meaning that a test particle experiences no force along $\hat{\mathbf{t}}$. At large $\mathbf{r}$, the system appears as a point so the equipotential surfaces are approximately circular, as if all of the mass were concentrated at the origin.

When $|\mathbf{r} - \mathbf{r}_i|$ is small, the potential is dominated by the mass M_i, and so the surfaces are again circular, though now centered on the star itself, rather than at the binary's center of mass.

Between these two limits, there exists a critical solution, let us call it Φ_{Rc}, for which the equipotential surface connects the two stars, making a figure eight enclosing the shaded region in figure 6.2. The two *lobes* connect at the inner Lagrangian point L_1, a saddle point of Φ_R. The relevance of this critical surface to our discussion is that matter rising through the potential well of the companion can traverse from one lobe to the other across this saddle point more easily than anywhere else on Φ_{Rc}. This happens because the potential decreases again across L_1, but rises in any direction away from the line joining the two stars.

Binary systems become particularly interesting, insofar as accretion is concerned, when either the binary separation has decreased, or the companion star M_2 has grown, to the point at which it *fills* its Roche lobe. Then matter can slide across L_1 into the Roche lobe of M_1, a compact star such as a white dwarf, a neutron star, or a stellar-size black hole.

For any binary system in which we wish to analyze the mass transfer, equation (6.84) may be solved completely, usually with numerical techniques. Often, however, it is more convenient and practical to use analytical relations that approximate these numerical results. For example, the average radius of M_2's Roche lobe is

$$\frac{R_2}{a} \approx 0.38 + 0.20 \log q \quad (0.5 \le q < 20), \tag{6.85}$$

where $q \equiv M_2/M_1$, and

$$\frac{R_2}{a} \approx 0.462 \left(\frac{q}{1+q}\right)^{1/3} \quad (0 < q < 0.5). \tag{6.86}$$

Obviously, the average radius of the primary star's Roche lobe is given by these same expressions with $q \to 1/q$, so

$$\frac{R_1}{a} \approx 0.38 - 0.20 \log q \quad (0.05 < q < 2) \tag{6.87}$$

and

$$\frac{R_1}{a} \approx 0.462 \left(\frac{1}{1+q}\right)^{1/3} \quad (2 \le q). \tag{6.88}$$

Another useful relation is the distance b_1 of L_1 from the center of M_1, which is given by

$$\frac{b_1}{a} = 0.500 - 0.277 \log q. \tag{6.89}$$

We will see in section 9.3 that binary systems evolve for several different reasons, including the release of gravitational radiation, the loss of angular momentum via a stellar wind, and, in some cases, even as a result of a self-sustaining mass transfer. Since mass transfer changes the values of q and a, it is worth examining under what conditions this self-induced mass exchange may occur.

Clearly, for the binary separation to continue shrinking, R_2 must decrease as mass is transferred across L_1 to keep the Roche lobe always in contact with the surface of M_2. Consider the following simplified picture. Assume (1) $M_1 + M_2 = $ constant, (2) $\mathbf{L}_{tot} = $ constant, and (3) $\mathbf{L}_{tot} = \mathbf{L}_1 + \mathbf{L}_2$, where $\mathbf{L}$ is an angular momentum (not to be confused with the inner Lagrangian point L_1) and the stars are treated simply as mass points.

From (1), we see that

$$M_1(1+q) = \text{constant}, \tag{6.90}$$

and (2) and (3) give

$$M_1 a_1^2 \omega + M_2 a_2^2 \omega = \text{constant}, \tag{6.91}$$

where $\omega = 2\pi/P_{orb}$ is the binary's angular frequency of rotation. But

$$a_1 = \frac{qa}{1+q} \tag{6.92}$$

and

$$a_2 = \frac{a}{1+q}. \tag{6.93}$$

Thus, equation (6.91) gives

$$\frac{q}{1+q} \frac{M_1 a^2}{P_{orb}} = \text{constant}. \tag{6.94}$$

In addition, Kepler's law (6.77) and equation (6.90) imply

$$a^2 \propto P_{orb}^{4/3}, \tag{6.95}$$

so in (6.94),

$$P_{orb} \propto \frac{(1+q)^6}{q^3} \propto \frac{1}{M_1^3 M_2^3}. \tag{6.96}$$

Returning to Kepler's law,

$$a \propto \frac{(1+q)^4}{q^2} \propto \frac{1}{M_1^2 M_2^2}. \tag{6.97}$$

Clearly, both P_{orb} and a have minima at $q = 1$. This may be seen easily, e.g., by differentiating a in equation (6.97) with respect to q and setting the result equal to zero. Thus, if $q > 1$, the reduction in q due to mass transfer causes a to decrease. The opposite occurs when $q < 1$.

Let us now write equations (6.85) and (6.86) as

$$R_2 = f(q)a,$$
(6.98)

where

$$f(q) = \begin{cases} 0.38 + 0.20 \log q & \text{if } (0.5 \le q < 20) \\ 0.462\, q^{1/3}(1+q)^{-1/3} & \text{if } (0 < q < 0.5). \end{cases}$$
(6.99)

Then

$$\ln R_2 = \ln f + \ln a,$$
(6.100)

so that

$$\frac{\Delta R_2}{R_2} = \left(\frac{f'}{f} + \frac{a'}{a} \right) \Delta q.$$
(6.101)

But it is clear from equations (6.97) and (6.99) that

$$\left| \frac{a'}{a} \right| > \left| \frac{f'}{f} \right|,$$
(6.102)

so

$$\frac{\Delta R_2}{R_2} \sim \frac{a'}{a} \Delta q = \frac{\Delta a}{a}.$$
(6.103)

Thus, as the binary separation decreases by an amount Δa (with $q > 1$), the Roche lobe around M_2 also shrinks, eating deeper into the envelope of the companion star. This forces more mass to transfer across L_1, decreasing q, and continuing this process until $q = 1$. The mass transfer is self-perpetuating as long as $M_2 > M_1$. Of course, in real systems other effects sustain this process even if $q < 1$. We will learn in section 9.3 that if the companion star is out of thermal equilibrium, as would occur after mass loss, its outer envelope swells and fills its Roche lobe, further driving the binary toward a more compact configuration.

6.4 FORMATION OF A DISK

Once matter flows across the inner Lagrangian point L_1, it proceeds to sweep ahead and past the compact object in the middle of the Roche lobe. Remember that we are looking at this system from within the rotating frame. Thus, plasma at L_1 has angular momentum with respect to the primary star, and therefore swings past the accretor to form an orbiting ring. We will study the results of a numerical simulation of this process shown in figure 7.5, but for now we will settle on the following heuristic approach to estimate where the *circularization* takes place, and how big the ring (and ultimately the disk) can be in these binary systems.

At L_1, thermal motion drives the expansion of the secondary's outer envelope. Therefore, the velocity of mass flowing across the saddle point has components

$$v_\perp \sim b_1 \omega$$
(6.104)

and

$$v_\parallel \lesssim c_s, \tag{6.105}$$

where c_s is the speed of sound we defined earlier in equation (6.49). Thus, for $b_1 \gtrsim 0.5a$ (see equation 6.89) and $\omega = 2\pi/P_{orb}$, and a given by Kepler's law (equation 6.77), we have

$$v_\perp \sim 100 \left(\frac{M_1}{M_\odot} \right)^{1/3} (1+q)^{1/3} \left(\frac{P_{orb}}{1\,\text{day}} \right)^{-1/3} \text{km s}^{-1} \tag{6.106}$$

and

$$v_\parallel \sim c_s \sim 10\,\text{km s}^{-1}, \tag{6.107}$$

for typical stellar envelope temperatures ($\lesssim 10^5$ K).

Within the primary's Roche lobe, the infalling plasma is controlled primarily by M_1's potential. It circularizes at a radius R_{circ} where the Keplerian specific angular momentum equals that of matter passing across L_1. That is,

$$v_\phi(R_{circ}) = \left(\frac{GM_1}{R_{circ}} \right)^{1/2}, \tag{6.108}$$

where

$$R_{circ} v_\phi(R_{circ}) = b_1^2 \omega. \tag{6.109}$$

Thus,

$$\frac{R_{circ}}{a} = \frac{4\pi^2}{GM_1 P_{orb}^2} a^3 \left(\frac{b_1}{a} \right)^4. \tag{6.110}$$

With Kepler's law, this simplifies to

$$\frac{R_{circ}}{a} = (1+q) \left(\frac{b_1}{a} \right)^4, \tag{6.111}$$

and with the fitting formula in equation (6.89), it reduces to the final form,

$$\frac{R_{circ}}{a} = (1+q)0.500 - 0.227 \log q^4. \tag{6.112}$$

In the case $q = 1$, we have $R_1 \approx b_1 = 0.38a$, and since $R_{circ} = 0.125a$, we see that $R_{circ} = 0.33R_1$. As it turns out, this result is actually generally true—the circularization ring in compact binaries tends to be roughly one-third to one-half the size of the primary Roche lobe.

Stellar mass compact objects have a radius R_* smaller than R_{circ} for any realistic binary parameters. For example, with $b_1/a \approx 0.5$ and $P_{orb} \sim 1$ h, $R_{circ} \sim 3.5 \times 10^9$ cm. By comparison, R_* cannot be larger than the radius of a low-mass white dwarf, typically $\sim 10^9$ cm. Plasma flowing from L_1 toward M_1 therefore misses the primary entirely and settles into a ring-like orbit with radius $\sim R_{circ}$. Viscosity then enters the picture and disperses the material to form a differentially rotating disk—whose physical properties we will explore in the next chapter.

SUGGESTED READING

An analytic treatment of accretion is invariably based on the so-called wind (or Parker) equation, first derived to account for the mass loss from the Sun. Invariant with respect to a reversal in time, this equation also describes mass flowing inward, i.e., mass gained by the central object. Its original derivation appeared in Parker (1960).

Several excellent review articles have been written on the theory of close binary evolution—on issues such as the origin of torques driving the loss of angular momentum, the consequence of mass transfer, the Roche lobe geometry, and the formation of compact accretion disks. The reader is encouraged to read some (or all) of the following: Paczyński (1971), Thomas (1977), Shu and Lubow (1981).

Chapter Seven

Accretion Disk Theory

Whether a compact object accretes stochastically from the surrounding medium, or preferentially from a binary companion, the accreting plasma invariably settles into a disk perpendicular to its net angular momentum vector **L**. Even if the accretion proceeds spherically at first, cooling processes eventually diminish the plasma's support parallel to **L** and the gas relaxes to a disk in which centrifugal support prevents further infall toward the origin (see figure 7.1). To fully understand a high-energy source, we must therefore have a viable theory of matter swirling on Keplerian orbits about a central accretor. Plasma on these trajectories rotates differentially with radially dependent azimuthal velocities. Any chaotic motion of gas elements about the circular streamlines gives rise to viscous forces, a process known as *shear viscosity*. In this chapter, we study the physics of thin-disk accretion driven by the transfer of angular momentum and energy via the dissipative effects of this internal "friction."

7.1 VISCOSITY AND RADIAL DISK STRUCTURE

Let us consider a segment of disk in cylindrical coordinates, as shown in figure 7.2. Because of chaotic motions, gas elements such as A and B' are constantly crossing a given surface (at R) with speeds $\tilde{v}$ over a length scale λ. But the angular momentum they carry is different. Element A takes a ϕ-velocity $R\Omega(R)$ to A', where it adds a specific angular momentum $(R+\lambda)R\Omega(R)$, in terms of the angular velocity Ω. Component B', on the other hand, takes a ϕ-velocity $(R+\lambda)\Omega(R+\lambda)$ to B, where it adds a specific angular momentum $R(R+\lambda)\Omega(R+\lambda)$.

These two transfers of specific angular momentum have different magnitudes; the net result is a viscous torque on the outer stream by the inner stream, which we may write as

$$\tau_{\text{out}} \sim \dot{M}_{A\to A'}(R+\lambda)R\Omega(R) - \dot{M}_{B'\to B}R(R+\lambda)\Omega(R+\lambda). \tag{7.1}$$

A disk of thickness H in steady state has

$$\dot{M}_{A\to A'} = \dot{M}_{B'\to B} = (2\pi RH)\rho(R)\tilde{v}. \tag{7.2}$$

Thus,

$$\tau_{\text{out}} \sim (2\pi RH)\rho(R)\tilde{v}R(R+\lambda)\left[\Omega(R) - \Omega(R+\lambda)\right]. \tag{7.3}$$

But

$$\left[\Omega(R) - \Omega(R+\lambda)\right] = -\lambda\Omega'(R), \tag{7.4}$$

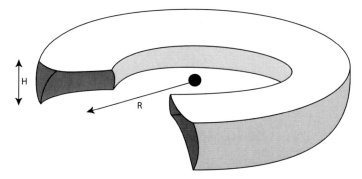

Figure 7.1 The motion of particles perpendicular to the angular momentum plane is suppressed by the cancellation of the z-components of velocity due to collisions. However, the motion in the ϕ-direction can never be suppressed since angular momentum is conserved. The inflowing matter therefore settles down toward the orbital plane and forms a disk, with a thickness H ($\ll R$) set by the thermal pressure gradients within the plasma (see figure 7.3).

so

$$\tau_{\text{out}} \sim -(2\pi R H)\rho(R)\tilde{v}R(R+\lambda)\lambda\Omega'(R). \tag{7.5}$$

We will now define two quantities used often in disk descriptions—the *surface density*

$$\Sigma \equiv \rho H \tag{7.6}$$

and the coefficient of *kinematic viscosity*

$$\nu \equiv \lambda\tilde{v}. \tag{7.7}$$

Thus, with $R + \lambda \approx R$,

$$\tau_{\text{out}} \sim -2\pi\nu\Sigma R^3\Omega'. \tag{7.8}$$

Notice that the torque goes to zero for rigid rotation, in which $\Omega' = 0$. If $\Omega(R)$ decreases outward, as it does for Keplerian motion, then $\tau_{\text{out}} > 0$, meaning that the outer ring gains angular momentum at the expense of the inner ring, so the gas slowly spirals in.

The major uncertainty in disk theory is the magnitude of the excursion length λ and the velocity $\tilde{v}$. In the most extreme case, we might expect that turbulent eddies giving rise to plasma elements, such as A and B' in figure 7.2, are limited in their extent by the actual dimensions of the disk, so that $\lambda \lesssim H$. In addition, it is difficult to envisage a situation in which the velocity of these plasma fluctuations is greater than the local sound speed c_s (see equation 6.49), so we expect that $\tilde{v} \lesssim c_s$. The now-famous α-prescription of viscosity[1] is the parametrization

$$\nu \equiv \alpha c_s H, \tag{7.9}$$

---

[1] Shakura and Sunyaev (1973) developed the α-prescription of viscosity in the early 1970s. Their foundational paper is now one of the most cited in astronomy and astrophysics.

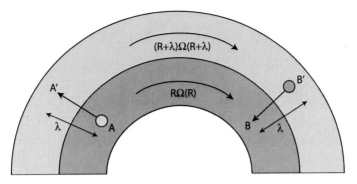

Figure 7.2 Two neighboring disk rings, at radius R and $R + \lambda$, respectively, exchange clumps of plasma A and B', which migrate over the characteristic length scale λ. These clumps, however, carry different azimuthal velocities, and therefore deposit unmatched specific angular momenta in their new locations. The clump $A \rightarrow A'$ contributes a specific angular momentum $(R + \lambda) R\Omega(R)$ to the outer ring, whereas $B' \rightarrow B$ returns an amount $R(R + \lambda)\Omega(R + \lambda)$ to the inner ring. The difference constitutes a net torque on the outer ring.

in which clearly the uncertainty in λ and $\tilde{v}$ is folded into a single parameter (α) of order one. Of course, there is no reason why α should be constant from one radius to the next, nor from one system to another. However, this formulation has become the most commonly used prescription in disk theory because of its simplicity and relevance to real systems.

7.2 STANDARD THIN-DISK THEORY

In attempting to describe the structure of our disk, we will find ourselves developing hydrodynamic equations not unlike those we already derived in section 6.1, though customized to make full use of the specific geometry we have in Keplerian motion.

At any radius R close to the orbital plane (see figure 7.1), the matter has a circular velocity

$$v_\phi(R) = R\Omega_K(R), \tag{7.10}$$

where

$$\Omega_K(R) \equiv \left(\frac{GM}{R^3}\right)^{1/2} \tag{7.11}$$

is the Keplerian angular velocity. Of course, due to the shear viscosity ν, the gas also has a small *drift* velocity v_R in the radial direction. The idea is for us to write conservation equations for mass and angular momentum corresponding to these two velocity components.

Between radii R and $R + \Delta R$, the mass and angular momentum contents are, respectively,

$$\text{mass} = 2\pi R \Delta R \Sigma, \tag{7.12}$$

$$\text{angular momentum} = (2\pi R \Delta R \Sigma) R^2 \Omega. \tag{7.13}$$

Thus,

$$\frac{\partial}{\partial t}(2\pi R \Delta R \Sigma) = v_R(R, t)2\pi R \Sigma(R, t)$$

$$- v_R(R + \Delta R, t)2\pi(R + \Delta R)\Sigma(R + \Delta R, t)$$

$$\approx -2\pi \Delta R \frac{\partial}{\partial R}(R\Sigma v_R), \tag{7.14}$$

or

$$R\frac{\partial \Sigma}{\partial t} + \frac{\partial}{\partial R}(R\Sigma v_R) = 0. \tag{7.15}$$

This is the mass conservation equation for a disk, analogous to equation (6.9) for a general hydrodynamic flow.

Similarly, for the angular momentum at R,

$$\frac{\partial}{\partial t}(2\pi R \Delta R \Sigma R^2 \Omega) = v_R(R, t)2\pi R \Sigma(R, t)R^2 \Omega(R)$$

$$- v_R(R + \Delta R, t)2\pi(R + \Delta R)\Sigma(R + \Delta R, t)$$

$$\times (R + \Delta R)^2 \Omega(R + \Delta R) + \tau_{\text{out}}(R) - \tau_{\text{out}}(R + \Delta R), \tag{7.16}$$

or

$$R\frac{\partial}{\partial t}(\Sigma R^2 \Omega) + \frac{\partial}{\partial R}(R\Sigma v_R R^2 \Omega) = -\frac{1}{2\pi}\frac{\partial \tau_{\text{out}}}{\partial R}. \tag{7.17}$$

But $\tau_{\text{out}}(R, t)$ is given by equation (7.8), so

$$R^3 \Omega \frac{\partial \Sigma}{\partial t} + \frac{\partial}{\partial R}\left(R\Sigma v_R R^2 \Omega\right) = \frac{\partial}{\partial R}\left(R\nu\Sigma R^2 \Omega'\right). \tag{7.18}$$

Using equation (7.15) to replace the partial derivative with respect to time, we get

$$R\Sigma v_R(R^2 \Omega)' = \frac{\partial}{\partial R}(R\nu\Sigma R^2 \Omega'). \tag{7.19}$$

If we now specialize to Keplerian flows, for which Ω is given by Ω_K in equation (7.11), we find

$$\frac{1}{2}R\Sigma v_R \left(\frac{GM}{R}\right)^{1/2} = \frac{\partial}{\partial R}\left[-\frac{3}{2}(GMR)^{1/2}\nu\Sigma\right], \tag{7.20}$$

or

$$v_R = -\frac{3}{\Sigma R^{1/2}}\frac{\partial}{\partial R}\left(\nu\Sigma R^{1/2}\right). \tag{7.21}$$

This is effectively our angular momentum equation. Together with equation (7.15), it specifies the dynamical structure of the disk.

In situations where ν is constant and the disk is in steady state, $R\Sigma v_R$ is constant, so the mass accretion rate may be written

$$\dot{M} = 2\pi R\Sigma(-v_R). \tag{7.22}$$

From equation (7.21), we infer that $v_R \sim O(\nu/R)$, and since $\nu = \alpha c_s H$, we gather that

$$\frac{v_R}{c_s} \sim \alpha \left(\frac{H}{R} \right). \tag{7.23}$$

As we shall soon see when we analyze their vertical structure, these disks are typically very thin, with $H/R \lesssim 0.01$, so

$$\frac{v_R}{c_s} \ll 1, \tag{7.24}$$

which validates our initial assumption that these disks are quasi-Keplerian. Though the shear viscosity leads to an inexorable transfer of angular momentum outward and a diffusion of mass inward, the dominant velocity at any given radius is always azimuthal (equation 7.10).

7.2.1 Energetics and Vertical Structure

The fact that $v_R \neq 0$ means, of course, that gravitational potential energy is being converted into kinetic and thermal energy. Some of this is radiated away; the rest is advected inward with the inflowing plasma. The structure of the disk, particularly in the vertical direction, can change dramatically depending on what fraction of the dissipated energy is radiated locally versus being transported somewhere else, e.g., through an event horizon. In thin-disk theory, one usually assumes that all of the energy is radiated away at the same location where it is liberated. In the next chapter, we will examine a particular example where this is not true, and we will see that the thin-disk approximation is then no longer valid.

For now, let us return for a moment to the expression we derived for the torque on a ring of material, and write

$$\tau_{\text{out}}(R) - \tau_{\text{out}}(R + dR) = -\frac{d\tau_{\text{out}}}{dR} dR. \tag{7.25}$$

Given our definition of τ_{out} (see equation 7.8), the left-hand side of this equation expresses the net torque acting on the ring at R. Thus, the power exerted on this ring by viscous dissipation is

$$P = -\Omega \frac{d\tau_{\text{out}}}{dR} dR. \tag{7.26}$$

That is,

$$P = -\left[\frac{d}{dR} (\tau_{\text{out}} \Omega) - \tau_{\text{out}} \Omega' \right] dR. \tag{7.27}$$

The first term simply gives a global transfer of rotational energy through the disk, and does not represent a local dissipation. We can understand this point by integrating this term over all radii,

$$\int_{R_{\text{in}}}^{R_{\text{out}}} \frac{d}{dR} (\tau_{\text{out}} \Omega) \, dR = \tau_{\text{out}} \Omega|_{R_{\text{out}}} - \tau_{\text{out}} \Omega|_{R_{\text{in}}}, \tag{7.28}$$

which demonstrates that this contribution to P is determined solely by the conditions at the edges of the disk.

The second term represents an actual local dissipation that produces heat. Thus, since each ring has a surface area $2(2\pi R)dR$, the dissipation rate per unit plane surface area is

$$D(R) = \frac{\tau_{\text{out}}\Omega'}{4\pi R}, \tag{7.29}$$

or

$$D(R) = -\tfrac{1}{2}\nu\Sigma\left(R\Omega'\right)^2. \tag{7.30}$$

This equation is correct as it stands, but it is not always practical because we don't necessarily know ν a priori. Instead, it would be more convenient to have an expression for $D(R)$ that depends on observables, such as $\dot{M}$. To that end, let us return to the mass conservation equation (7.22) and work with the angular momentum equation (7.17) in steady state. With the time derivative in this expression set to zero,

$$R\Sigma v_R R^2 \Omega = -\frac{1}{2\pi}\tau_{\text{out}} + \text{constant}, \tag{7.31}$$

and substituting for τ_{out}, we obtain

$$-\nu\Sigma\Omega' = \Sigma(-v_R)\Omega + \frac{\text{constant}}{R^3}. \tag{7.32}$$

Near the surface of the star (other than a black hole, of course), Ω must eventually merge with the stellar rotation rate, so $\Omega' \to 0$. And setting $\Omega = \Omega_K(R_*)$, where R_* is the stellar radius, we get

$$\text{constant} = -\frac{\dot{M}}{2\pi}(GMR_*)^{1/2}. \tag{7.33}$$

Thus, returning to equation (7.32), and rearranging terms, we arrive at an expression for ν as a function of $\dot{M}$ (using $\Omega = \Omega_K$),

$$\nu\Sigma = \frac{\dot{M}}{3\pi}\left[1 - \left(\frac{R_*}{R}\right)^{1/2}\right]. \tag{7.34}$$

And with this, the dissipation rate (equation 7.30) is completely expressible in terms of observables,

$$D(R) = \frac{3GM\dot{M}}{8\pi R^3}\left[1 - \left(\frac{R_*}{R}\right)^{1/2}\right]. \tag{7.35}$$

In the next section, we analyze the spectrum produced by the disk as a result of this dissipation of gravitational energy, but to do so, we first need to know something about its vertical structure to assess the importance of radiative diffusion.

As a first approximation, we may assume that the gas is in local hydrostatic equilibrium, as illustrated schematically in figure 7.3. The two forces contributing to this equilibrium in the vertical direction are gravity and a thermal pressure gradient.

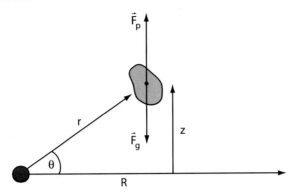

Figure 7.3 Assuming that the gas within the disk (see figure 7.1) is in hydrostatic equilibrium, we may consider all the forces acting on a typical clump at radius r, making an angle θ with respect to the equatorial plane, to determine its equilibrium height z. In this fashion, balancing the vertical component of gravity $\mathbf{F}_g$ with the z-gradient of pressure (producing a counter force $\mathbf{F}_p$), we can ascertain the vertical distribution of density once the equation of state is known (or assumed).

The gravitational force per unit volume is

$$\mathbf{F}_g = -\frac{GM\rho}{r^2} \sin\theta \, \hat{\mathbf{z}}, \tag{7.36}$$

or

$$\mathbf{F}_g = -\frac{GM\rho z}{r^3} \hat{\mathbf{z}} \approx -\frac{GM\rho z}{R^3} \hat{\mathbf{z}}, \tag{7.37}$$

with the supposition that $r \approx R$ for a thin disk. The pressure force per unit volume is

$$\mathbf{F}_p = -\frac{\partial P}{\partial z} \hat{\mathbf{z}}, \tag{7.38}$$

so together, these equations yield the hydrostatic equilibrium condition

$$\frac{1}{\rho}\frac{dP}{dz} = -\frac{GMz}{R^3}. \tag{7.39}$$

It is often a reasonable approximation to assume that the gas is locally isothermal in the vertical direction, so that T varies only with R. Of course, this cannot be strictly correct, especially in optically thick environments, where the diffusion of radiation relies on a nonzero gradient in temperature (see equation 6.42). However, even a small gradient in T can produce a large radiative flux if T itself is large, and our analysis will show that disks in compact objects are characterized by temperatures as high as $\sim 10^6$ K or more.

Putting $T = T(R)$ in equation (7.39) gives

$$\frac{\partial P}{\partial z} = \frac{R_g}{\mu} T(R) \frac{\partial \rho}{\partial z} = \frac{P}{\rho} \frac{\partial \rho}{\partial z} = -\frac{GMz}{R^3}, \tag{7.40}$$

where R_g is the gas constant and μ is the mean molecular weight per particle. Therefore,

$$\frac{1}{\rho}\frac{\partial \rho}{\partial z} = -\left(\frac{GM}{R^3}\frac{\mu}{R_g T}\right)z, \qquad (7.41)$$

with solution

$$\rho(R, z) = \rho_c(R) \exp\left[-\frac{z^2}{2\zeta^2}\right]. \qquad (7.42)$$

The newly defined quantities are the central (or midplane) density $\rho_c(R)$, and the scale height

$$\zeta \equiv \left(\frac{R^3 R_g T}{GM\mu}\right)^{1/2}, \qquad (7.43)$$

which is $H/2$—half the disk thickness introduced in figure 7.1.

As a concrete example, let us consider the structure of a disk orbiting about a one-solar mass object. As we shall see shortly, the midplane temperature in such a configuration is $\sim 10^4$–10^5 K, so from equation (7.43), we get $\zeta/R \sim 10^{-7} R^{1/2}$. This ratio is therefore $\sim 10^{-4}$ at 10 km, appropriate for a neutron star, and $\sim 10^{-2}$ at 10^5 km, as one might encounter around a white dwarf. In any case, ζ is never more than a few percent of R, so our assumption of a thin-disk geometry is often well justified.

7.2.2 Thin-Disk Spectrum

In deciding how to approach the calculation of a disk spectrum, we first need to know whether or not the emitting region is optically thick. An optically thin medium may produce photons by various means, as we learned in chapter 5, sometimes following a combination of several processes. For example, in high-energy environments, radio photons emitted via synchrotron are often Comptonized by the same particles that produced them, giving rise to a phenomenon known as synchrotron self-Comptonization (SSC). However, when the optical depth is much larger than one, all memory of how the radiation was produced is lost and the emerging spectrum is always a Planck function.

Since the optical depth is proportional to the column density, a good place to start with our exploration of the disk's emissivity is equation (7.34), where we learn that

$$\nu\Sigma \sim \frac{\dot{M}}{3\pi}. \qquad (7.44)$$

Since

$$\nu = \alpha c_s H = 2\alpha \left(\frac{R_g T}{\mu}\right)\left(\frac{R^3}{GM}\right)^{1/2}, \qquad (7.45)$$

we have

$$\Sigma \sim \alpha^{-1}\frac{\dot{M}}{6\pi}\left(\frac{\mu}{R_g T}\right)\left(\frac{GM}{R^3}\right)^{1/2}. \qquad (7.46)$$

Thus,

$$\Sigma \sim 10^3 \left(\frac{\dot{M}}{10^{16} \text{ g s}^{-1}}\right) \left(\frac{0.1}{\alpha}\right) \left(\frac{10^4 \text{ K}}{T}\right) \left(\frac{M}{M_\odot}\right)^{1/2} \left(\frac{10^9 \text{ cm}}{R}\right)^{3/2} \text{ g cm}^{-2}.$$

$$(7.47)$$

Now, the optical depth perpendicular to the disk is

$$\tau_\perp = \int_{-\infty}^{\infty} n(z)\sigma_T \, dz, \tag{7.48}$$

where $n(z)$ is the z-dependent particle number density and σ_T is the Thomson cross section. We can write this as

$$\tau_\perp = \frac{\sigma_T}{\mu m_H} \int_{-\infty}^{\infty} \rho(z) \, dz = \frac{\sigma_T}{\mu m_H} \Sigma, \tag{7.49}$$

where m_H is the proton mass.[2] Hence, with the value of Σ in equation (7.47), we estimate that $\tau_\perp \approx 0.8\Sigma \sim 10^3 \gg 1$. There is no question, therefore, that geometrically thin disks are optically thick, and that the emission must be blackbody limited.

But though thin disks evidently radiate as a blackbody, the temperature $T(R)$ changes with radius. By conservation of energy,

$$\sigma_B T_{\text{eff}}^4(R) = D(R), \tag{7.50}$$

where σ_B is the Boltzmann constant and $T_{\text{eff}}(R) \sim T(R)$ is the effective (blackbody) temperature (see also equation 1.20). Thus, with equation (7.35), we derive one of the most important equations in accretion-disk theory,

$$T_{\text{eff}}(R) = \left\{\frac{3GM\dot{M}}{8\pi R^3 \sigma_B}\left[1 - \left(\frac{R_*}{R}\right)^{1/2}\right]\right\}^{1/4}. \tag{7.51}$$

Notice that over most of the disk,

$$T_{\text{eff}}(R) \sim R^{-3/4}, \tag{7.52}$$

an easily testable theoretical prediction that we will compare with observations shortly (see figure 7.7).

At each radius R, the intensity of emerging radiation is therefore

$$I_\nu(R) = B_\nu \left[T(R)\right], \tag{7.53}$$

where B_ν is the Planck (blackbody) function

$$B_\nu = \frac{2h\nu^3/c^2}{\exp h\nu/kT(R) - 1}, \tag{7.54}$$

in units of energy per unit area, per unit time, per unit frequency, per unit steradian. The quantity h in this expression is Planck's constant. Thus, an observer a distance

---

[2] Earlier, we had approximated the optical depth as ρH in equation (7.6). Here we are using the more formal definition $\Sigma = \int_{-\infty}^{\infty} \rho(z) \, dz$.

D away, with a line of sight making an angle i relative to the disk's symmetry axis, will measure a flux

$$F_\nu = \int_{R_*}^{R_{out}} I_\nu \, d\Omega(R),$$ (7.55)

where R_{out} is the disk's outer radius, and $d\Omega(R)$ is the solid angle subtended at the observer's position by the emitting annulus at radius R,

$$d\Omega(R) = \frac{2\pi R \, dR \, \cos i}{D^2}.$$ (7.56)

Thus,

$$F_\nu = \frac{4\pi h \nu^3 \cos i}{c^2 D^2} \int_{R_*}^{R_{out}} \frac{R \, dR}{\exp h\nu/kT(R) - 1}.$$ (7.57)

In the Rayleigh-Jeans limit, here defined by the condition $\nu \ll kT(R_{out})/h$,

$$B_\nu^{RJ} \approx \frac{2kT(R)\nu^2}{c^2},$$ (7.58)

and therefore

$$F_\nu^{RJ} \propto \nu^2.$$ (7.59)

In the opposite limit, $\nu \gg kT(R_*)/h$, the Planck function has the Wien shape,

$$B_\nu^{W} \approx \frac{2h\nu^3}{c^2} e^{-h\nu/kT},$$ (7.60)

and the flux is dominated by the hottest portion of the disk near R_*, with essentially an exponential dependence on frequency,

$$F_\nu^{W} \propto \nu^3 e^{-h\nu/kT}.$$ (7.61)

Both of these limits are highlighted in figure 7.4, which shows an idealized thin-disk spectrum.

In between these two limits, the thin-disk spectrum is a sum of blackbodies. We will make use of the approximation in equation (7.52), and define the variable

$$\eta \equiv \frac{h\nu}{kT(R)} \approx \frac{h\nu}{kT(R_*)} \left(\frac{R}{R_*}\right)^{-3/4}.$$ (7.62)

Then, in (7.57),

$$F_\nu \propto \nu^{1/3} \int_0^\infty \frac{\eta^{5/3}}{\exp(\eta) - 1} d\eta$$
$$\propto \nu^{1/3}.$$ (7.63)

The thin-disk spectrum may be distinguished from those of other thermal sources by its stretched out appearance. If it weren't for this $\sim \nu^{1/3}$ (so-called flat) segment, one might mistake it for a regular blackbody shape. Of course, the other feature that makes it unique is that the temperatures associated with the Rayleigh-Jeans

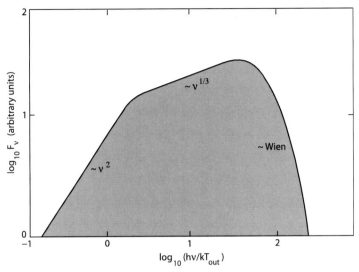

Figure 7.4 The disk spectrum may be thought of as a sum of blackbodies, each produced by a particular ring with its own characteristic temperature. As such, the low-frequency portion is dominated by the lowest-temperature ring (typically the outer one), and is characterized by the Rayleigh-Jeans law ($\propto \nu^2$). At the highest energies, the disk spectrum is dominated by the hottest ring, usually near the inner edge, and falls off according to the Wien law for that particular temperature. In between, the spectrum is a sum of Planck functions, each at progressively higher frequency as the temperature increases from the outer to the inner rings, and is characterized by a relatively flat shape going as $\sim \nu^{1/3}$. On the abscissa, $T_{\text{out}} \equiv T(R_{\text{out}})$.

and Wien regions are very different. So it is usually not difficult to identify a spectrum as belonging to a thin disk once it has been assembled from multifrequency observations.

Now that we have studied accretion by compact objects, and the properties of accretion disks that often form around them, we are in a position to put this theory to a more stringent test by comparing these results with the observations. Cataclysmic variables, which we will study as a class in section 9.3, are binaries that contain a white dwarf primary accreting from a low-mass main-sequence (or sometimes degenerate) companion star. Because many are nearby, and thus relatively easy to observe, they provide us with the best evidence for the presence of accretion disks around high-energy sources.

In section 6.4 we took a heuristic approach to argue that matter crossing the inner Lagrangian point L_1 has too much angular momentum to fall directly into the compact object, and that it settles instead onto a disk that subsequently decays through the dissipational action of shear viscosity. By now, several detailed numerical simulations of this process have been carried out, including those shown in figure 7.5. What emerges from these realistic calculations is that the disk deviates significantly from azimuthal symmetry. In fact, once the disk has spread out to fill roughly half of the primary's Roche lobe, the incoming stream impacts its outer edge and helps to energize a spiral pattern that orbits the accretor.

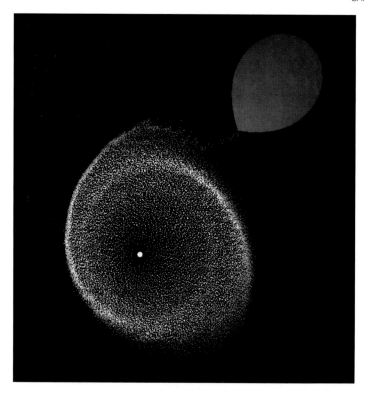

Figure 7.5 When the companion star (upper right) fills its Roche lobe (see figure 6.2), particles stream across the inner Lagrangian point L_1 and impact the accretion disk inside the primary's own Roche lobe. The latter has a size approximately twice that of the disk forming within it. But note that due to the stream–disk interaction and tidal forces within the Roche lobe, the accretion disk deviates noticeably from perfect azimuthal symmetry, and acquires a spiral structure illustrated with better contrast in figure 7.6. (Image courtesy of Matt A. Wood)

These effects have been beautifully confirmed in Cataclysmic Variables, such as V1494 Aql (figure 7.6). In this particular case, the irradiation of the companion star by the primary is modulated by features rising in the accretion disk. Both the period and amplitude of the modulation are consistent with the basic theory, as illustrated in figure 7.5. Notice also that the disk is thin (confirmed even more compellingly by the excellent spectral fits to the data in figure 7.7), and that its size complies with the volume limitations of the critical Roche lobe.

But the true test is in the spectrum, an example of which is shown in figure 7.7. The data included in this plot correspond primarily to the $\sim \nu^{1/3}$ part of the curve (figure 7.4). On the vertical scale appears the so-called brightness temperature, as a function of the distance from the primary star in units of the distance between the white dwarf and the inner Lagrangian point (which we called b_1 earlier; see equation 6.89). The brightness temperature is the temperature of a blackbody whose radiant intensity (given in equation 7.54) in the range of frequencies between ν and $\nu + d\nu$ is the same as that of the observed source. If the source is a

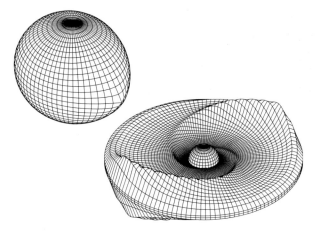

Figure 7.6 Observational evidence suggests that the two-armed spiral structure predicted by numerical simulations of accretion disks in Cataclysmic Variables (see figure 7.5) influences the irradiation of the companion star by the primary, producing a detectable signature in the lightcurve of such a system. This particular schematic diagram describes the eclipsing fast nova V1494 Aql, in which the cool companion (far left side) is a main sequence star ($\sim$0.3 $M_{\odot}$) filling its Roche lobe. Its north and south poles are irradiated by the hot ($\sim$1 $M_{\odot}$) white dwarf primary (near right side), though the irradiation is modulated by the spiral disk structure. The inferred separation of the two stars is 1.21 $R_{\odot}$, and the effective radii of the Roche lobes are $R_1 = 0.59 R_{\odot}$ and $R_2 = 0.34 R_{\odot}$ (compare with figure 6.2). Note that the white dwarf surface is artificially enlarged in order for us to see it more easily. (From Hachisu et al. 2004)

true blackbody, this brightness temperature is simply the effective temperature we defined in equation (7.50).

A comparison between the data and model fits (obtained by numerically solving equation 7.57) shows that the radial temperature profiles of the continuum maps are well described by steady-state models. The mass accretion rate ($\dot{M}$) required in the spectral fitting seems to have increased from August to November, in accordance with the observed overall increase in brightness of the source during that time. This result in turn suggests that the mass transfer rate in UX UMa can vary by a substantial amount—50 percent, or so—on a timescale of only a few months.

Sometimes a third characteristic of disk emission is observed when emission lines are quite prominent in the spectrum. The additional features are associated with the fact that the disk material is in circular motion, so that emission lines exhibit a double-humped profile, consistent with a blueward and redward Doppler shift relative to the line center. In some systems that are observed edge-on, shadowing by the companion can lead to the alternate appearance and disappearance of these two components, consistent with the additional rotation in the disk.

7.2.3 Boundary Layers

From this point on, the theory of accretion disks evolves in one of several directions, depending on which features are important for interpreting a given set of

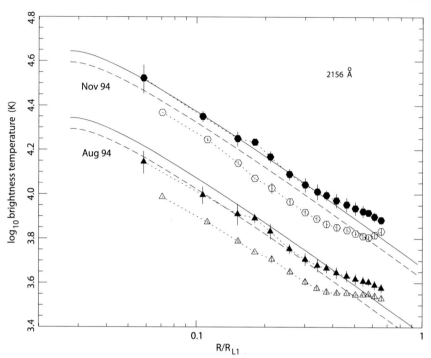

Figure 7.7 Accretion disks in Cataclysmic Variables may be "imaged" using the so-called eclipse method, which translates the eclipse shape into a map of the disk surface brightness distribution (Baptista and Steiner 1993). In the case of UX UMa shown here, the intensities were further converted into blackbody brightness temperatures (see text) for comparison with the radial profile of effective temperature predicted by steady-state disk models. The data (for two different observations) correspond to a passband centered at λ2156 Å. A distance of 345 parsecs is assumed for the source. The steady-state disk models are for mass accretion rates of $10^{-8.1}$ (solid) and $10^{-8.3}$ (dashed) $M_\odot$ yr^{-1}, assuming a white dwarf mass $M_1 = 0.47\,M_\odot$ and radius $R_1 = 0.014\,R_\odot$. The lowest curves are on the true temperature scale. The other diagrams are vertically displaced by 0.3 on the log scale for clarity. The horizontal scale is in units of the distance from the disk center to the inner Lagrangian point L_1. Filled symbols correspond to the profile for the back of the disk; open symbols correspond to the front of the disk. (From Baptista et al. 1998)

observations. For example, a disk with a large $\dot{M}$ may develop unstable two-temperature regions, as we shall see in section 7.4. This type of environment may be relevant to binaries containing stellar-size black holes, such as Cygnus X-1 (see section 10.1). Very hot disks swell and thicken, as apparently happens in some AGNs (see chapter 8). In some cases, the accreting plasma may be so magnetized and tenuous that its Keplerian geometry has only a marginal impact on its physical properties. Low-activity AGNs, such as our galactic center, may be members of this class (see section 12.2). In this section, we begin our extension of the "standard" thin-disk theory with a more modest addition, but one that nonetheless has significant observational consequences.

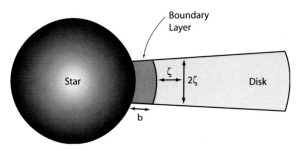

Figure 7.8 When the central star has a "hard" surface (as opposed to a black hole, whose outer edge is an event horizon), matter spiraling inward through the accretion disk has an azimuthal velocity at the disk's terminus greater than that of the stellar material with which it collides. The dissipation of energy at the star's equator produces a boundary layer, shown here as a sheath of width b, where essentially half of the accretion energy is released. In some high-energy sources, the spectral signature of this region may be as significant as that of the disk itself (see figure 7.9).

The feature we wish to study at greater depth is the role played in high-energy astrophysics by the transition region between the inner edge of the disk and the stellar surface. Since $v_R \ll v_\phi$, we expect that the angular velocity $\Omega(R)$ in the disk remains very close to the Keplerian value (equation 7.11) until the accreting matter enters a boundary layer of radial extent b, just outside the surface of the star at $R = R_*$ (see figure 7.8). Within this region, Ω must decrease from a value $\Omega(R_* + b) \approx \Omega_K(R_* + b)$ to the surface angular velocity $\Omega_* < \Omega_K(R_*)$.

Since by definition the gas at $R_* + b$ "knows" about the presence of the hard surface at R_* (otherwise Ω would not stop increasing there), $v_R < c_s$ in this region. This is always true in thin disks anyway, but it must be an even stronger inequality here where the matter is stopped in its motion once it impacts the surface. Thus, the boundary layer must be in approximate hydrostatic equilibrium in the radial direction, with

$$\frac{1}{\rho}\frac{\partial P}{\partial R} \approx -\frac{GM}{R^2}, \tag{7.64}$$

where $P \sim c_s^2 \rho$. Dimensionally, this means that

$$\frac{c_s^2}{b} \sim \frac{GM}{R_*^2}. \tag{7.65}$$

However, just outside the boundary layer, we also have hydrostatic equilibrium in the vertical direction,

$$\frac{1}{\rho}\frac{\partial P}{\partial z} \approx -\frac{GMz}{R^3} \tag{7.66}$$

or

$$\frac{P}{\rho\zeta} \approx \frac{GM\zeta}{R^3}, \tag{7.67}$$

so that

$$\zeta \approx c_s R \left(\frac{R}{GM} \right)^{1/2}. \tag{7.68}$$

Assuming that ζ and c_s just inside $R_* + b$ are similar to their values just outside, we therefore conclude that

$$b \sim \frac{R_*^2}{GM} c_s^2 \sim \frac{\zeta^2}{R_*}, \tag{7.69}$$

or

$$b \sim \left(\frac{\zeta}{R_*} \right) \zeta \ll R_*. \tag{7.70}$$

The radiation (usually X-rays) emitted by the boundary later emerges through a region of radial extent $\sim \zeta$ on the two disk faces. Normally, this region is optically thick (see equation 7.49) and therefore radiates as a blackbody with area $\sim 2(2\pi R_* \zeta)$. Its luminosity clearly depends on how much gravitational energy is expended in this region. Returning for a moment to the dissipation rate $D(R)$ (equation 7.35), we calculate that the total luminosity produced by the disk is

$$\begin{aligned}
L_d &= 2 \int_{R_*}^{\infty} D(R) 2\pi R \, dR \\
&= \frac{3GM\dot{M}}{2} \int_{R_*}^{\infty} \left[1 - \left(\frac{R_*}{R} \right)^{1/2} \right] \frac{dR}{R^2} \\
&= \frac{GM\dot{M}}{2R_*}. \tag{7.71}
\end{aligned}$$

But by the time the plasma has reached R_*, the total gravitational energy released is

$$L_{\text{acc}} = \frac{GM\dot{M}}{R_*}. \tag{7.72}$$

Thus, the power emitted within the boundary layer is actually *half* of the total available power,

$$L_b = \frac{GM\dot{M}}{2R_*}. \tag{7.73}$$

The boundary layer therefore radiates as a blackbody with temperature T_b, where

$$4\pi R_* \zeta \sigma_B T_b^4 \approx \frac{GM\dot{M}}{2R_*}, \tag{7.74}$$

so that

$$T_b \approx \left(\frac{GM\dot{M}}{8\pi R_*^2 \zeta \sigma_B} \right)^{1/4}. \tag{7.75}$$

A direct comparison between T_b and the disk temperature (equation 7.51) shows that

$$T_b \approx \left(\frac{R_*}{\zeta} \right)^{1/4} T(R_*). \tag{7.76}$$

From equation (7.68), we infer that

$$\zeta \approx R_* \left[\frac{kT(R_*)}{\mu m_H}\right]^{1/2} \left(\frac{R_*}{GM}\right)^{1/2}, \qquad (7.77)$$

and therefore

$$T_b \approx \left[\frac{T_0}{T(R_*)}\right]^{1/8} T(R_*), \qquad (7.78)$$

where

$$T_0 \equiv \frac{3}{8} \frac{GM\mu m_H}{k R_*}. \qquad (7.79)$$

In a neutron star, with mass $M \approx 1\, M_\odot$ and radius $R_* = 10\,\mathrm{km}$, $T_0 \approx 3 \times 10^{11}$ K, whereas $T(R_* + b) \approx 10^7$ K. Therefore, $[T_0/T(R_*)]^{1/8} \approx 3.6$ and $T_b \approx 3 \times 10^7$ K. At this temperature, the boundary layer can produce a significant blackbody component above the regular disk spectrum shown schematically in figure 7.4. Indeed, this combination of a stretched out thin-disk spectrum plus a single (hotter) blackbody component is seen often in low-mass X-ray binaries, as illustrated in figure 7.9. One clearly sees in this figure that the boundary layer contribution extracted from the total spectrum lies at much higher energy than the rest of the disk component, consistent with the idea that a significant fraction of the total gravitational energy is dissipated very close to the accretor.

7.3 ACCRETION COLUMNS

Though boundary layers are prevalent in high-energy astrophysics, they are not always present in accreting compact objects. Sometimes, the disk does not extend all the way down to the stellar surface, as may happen when the central object is so strongly magnetized that it disrupts the flow some distance away from the star.

When present, the magnetic field may be quite complex near the surface of the compact object, with several multipole moments contributing to its overall configuration. However, all of the moments higher than dipole die off too quickly to be of dynamical relevance away from the star. It usually suffices for us to assume that the primary's magnetic field is dominated by a dipole moment $\mathbf{m}$, whose field may be written in polar coordinates as

$$\mathbf{B(r)} = \frac{3\hat{\mathbf{n}}(\hat{\mathbf{n}} \cdot \mathbf{m}) - \mathbf{m}}{r^3}, \qquad (7.80)$$

where $\hat{\mathbf{n}}$ is a unit vector pointing from the center of the dipole to the observation point at $\mathbf{r}$.

The inner edge of the disk is located at a radius R_m, where the magnetic field pressure P_{mag} begins to dominate over the matter pressure P_{gas}. Now

$$P_{\mathrm{mag}} = \frac{B^2}{8\pi}, \qquad (7.81)$$

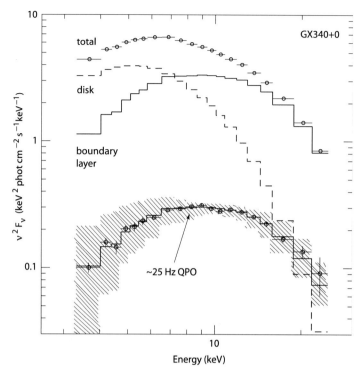

Figure 7.9 In some X-ray binaries, the spectral signature of the boundary layer can be as significant as those of the star itself and the accretion disk. In this figure, we see the decomposition of GX340+0's X-ray spectrum, showing the boundary layer component as distinct from the disk emission. The shaded area shows a plausible range of the boundary layer spectral shape calculated subtracting the predicted disk spectrum from the total spectrum and renormalizing the residual to the total energy flux of the frequency resolved spectrum. The dashed histogram shows the accretion-disk spectrum. The upper solid histogram shows the difference between the total and accretion-disk spectrum, i.e., roughly the boundary layer spectrum. The lower solid histogram is the same but scaled to the total energy flux of the frequency resolved spectrum. (From Gilfanov et al. 2003)

where for a dipole field (using R to denote radii in the equatorial plane)

$$B(R) = B_* \left(\frac{R_*}{R}\right)^3, \tag{7.82}$$

in terms of the field intensity B_* at the polar cap. The gas is supersonic since $v_\phi \gg c_s$, so it is the ram pressure that dominates over thermal pressure,[3]

$$P_{\text{gas}} \approx \rho \mathbf{v}_\phi^2 = \frac{GM\rho}{R}. \tag{7.83}$$

---

[3]By ram pressure, we mean the force per unit area applied to an element of gas by the transfer of momentum arising from the motion of another element. The momentum density in the plasma is ρv_ϕ, so $(\rho v_\phi)v_\phi$ is its momentum flux, which has the same units as pressure.

Thus, with Σ given in equation (7.46) and $\rho \sim \Sigma/2\zeta$, we have

$$\rho \sim \frac{\dot{M}}{12\pi} \frac{1}{\zeta\alpha} \left(\frac{\mu}{R_g T}\right) \left(\frac{GM}{R^3}\right)^{1/2}. \tag{7.84}$$

So in equation (7.83),

$$P_{\text{gas}} \sim \frac{\dot{M}}{12\pi} \frac{1}{\zeta\alpha} \left(\frac{\mu}{R_g T}\right) \frac{(GM)^{3/2}}{R^{5/2}}. \tag{7.85}$$

But the scale height ζ is also known in terms of M, R, and c_s (see equation 7.68), so we can write

$$P_{\text{gas}} \sim \frac{\dot{M}}{12\pi} \frac{1}{\alpha} \left(\frac{\mu}{R_g T}\right) \left(\frac{GM}{R}\right)^2 \frac{1}{R^2}. \tag{7.86}$$

The magnetic radius R_m occurs when $P_{\text{mag}} = P_{\text{gas}}$, which gives (with equations 7.81, 7.82, and 7.86)

$$R_m \approx B_* R_*^3 \left(\frac{R_g T}{\mu}\right)^{3/4} \frac{1}{GM} \left(\frac{3}{4}\alpha\right)^{1/2} \dot{M}^{-1/2}. \tag{7.87}$$

Numerically, this is

$$R_m \approx (30\,\text{km})\, \alpha^{1/2} \left(\frac{B_*}{10^{12}\,\text{G}}\right) \left(\frac{R_*}{10^6\,\text{cm}}\right)^3 \left(\frac{T}{10^6\,\text{K}}\right)^{3/4}$$

$$\times \left(\frac{M}{M_\odot}\right)^{-1} \left(\frac{\dot{M}}{10^{16}\,\text{g s}^{-1}}\right)^{-1/2}. \tag{7.88}$$

For a neutron star, $R_m \sim 3R_*$ when the parameters take on their "natural" values used in the above scaling relation.

7.3.1 Accretion Columns in Magnetic Cataclysmic Variables

At R_m, the ionized plasma in the disk flows onto the dominant magnetic field lines and is funneled toward the stellar polar caps. This cropping of the disk before it reaches R_* is seen in both white-dwarf (Cataclysmic Variables; see figure 7.10) and neutron-star (pulsar) systems, though, of course, the physical conditions in the latter (with $B_* \sim 10^{12}$ G and a free-fall velocity $\sim c/2$ at R_*) are much more extreme and therefore more difficult to model. We'll return to these objects in the next section.

Magnetic Cataclysmic Variables (mCVs), which we will study as a class in section 9.3, are better understood than their neutron-star counterparts, not only because the physical modeling is more straightforward, but also because there is correspondingly more observational evidence available for comparison with theory. There are many issues to consider, but there is one major characteristic that is common to all of these systems—the structure of the termination region at the white dwarf's surface.

A dipole field line is described (in polar coordinates) by the relation

$$r = \text{constant} \times \sin^2 \theta. \tag{7.89}$$

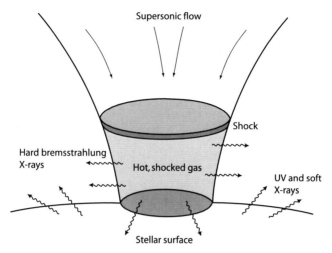

Figure 7.10 When the primary's magnetic field is strong enough to disrupt the inner portion of the disk, plasma flows along the dipole field lines onto the polar caps. Approaching the surface at supersonic speeds, the infalling matter shocks and converts the upwind kinetic energy into a downwind heat. The radiation field produced by the accretion column is a composite of UV cyclotron emission, hard, optically thin bremsstrahlung X-rays, and a soft, thermal X-ray component from the reprocessing of hard X-rays that have penetrated below the stellar surface.

Thus, since R_m corresponds to the radius at $\theta = \pi/2$, we estimate the outer angle of impact on the white dwarf's surface to be

$$\delta = \sin^{-1} \left(\frac{R_*}{R_m} \right)^{1/2}. \tag{7.90}$$

Other magnetic field lines, threading the disk beyond R_m, will feed the termination funnel at smaller angles. We therefore expect the shocked region shown in figure 7.10 to be full. In the case of a white dwarf, with $B_* \sim 2 \times 10^7$ G, $T \sim 10^5$ K, and $R_* \sim 10^9$ cm, we have $R_m \sim 10^{10}$ cm (though this also depends on the actual value of $\dot{M}$). Therefore, $\delta \sim 20°$. With reference to figure 7.10, this means that the polar cap area beneath the termination funnel is $\approx \frac{1}{40}$ of the total stellar surface area or, since, in principle, we have two active polar caps, roughly a fraction $f = \frac{1}{20}$ of the entire white dwarf's surface participates in this accretion process. This fraction can be even smaller for neutron stars, since their magnetic fields are stronger, which means the field lines transferring plasma to the polar caps are pinned on the surface closer to the magnetic dipole axis, i.e., at even smaller values of the angle δ.

As we shall see shortly, the gas within the termination funnel has a density $\rho_{shock} \sim 10^{-10}$ g cm^{-3} and a temperature $T_{shock} \lesssim 10^9$ K. Thus, the gas pressure there is $P_{gas} = R_g \, \rho_{shock} T_{shock} / \mu \sim 10^8$ dyne cm^{-2}. By comparison, the magnetic field pressure is $P_{mag} \sim 10^{13}$ dyne cm^{-2} for a field of 2×10^7 G. So the magnetic field easily dominates over the incoming plasma, which is forced to flow strictly along the dipole structure.

Inside the funnel, the falling matter reaches a velocity $v_{ff} = (2GM/R_*)^{1/2} \sim 5 \times 10^8$ cm s^{-1} (for standard white dwarf parameters). By comparison, the thermal velocity in the preshocked material leaving the disk is $v_{th} = (kT/m_e)^{1/2} \sim 10^8$ cm s^{-1} for electrons, and even lower for ions since their mass is greater than m_e. Thus, somewhere near the stellar surface, this supersonic inflow must produce a shock, where most of the directed kinetic energy is converted into (a randomized) internal energy of the plasma within the termination funnel.

Across a strong shock, the velocity drops by a factor four (from $\sim v_{ff}$ to $v_{ff}/4$) and, given mass flux conservation, the density must increase by the same factor (see the discussion following equation 4.61). That is, the density in the shocked region is

$$\rho_{shock} \approx 4 \frac{\dot{M}}{4\pi f R_*^2} \left(\frac{2GM}{R_*} \right)^{-1/2}, \tag{7.91}$$

which is $\rho_{shock} \sim 10^{-10}$ g cm^{-3} for typical parameters.

Correspondingly, the temperature in the postshock region is

$$T_{shock} \sim \frac{m_H}{3k} \left(\frac{3v_{ff}}{4} \right)^2 \tag{7.92}$$

(since the protons carry most of the kinetic energy). This gives $T_{shock} \sim 6 \times 10^8$ K. However, the hot plasma cools quickly because the radiated photons leave the system with little or no trapping, given that the optical depth transverse to the column is

$$\tau_{\perp} \sim \left(\frac{\rho_{shock}}{\mu m_H} \right) \sigma_T f^{1/2} R_* \approx 0.03. \tag{7.93}$$

The radiative cooling is therefore due to cyclotron and (primarily) optically thin bremsstrahlung emission. Some of the hard X-radiation produced in this fashion penetrates into the stellar surface and is reprocessed into a soft blackbody component. As figure 7.10 indicates, the overall spectrum from a white-dwarf accretion column therefore includes a cyclotron component, hard, optically thin bremsstrahlung emission, and a soft blackbody contribution.

7.3.2 Accretion Columns in X-ray Pulsars

The main complication in moving from accretion columns in mCVs to pulsars (see section 9.2) is that the radiation pressure due to the accretion luminosity in the latter becomes dynamically important. To see this, recall that in magnetically dominated systems, all of the accreted plasma is funneled onto small polar cap regions. Within these columns, the effective enhancement in accretion power may be expressed as

$$L_{acc}^{eff} \sim \frac{1}{f} L_{acc}, \tag{7.94}$$

which is not to say that the actual total power is larger by a factor $1/f$, but rather that the flux is higher by that factor than it would be in an isotropically emitting source. In other words, even with an accretion luminosity $L_{acc} \sim 0.1 L_{edd}$ (see equation 1.19), the effective luminosity out of the funnel can be comparable to the Eddington limit.

A second important complication is that the transverse optical depth $\tau_\perp$ is no longer less than one, an effect due to the enhanced confinement of the accreting plasma in the much smaller neutron-star environment. The density scales inversely with area in the funnel (so $n_e \propto R_*^{-2}$), whereas $\tau_\perp \propto n_e R_*$. Thus, $\tau_\perp \propto R_*^{-1}$, and therefore $\tau_\perp(ns) \sim (R_{\rm wd}/R_{\rm ns})\tau_\perp(wd) \approx 30 \gg 1$.

So now consider this, we have an Eddington flux being produced within the neutron-star funnel, yet the medium is optically thick. Numerical simulations confirm the view that the accretion column in these circumstances must be unstable.[4] Instead of forming an ordered shock, the infalling plasma breaks up into clumps and relative voids. It becomes efficient for local radiation to pass into regions of lower opacity. In effect, the radiation forms *photon bubbles* trapped by the high-opacity gas surrounding them. The simulations show that the presence of the low-density regions helps to increase the photon diffusion speed by a factor of several, while the buoyancy of the bubbles serves to transport energy via advection.

7.4 TWO-TEMPERATURE THIN DISKS

Thus far in our discussion of thin disks, we have assumed that the structure, once defined, remains in stable equilibrium. However, this is not always the case, and, in fact, there is some evidence that the inner portions of disks around stellar-size black holes may be extended. Having already learned from our study of accretion columns in pulsars that high local accretion rates and large optical depths can lead to instabilities, we would not be surprised now to learn that primaries more massive than white dwarfs and neutron stars can create physical conditions conducive to instabilities in the disks themselves.

We can make significant progress in trying to understand the behavior of accretion disks in black-hole binaries by comparing the various timescales of interest:

$$t_\phi \equiv \frac{R}{v_\phi} = \Omega_K^{-1} \quad \text{(dynamical)}$$

$$t_{\rm visc} \equiv \frac{R}{v_R} \quad \text{(viscous)}$$

$$t_z \equiv \frac{\zeta}{c_s} \quad \text{(hydrodynamic)}$$

$$t_{\rm th} \equiv \frac{\text{heat content/area}}{\text{dissipation rate/area}} \quad \text{(thermal).} \tag{7.95}$$

In this endeavor, we can use the full expression for v_R in equation (7.21) to write

$$v_R \sim \frac{\nu}{R}, \tag{7.96}$$

---

[4] Some of the most detailed simulations of these systems have been carried out by Klein, Arons, and their colleagues. See, e.g., Hsu et al. (1997).

so that

$$t_{\text{visc}} \sim \frac{R^2}{\nu}. \tag{7.97}$$

And since $\nu \equiv \alpha H c_s$,

$$t_{\text{visc}} \sim \frac{1}{\alpha} \frac{R}{H} \frac{R}{c_s}. \tag{7.98}$$

But from the hydrostatic equilibrium equation (7.39), we know that the scale height $\zeta = H/2$ may be expressed as

$$\zeta \sim c_s \left(\frac{R}{GM} \right)^{1/2} R \tag{7.99}$$

(see also equation 7.43), so that

$$\frac{H}{R} \sim \frac{2c_s}{v_\phi}. \tag{7.100}$$

Therefore,

$$t_{\text{visc}} \sim \frac{1}{\alpha} \frac{v_\phi}{2c_s} \frac{R}{c_s}, \tag{7.101}$$

or

$$t_{\text{visc}} = \frac{1}{2\alpha} \left(\frac{v_\phi}{c_s} \right)^2 t_\phi. \tag{7.102}$$

To derive an analogous expression for the thermal timescale, we need to use equation (7.30) for the local dissipation rate, which gives

$$t_{\text{th}} \sim \int_{-\infty}^{\infty} \rho c_s^2 \, dz \div D(R). \tag{7.103}$$

With the additional assumption that $\Omega = \Omega_K$,

$$t_{\text{th}} \sim \frac{\Sigma c_s^2}{D(R)} \sim \frac{R^3 c_s^2}{GM \nu}, \tag{7.104}$$

and therefore

$$t_{\text{th}} \sim \left(\frac{c_s}{v_\phi} \right)^2 \frac{R^2}{\nu}. \tag{7.105}$$

A comparison between equations (7.97) and (7.105) shows that

$$t_{\text{th}} \sim \left(\frac{c_s}{v_\phi} \right)^2 t_{\text{visc}}. \tag{7.106}$$

Finally, for the remaining timescale, we have

$$t_z \sim \frac{\zeta}{c_s} = \frac{\zeta}{R} \frac{R}{c_s} = \frac{c_s}{v_\phi} \frac{R}{c_s} = t_\phi. \tag{7.107}$$

We now have enough information to develop a timescale hierarchy. Noting that $v_\phi \gg c_s$ and $\alpha \lesssim 1$, we find that a direct comparison of all the equalities and inequalities we have derived shows that

$$t_\phi \sim t_z \lesssim t_{th} \ll t_{visc}. \tag{7.108}$$

Numerically,

$$t_\phi \sim t_z \sim (100\,\text{s}) \left(\frac{M}{M_\odot}\right)^{-1/2} \left(\frac{R}{10^{10}\,\text{cm}}\right)^{3/2}, \tag{7.109}$$

$$t_{th} \sim \frac{1}{2\alpha} t_\phi \sim (500\,\text{s}) \left(\frac{\alpha}{0.1}\right)^{-1} \left(\frac{M}{M_\odot}\right)^{-1/2} \left(\frac{R}{10^{10}\,\text{cm}}\right)^{3/2}, \tag{7.110}$$

and using equation (7.23),

$$t_{visc} \sim \frac{R}{c_s} \frac{c_s}{v_R} \sim \frac{R^2}{\alpha c_s H} \sim 10^5\,\text{s}, \tag{7.111}$$

for typical parameters.

Mass transfer instabilities thus propagate through the disk on a timescale of days to weeks, whereas dynamical and thermal instabilities manifest themselves on a timescale of only minutes. The question now is, what happens when the cooling rate in the disk cannot keep up with the heating rate?

An example of where this situation might arise is the inner region of a strongly accreting disk, where the dissipation rate $D(R)$ is largest and the cooling is dominated by two-body processes, such as bremsstrahlung. The frequency-integrated emissivity, j, is then given by

$$4\pi j \propto n_e n_i \Lambda(T), \tag{7.112}$$

where n_e and n_i are, respectively, the electron and ion number densities, and Λ is the cooling function, in units of energy times volume, per unit time (see figure 7.11). If the heating and cooling rates are not balanced, then adjustments must occur on a thermal timescale t_{th}. And since $t_{visc} \gg t_{th}$, the column density Σ cannot keep up with the immediate changes in the internal energy content, and remains roughly constant. However, for $\alpha < 1$, we have $t_\phi \sim t_z < t_{th}$, so the vertical structure does adjust more quickly than the internal energy profile does, and we may assume that (quasi) hydrostatic equilibrium is maintained throughout this process.

For simplicity, let us take $n_e = n_i$. The cooling rate may then be written as

$$Q^- = 4\pi j \propto n_e^2 \Lambda(T) \sim \left(\frac{\Sigma}{H}\right)^2 \Lambda(T). \tag{7.113}$$

Given that $\Sigma \approx$ constant, and that $H \sim c_s$ (since R and v_ϕ are also constant; see equation 7.100), we see that

$$Q^- \sim c_s^{-2} \Lambda(T) \sim T^{-1} \Lambda(T). \tag{7.114}$$

On the other hand, the heating rate (with the same units as Q^-) may be written

$$Q^+ \sim \frac{D(R)}{H}. \tag{7.115}$$

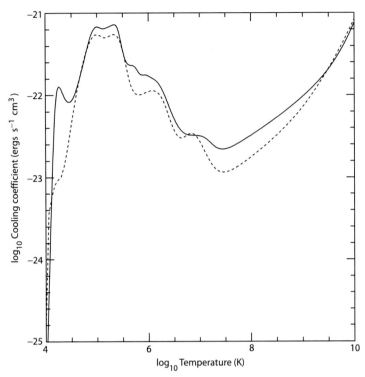

Figure 7.11 Temperature dependence of the cooling coefficient, defined to be the cooling rate divided by the baryon number density squared, for an optically thin plasma with cosmic abundances. The thick solid curve is the coefficient updated from the first usage of this function (dashed curve) in Gehrels and Williams (1993). The thick solid curve is based on detailed modeling with CLOUDY (Ferland 1996). Line emission, recombination, and two-photon continuum emission dominate the emissivity below about 10^6 K. Thermal bremsstrahlung and inverse Compton scattering take over at higher temperatures, leading to the inversion of the cooling coefficient's dependence on temperature. Note that the plasma is located in an unstable portion of this curve when its temperature lies between 10^5 and 10^7 K. Thus, depending on the gas density, which determines the cooling rate, the plasma may either heat up to about 10^{10} K, or cool down to around 10^4 K. (From Liu, Fromerth, and Melia 2002)

Notice from the functional dependence of $D(R)$ in equation (7.30) that the only quantity possibly changing on a thermal timescale is ν. Therefore,

$$Q^+ \sim \frac{\alpha c_s H}{H} = \alpha c_s \sim \alpha T^{1/2}. \tag{7.116}$$

A thermal instability will grow if the heating rate exceeds the cooling rate, i.e., if

$$T^{-1}\Lambda(T) < \text{constant} \times \alpha T^{1/2}, \tag{7.117}$$

or

$$\frac{d \ln(\Lambda/\alpha)}{d \ln T} < \frac{3}{2}. \tag{7.118}$$

A close inspection of the cooling function in figure 7.11 shows that this condition is not met—and therefore the disk is stable—for temperatures below $\sim 10^5$ K and greater than $\sim 10^7$ K. However, plasma between these two limits has a negative value of $d \ln(\Lambda/\alpha)/d \ln T$, and is therefore subject to the thermal instability we have been discussing above.

One can understand this behavior by realizing that, whereas line emission dominates at low temperature and inverse Comptonization is most active at high temperature, bremsstrahlung is the most prominent emission mechanism between these limits. The first two processes become more efficient with increasing temperature. However, when the plasma is fully ionized (around 10^5 K), line emission is suppressed relative to bremsstrahlung (a two-body mechanism). But as the temperature rises, the disk expands in the vertical direction (though always maintaining quasi hydrostatic equilibrium) and the density drops. This causes the inversion in Λ, which now decreases with increasing T. The process runs away in this temperature regime because Q^+ pumps in more energy than Q^- can take out. The instability is quenched only when inverse Comptonization re-emerges as the dominant emission process above $T \sim 10^7$ K.

The reason this instability may be an important issue in some binaries is most easily seen in equation (7.51), which shows that the temperature in the disk depends (though weakly) on the primary's mass and the rate at which it accretes from its companion. Systems such as Cygnus X-1, which will be featured in section 10.1, are likely candidates where the disk may become unstable and puff up.[5] The consequences of this eventuality are many and varied, including the possibility that that the plasma may separate into two species, each characterized by its own temperature.

The reason a single-temperature profile is difficult to maintain when the disk becomes unstable is that most of the energy is carried by the protons, whereas most of the cooling is mediated by the electrons. For the disk to lose energy, a net transfer of heat must occur from the protons to the electrons, which, based on simple thermodynamic principles, means that $T_i > T_e$, where T_i and T_e are, respectively, the ion and electron temperatures.

The equations that describe a two-temperature region (presumably close to the black hole where $D(R)$ is largest), now include the condition for hydrostatic equilibrium (equation 7.39), the equation of state

$$P = kn_e(T_i + T_e), \tag{7.119}$$

and an expression formalizing the idea that the net dissipation rate $D(R)$ must be balanced by the transfer of energy from the protons to the electrons,

$$D(R) = \tfrac{3}{2} \nu_E \, n_e k(T_i - T_e), \tag{7.120}$$

where ν_E is the collision frequency and the right-hand side of this equation represents the electron–ion Coulombic energy exchange rate.

The electrons cool via bremsstrahlung, bremsstrahlung-self-Compton (analogous to synchrotron-self-Compton, though seed photons are emitted through brems-

---

[5]This possibility was first explored by Shapiro, Lightman, and Eardley (1976).

strahlung), and inverse Comptonization of ambient low-energy photons, e.g., from the outer disk. Recent modeling of sources such as Cygnus X-1 has had to contend with many observational constraints at energies from ~1 keV all the way up to 10 MeV (see, e.g., figures 10.1 and 10.2). It now looks like a simple thermal picture is untenable; the observational evidence clearly points to the presence of nonthermal particles in these systems, along with a thermal population (or populations, if the species split into two temperatures). Thus, although the simple extension to the thin-disk theory we have outlined above provides an adequate beginning, realistic simulations are far more complicated and include nonlocal transfer effects, such as the disk's reflection of spectral components originating above it. We shall return to these very interesting objects in chapter 10.

SUGGESTED READING

Almost every compact object imaginable is surrounded by an accretion disk. But accretion disks come in various sizes (and shapes), fashioned by the depth of the gravitational potential and the source of infalling matter, e.g., the interstellar medium versus Roche lobe overflow from a stellar companion. An excellent review on this broad subject is the two-part series Papaloizou and Lin (1995, 1996).

An expanded discussion of the role of accretion disks in astrophysics, far beyond that attempted here, may be found in the well-written monograph Frank, King, and Raine (2002).

Chapter Eight

Thick Accretion Disks

By now, we've covered almost every conceivable accretion scenario but one, and this last one is likely to apply to AGNs and quasars. In chapter 6, we considered situations with low accretion rates and very low specific angular momentum, and found that the medium surrounding a compact object may be adequately described by quasi-spherical hydrodynamics. The outgoing radiation pressure in that case was negligible. In chapter 7, we studied scenarios in which the infalling angular momentum could not be ignored, and learned a great deal about geometrically thin disks. But these disks were thin (for the most part) because the gravitational energy released during the accretion process was not sufficient to puff them up and produce a more extended structure, except possibly in the case of black-hole binaries, in which M and/or $\dot{M}$ could be high enough to heat the disk material and provide sufficient pressure in the vertical direction to partially offset the effects of gravity.

In AGNs, the accretion disk may be geometrically thick for at least two reasons. It may possess a significant amount of angular momentum, but radiate inefficiently, in which case the orbiting plasma retains much of the dissipated gravitational energy. Aside from advecting most of the gravitational power through its event horizon, the black hole in these systems is surrounded by a vertically extended structure, supported by the disk's internal pressure. We will return to these objects later in this chapter. In the opposite limit, an AGN disk may be accreting at such a greatly super-Eddington rate that the transfer of radiation through the optically thick medium cannot keep up with the heating rate. Again, the disk is prevented from collapsing to the equatorial plane.

But none of the disk properties we've examined thus far can accommodate either of these two scenarios. We learned at the very beginning of this book that accretion is heavily suppressed when the outward radiation pressure matches the inward force per unit area supplied by gravity. Note, however, that this is true only in the case of spherical symmetry (see equation 1.17). Still, we saw in the previous chapter that even with thin disks, half of the accretion luminosity is dissipated at the stellar surface (for neutron stars and white dwarfs), and this radiation comes out more or less isotropically, at least as viewed from large distances. Again, it would appear that no possibility exists for observing accretion rates well beyond the Eddington limit. But we have not yet exhausted all possible disk configurations, and understanding how these limitations may be circumvented is a primary focus of the present chapter.

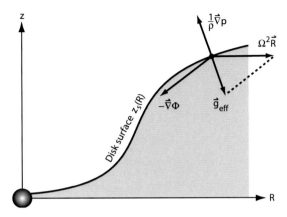

Figure 8.1 When the accreting plasma has sufficient heat to partially offset the vertical component of gravity and expand the disk in the vertical direction, the corresponding pressure gradients in the radial direction reduce the angular velocity Ω (required to maintain a circular orbit) below its Keplerian value. In that case, one may infer the vertical structure of the disk by finding the isobaric surfaces $z_s(R)$, along which $\vec{\nabla} P|_s = 0$ (see text). In this figure, $\Omega^2 \mathbf{R}$ is the centrifugal force, $-\vec{\nabla}\Phi$ is the gravitational force, and $\mathbf{g}_{\text{eff}}$ (the sum of these two) is the effective gravity.

8.1 THICK-DISK STRUCTURE

As we shall see, thick disks require non-Keplerian rotation laws, and hence radial pressure gradients and unfamiliar viscosities. So the theory of thick-disk accretion is not as well understood—perhaps one should say, not as widely agreed upon—as that of thin disks.[1]

Let us begin by considering the equations describing a rotating thick disk, which are quite different from those of chapter 7, in that we will not be as concerned here with the microphysics of these structures as we were in the case of thin disks. A sketch of the thick-disk geometry is shown in figure 8.1.

In these systems, the vertical structure becomes important, and in general

$$\Omega = \Omega(R, z). \tag{8.1}$$

For simplicity (and since $v_R \ll v_\phi$ anyway), let's take

$$v_R \approx 0$$
$$v_\phi \approx R\Omega$$
$$v_z \approx 0. \tag{8.2}$$

---

[1]The theory of thick disks was first introduced in the early 1980s (see, e.g., Jaroszyński, Abramowicz, and Paczyński 1980, Abramowicz, Calvani, and Nobili 1980, and Paczyński 1998). It has been argued, however, that such structures are unstable, though stability may be regained under some circumstances (Blaes 1987). More recent numerical simulations appear to have rediscovered many of the results first seen in the early analytical treatments of these disks (e.g., Igumenshchev, Chen, and Abramowicz 1996), which therefore still have a pedagogical value in demonstrating the astrophysically relevant features of these sources.

We still assume hydrostatic equilibrium, but whereas before we could simply balance the vertical pressure gradient with the corresponding component of gravity, here we must also include the centrifugal force in the radial component of the forces, which may be significant compared with the previous two. Formally,

$$\frac{1}{\rho}\frac{\partial P}{\partial R} = -\frac{\partial \Phi}{\partial R} + \Omega^2 R$$

$$\frac{1}{\rho}\frac{\partial P}{\partial z} = -\frac{\partial \Phi}{\partial z}, \tag{8.3}$$

where Φ is the gravitational potential and P is the pressure. These are usually written in vector form as

$$\frac{1}{\rho}\vec{\nabla} P = -\vec{\nabla}\Phi + \Omega^2 \mathbf{R} \equiv \mathbf{g}_{\text{eff}}, \tag{8.4}$$

where $\mathbf{g}_{\text{eff}}$ is defined to be the *effective* gravity, representing the sum of gravitational and centrifugal accelerations.

We can use this equation to determine the *isobaric* surfaces, along which $\vec{\nabla} P|_s = 0$, for any given $\mathbf{g}_{\text{eff}}$, i.e., for any given rotation law $\Omega(R)$. Along the surface $z_s(R)$, we have

$$0 = -\vec{\nabla}\Phi|_s + \Omega^2 R|_s, \tag{8.5}$$

or

$$(\Omega^2 R)_s = \left(\frac{\partial \Phi}{\partial z}\right)_s \frac{\partial z_s}{\partial R_s} + \left(\frac{\partial \Phi}{\partial R}\right)_s. \tag{8.6}$$

Note that, in general, surfaces of constant Ω are not cylinders, so isobaric surfaces do not necessarily coincide with isopicnic (i.e., constant density) surfaces. This is not a problem, however, since it is, in fact, the isobaric surfaces that determine the flux (see below).

To see why a thick disk such as this can admit a luminosity in excess of the Eddington value, let us consider a highly simplified geometry in which the disk's boundary is a surface of revolution generated by two straight lines making an angle $\pm\alpha$ relative to the equatorial plane. (In figure 8.1, this would correspond to approximating the inner funnel region with a straight line directed outward from the origin.) The equation describing this particular surface is

$$z_s(R) = \pm(\tan \alpha) R. \tag{8.7}$$

Let us also assume that the point mass M at the origin dominates the gravitational potential, so that

$$\Phi = -\frac{GM}{(R^2 + z^2)^{1/2}}. \tag{8.8}$$

Then,

$$\frac{\partial \Phi}{\partial z} = \frac{GMz}{(R^2 + z^2)^{3/2}}, \tag{8.9}$$

so

$$\frac{\partial \Phi}{\partial z}\bigg|_s = \frac{GM (\tan \alpha) \cos^3 \alpha}{R^2}. \tag{8.10}$$

Thus,

$$\frac{\partial \Phi}{\partial z}\bigg|_s \frac{\partial z_s}{\partial R_s} = \frac{GM (\tan^2 \alpha) \cos^3 \alpha}{R^2} \tag{8.11}$$

and

$$\frac{\partial \Phi}{\partial R}\bigg|_s = \frac{GM \cos^3 \alpha}{R^2}. \tag{8.12}$$

Putting all of this together, we arrive at a "solution" for $\Omega^2 R$ in equation (8.6),

$$(\Omega^2 R)_s = \frac{GM \cos \alpha}{R^2}, \tag{8.13}$$

so in this case Ω is only a function of R and, indeed,

$$\Omega^2(R) = \frac{GM \cos \alpha}{R^3}. \tag{8.14}$$

Remarkably, this type of disk is still Keplerian, but notice that the effective mass has been reduced to $M \cos \alpha < M$. The velocities required to maintain equilibrium are sub-Keplerian because the pressure distribution necessary to support such a disk vertically also has a radial gradient and is able to partially counteract the inward pull of gravity. The angular frequency Ω approaches Ω_K asymptotically as $\alpha \to 0$, which recovers the thin-disk configuration we developed previously.

Let us now see how this new disk shape affects the observable luminosity. As one may already intuit, because these disks are thick, the emitting surface area is significantly larger than that of a thin disk. Since the internal energy is so high, the pressure tends to be dominated by radiation, so $P \approx aT^4/3$. Thus, from equation (6.42), we gather that the radiative flux may be expressed in the alternative form

$$\mathbf{F}_{\text{rad}} = -\frac{c}{\kappa \rho} \vec{\nabla} P, \tag{8.15}$$

where (again) κ is the opacity, in units of area per unit mass of material. Thus, using equation (8.4) for the effective gravity, we have

$$\mathbf{F}_{\text{rad}} = -\frac{c}{\kappa} \mathbf{g}_{\text{eff}} = -\frac{c}{\kappa} \vec{\nabla} \Phi - \frac{c}{\kappa} \Omega^2(R, z) \mathbf{R}. \tag{8.16}$$

The total luminosity is just the integral of this expression over the whole surface of the disk, so

$$L = \frac{c}{\kappa} \int_S \vec{\nabla}\Phi \cdot d\mathbf{s} - \frac{c}{\kappa} \int_S \Omega^2 \mathbf{R} \cdot d\mathbf{s}, \tag{8.17}$$

where for simplicity we have also assumed that κ is constant.

Using Gauss's theorem, we get

$$L = \frac{c}{\kappa} \int_V \vec{\nabla}^2\Phi \, dV - \frac{c}{\kappa} \int_V \vec{\nabla} \cdot (\Omega^2 \mathbf{R}) \, dV, \tag{8.18}$$

where V is the volume enclosed by the surface S. But Poisson's equation for the gravitational field says that

$$\vec{\nabla}^2\Phi = 4\pi G\rho. \tag{8.19}$$

Therefore,

$$L = \frac{c}{\kappa} \int_V 4\pi G\rho \, dV - \frac{c}{\kappa} \int_V \vec{\nabla} \cdot (\Omega^2 \mathbf{R}) \, dV, \tag{8.20}$$

or more simply,

$$L = \frac{4\pi cGM}{\kappa} - \frac{c}{\kappa} \int_V \vec{\nabla} \cdot (\Omega^2 \mathbf{R}) \, dV. \tag{8.21}$$

The first term is the usual Eddington luminosity for mass M. It is the second term that can make L exceed L_{edd}.

To see this, let us write

$$\vec{\nabla} \cdot (\Omega^2 \mathbf{R}) = 2R\Omega \frac{\partial \Omega}{\partial R} + 2\Omega^2. \tag{8.22}$$

That is,

$$\vec{\nabla} \cdot (\Omega^2 \mathbf{R}) = \frac{1}{2}\left[\frac{1}{R}\frac{\partial}{\partial R}(R^2\Omega)\right]^2 - \frac{1}{2}\left(R\frac{\partial \Omega}{\partial R}\right)^2. \tag{8.23}$$

But

$$\frac{\partial \Omega}{\partial R} = -\frac{3}{2}\frac{(GM\cos\alpha)^{1/2}}{R^{5/2}}, \tag{8.24}$$

so

$$\vec{\nabla} \cdot (\Omega^2 \mathbf{R}) = -\Omega^2. \tag{8.25}$$

Thus, in equation (8.21),

$$L \approx L_{\text{edd}} + \frac{cGM\cos\alpha}{\kappa} \int_V \frac{1}{R^3} \, dV > L_{\text{edd}}. \tag{8.26}$$

Further, since $dV = 4\pi R^2 \tan\alpha \, dR$, we get, finally,

$$L = L_{\text{edd}}\left[1 + \sin\alpha \ln\left(\frac{R_2}{R_1}\right)\right], \tag{8.27}$$

where R_1 and R_2 are, respectively, the inner and outer radii of the *funnel* region in the thick disk (see figure 8.1). For example, with $\alpha = 45°$ and $R_2 = 100R_1$, an AGN can emit at over four times the Eddington limit. And though it may not have been obvious, we actually incorporated into the analysis the radiative transfer resulting from this prodigious $\dot{M}$ through the use of equation (8.16). The vertical extension in these structures is a direct result of the gravitational dissipation associated with such large accretion rates.

8.2 RADIATIVELY INEFFICIENT FLOWS

The nature of disk accretion onto compact objects changes considerably once $\dot{M}$ drops well below the Eddington value. Instead of an optically thick medium, in which radiative transfer is an essential component in supporting the disk internally, these systems tend to be filled with an optically thin plasma radiating inefficiently. If we stop and think about this for a moment, we also realize that the principles of viscous hydrodynamics we have espoused so far probably don't work in this context either. Indeed, attempts at understanding the nature of high angular momentum disks accreting at very low rates fail without the introduction of a magnetic field. This new agent is invoked to provide an *anomalous* viscosity via the magneto-rotational instability (see below).

The presence of a magnetic field, however, also means that the plasma can radiate via synchrotron emission (see section 5.5), and if one is attempting to model a faint source with such a disk, it matters how one chooses the proper balance between viscosity and emissivity. Of all the disk scenarios we have considered thus far, this last one is perhaps the most challenging to handle theoretically, not only because of the various physical effects that must be included, but also because some of the physics is actually not well understood yet. As we shall see below, the magnetic field invariably intensifies to superequipartition values as the plasma spirals in toward smaller radii, and one cannot avoid having to deal with the effects of magnetic reconnection. Unfortunately, this is one of those processes that still requires prolonged study.

The importance of magnetic field dissipation within the accreting gas was recognized from the earliest thinking on this subject.[2] Simply stated, we have a situation in which the highly ionized plasma "freezes" the magnetic field and intensifies it due to flux conservation in the flow converging toward the black hole.

Consider what happens in the case of an AGN accreting from its environment. The plasma is highly ionized even at the capture radius r_{acc} (equation 6.73), with a temperature typically well above 10^4 K, and the implied high conductivity therefore prevents any diffusion of magnetic field lines across the gas. As the accreting matter converges toward the black hole, the area element of any given flux tube decreases as $\sim r^2$, so flux conservation implies that the magnetic field intensity must go as

$$B(r) \propto r^{-2}, \tag{8.28}$$

---

[2]The very first paper on this topic (Shvartsman 1971) already foresaw the consequence of this process on the energetics, though its influence on the dynamics has become more apparent in recent years.

and hence the energy density goes as r^{-4}. By comparison, the specific kinetic energy density $v_r^2 n_e/2$ is proportional to $r^{-5/2}$.

Evidently, the magnetic field intensity becomes divergent very quickly as the radius shrinks. But this cannot go on indefinitely since eventually the magnetic pressure would overwhelm that of the gas and any further constriction would cease. The crossover point occurs at equipartition, where all the specific energies are equal, including that of the magnetic field:

$$\frac{B_{\text{eq}}^2}{8\pi} \equiv \frac{1}{2} v_r^2 \, n_e(r) \, m_H. \tag{8.29}$$

As the dominant agent, any tangled superequipartition field would realign itself, possibly even annihilate adjacent sheared field components until the intensity again falls below the equipartition value. However, as we have already stated, the physics of magnetic reconnection is poorly known. Some attempts have been made to incorporate the magnetic tearing mode instability into a model of superequipartition field annihilation (e.g., in the supermassive black hole at the galactic center; see chapter 12), but there is still precious little agreement between our understanding of plasma processes such as this and what is actually observed in real systems, such as the Sun.[3] We shall return to this issue below, when we will be forced to deal with the impact of equation (8.29) and superequipartition magnetic fields.

Numerical simulations of magnetized disk accretion solve a set of coupled hydrodynamic and magnetic field equations, several of which we have already encountered in one guise or another. The first is simply the conservation of mass (or continuity) equation (6.9). The second expression is a generalization of the Lorentz force equation (4.23), in which we add pressure gradients and gravity,

$$\rho\frac{\partial \mathbf{v}}{\partial t} + \rho(\mathbf{v} \cdot \vec{\nabla})\mathbf{v} = -\vec{\nabla}(P + Q) - \rho\vec{\nabla}\Phi + \frac{1}{4\pi}(\vec{\nabla} \times \mathbf{B}) \times \mathbf{B}. \tag{8.30}$$

The symbols here have their usual meaning. In addition, the appearance of Q corresponds to an artificial (small) viscosity introduced to correctly treat shocks. For the energy equation, we take the expression for internal energy in (6.23), and replace the heat transfer term ($\rho T \, ds/dt$) with the more specific formulation arising from resistive heating,

$$\frac{d}{dt}(\rho u) + \rho u(\vec{\nabla} \cdot \mathbf{v}) = -(P + Q)(\vec{\nabla} \cdot \mathbf{v}) + \eta \mathbf{J}^2, \tag{8.31}$$

where $\mathbf{J} = (c/4\pi)\vec{\nabla} \times \mathbf{B}$ is the current density, and η is the resistivity. Notice that radiative cooling is explicitly omitted from this set of equations, imposing from the beginning the restriction that the disks modeled here are inefficient radiators. Finally, we also need to include an equation describing the evolution of the magnetic field itself,

$$\frac{\partial \mathbf{B}}{\partial t} = \vec{\nabla} \times (\mathbf{v} \times \mathbf{B} - c\eta\mathbf{J}). \tag{8.32}$$

---

[3]The reader interested in learning more about magnetic dissipation in converging flows may want to read Kowalenko and Melia (1999) and Melia and Kowalenko (2001).

The last term in equation (8.30) is the Lorentz force density with $\mathbf{J}$ replaced with Ampère's law (ignoring the displacement current). Equation (8.31) is the conservation of energy equation, in which the internal energy density of the gas changes through $P\,dV$ work (first term) or resistive dissipation (second term), and (8.32) is the equation of resistive MHD.

To understand the origin of this expression, consider that in the rest (primed) frame of the fluid, Ohm's law says

$$\mathbf{J}' = \frac{1}{\eta}\,\mathbf{E}', \tag{8.33}$$

where $1/\eta$ is the conductivity. From the Lorentz transformations derived in section 5.3, we know that

$$\mathbf{E}' = \gamma\left(\mathbf{E} + \frac{\mathbf{v}}{c} \times \mathbf{B}\right), \tag{8.34}$$

so that for nonrelativistic motion,

$$\mathbf{J}' \approx \frac{1}{\eta}\left(\mathbf{E} + \frac{\mathbf{v}}{c} \times \mathbf{B}\right). \tag{8.35}$$

Thus, if the medium is neutral, for which there are no advected currents,

$$\mathbf{J} = \mathbf{J}' \approx \frac{1}{\eta}\left(\mathbf{E} + \frac{\mathbf{v}}{c} \times \mathbf{B}\right). \tag{8.36}$$

In *ideal* MHD, the conductivity goes to infinity (or equivalently, η goes to zero), whereupon

$$\mathbf{E} \approx -\frac{\mathbf{v}}{c} \times \mathbf{B}. \tag{8.37}$$

In this limit, equation (8.32) is then simply a restatement of Faraday's law with $\mathbf{E}$ replaced with this expression (and $\eta = 0$). In *resistive* MHD, η is finite and cannot be ignored. Equation (8.32) is therefore the more general application of Faraday's law with $\mathbf{E}$ replaced with equation (8.36).

Returning now to the question of magnetic reconnection, we note that rapid annihilation of the field occurs wherever any component of the magnetic field changes sign and is not held apart by suitable fluid pressure. In the Petschek mechanism,[4] dissipation of the sheared magnetic field occurs in the form of shock waves surrounding special neutral points in the current sheets, and thus nearly all the dissipated magnetic energy is converted into the magnetic energy carried by the emergent shocks. Rapid reconnection may occur at speeds 10^{-2}–10^{-1} times the Alfvén velocity, $v_A = B/\sqrt{4\pi\rho}$, and causes vigorous dissipation of the magnetic field.[5]

In an alternative picture, the reconnection process develops as a result of resistive diffusion of the magnetic field lines through the plasma. It has been pointed out[6] that, aside from the fact that the Petschek mechanism suffers conceptually from

---

[4]This subject was first broached in connection with solar flares. See Petschek (1964).

[5]See also Parker (1979).

[6]A leading proponent of an alternative to the Petschek mechanism has been Van Hoven (1976), who also carried out laboratory experiments to measure the speed of reconnection in sheared magnetic fields.

the lack of an observable time scale and a predictable energy output, there is also a question as to whether it exists at all, since it has never been observed in either laboratory or astrophysical applications.[7] In the alternative picture, referred to as resistive magnetic tearing (or the tearing instability), the instability grows temporally and is driven by the free energy of a sheared magnetic field.

Another complication is that we have little guidance on what to take for the geometry of the magnetic field itself. Since the dissipation presumably occurs only for counteraligned field components, it clearly matters whether the field is completely turbulent or whether it acquires a partial coordination with the flow direction.

In the end, even tearing-mode instabilities may be ineffective at spreading the reconnection layer, since they may be stabilized by shear and grow slowly past the linear stage. But there is no question that rapid reconnection does take place (e.g., in the Sun, where observations point to reconnection at the Alfvén speed, at least when the magnetic field is dynamically important), so the growth rate calculated from either of these two mechanisms may simply be underestimates of the actual value.

The approach often taken in numerical simulations of optically thin, accreting plasmas is a rather simple one. An explicit artificial resistivity

$$\eta = \eta_0 \frac{|\vec{\nabla} \times \mathbf{B}|}{\sqrt{4\pi\rho}} \Delta^2 \qquad (8.38)$$

is introduced with a magnitude set to be larger than the effective numerical resistivity one would achieve with numerical reconnection.[8] Here, Δ is the grid spacing, and η_0 is a dimensionless parameter. In essence, this prescription guarantees that the magnetic field reconnects vigorously once it exceeds its equipartition value specified in equation (8.29).

Within the Keplerian structure, the captured plasma falls prey to the action of a magnetic dynamo fed by the differential rotation; this overwhelms the field annihilation and leads to a saturated field intensity (at a fraction of the equipartition value). The discovery of this magneto-rotational instability (MRI) in weakly magnetized disks[9] spawned several numerical simulations that confirmed the crucial role played by this process in rotational accretion systems.[10] The magnetic field generated in this fashion produces a significant viscosity via the Maxwell stress, dominating the transfer of angular momentum across the disk.

To appreciate how this instability develops in a weakly magnetized Keplerian flow, let us revist the the basic dynamical equations (6.9), (8.30), and (8.32). For simplicity, and given that we will find the MRI to be very efficient compared to magnetic field annihilation due to resistive diffusion in the inner region of the disk,

---

[7]However, more recent *Yohkoh* observations may have provided an indication that Petschek-like reconnection may be taking place in the Sun after all. See Blackman (1997). Theoretically, Petschek reconnection is perceived to be possible as long as some enhanced anomalous resistivity is active at the origin.

[8]This specification was first made by Stone and Pringle (2001), and is designed to mimic the magnetic Reynolds number one would get from the resistivity associated with simple, local numerical fluctuations.

[9]See Balbus and Hawley (1991).

[10]Some of these simulations have been reported in Hawley and Balbus (1991, 2002), Hawley, Gammie, and Balbus (1995), and Stone, Hawley, Gammie, and Balbus (1996).

we will set $\eta = 0$ for this particular analysis. We will adopt standard cylindrical coordinates $\mathbf{r} = (R, \phi, z)$, where R is the perpendicular distance from the z-axis.

Again, for simplicity, we will consider perturbations of an initial field $\mathbf{B} = (0, 0, B_z)$. It turns out that the maximal growth rate is reached in the axisymmetric case with a weak B_z, so this is a good initial configuration for pedagogical purposes as well. We will denote the Eulerian perturbations by δv, δB, and so forth, each modulated by the function $e^{i(k_R R + k_z z - \omega t)}$, where k_R and k_z are, respectively, the radial and vertical components of the wavevector.

The numerical simulations show that buoyancy is not a significant factor influencing the instability, nor is the compressibility of the fluid. By neglecting these terms and assuming incompressibility (so that $\delta \rho = 0$ in all the equations other than the equation of motion and the equation of state) and that $\delta P = 0$ in the equation of state, one obtains[11] the following linearized dynamical equations:

$$k_R \delta v_R + k_z \delta v_z = 0, \tag{8.39}$$

$$\frac{\partial \delta v_R}{\partial t} - 2\Omega \delta v_\phi = i \frac{k_z B_z}{4\pi \rho} \delta B_R - i k_R \left(\frac{\delta P}{\rho} + \frac{B_z \delta B_z}{4\pi \rho} \right), \tag{8.40}$$

$$\frac{\partial \delta v_z}{\partial t} = -i k_z \frac{\delta P}{\rho}, \tag{8.41}$$

$$\frac{\partial \delta v_\phi}{\partial t} + \frac{\kappa^2}{2\Omega} \delta v_R = i \frac{k_z B_z}{4\pi \rho} \delta B_\phi, \tag{8.42}$$

$$\frac{\partial \delta B_R}{\partial t} = i k_z B_z \delta v_R, \tag{8.43}$$

$$\frac{\partial \delta B_z}{\partial t} = i k_z B_z \delta v_z, \tag{8.44}$$

$$\frac{\partial \delta B_\phi}{\partial t} = \frac{R \, d\Omega}{dR} \delta B_R + i k_z B_z \delta v_\phi, \tag{8.45}$$

where Ω is the angular velocity in the circularized flow, and $\kappa^2 = (2\Omega/R) \, d(R^2\Omega)/dR$ is the square of the epicyclic frequency.

Replacing the Lagrangian time derivatives with $-i\omega$ in the linearized equations and eliminating the Eulerian perturbations, one obtains the dispersion relation

$$(\omega^2 - k_z^2 v_{A,z}^2)^2 - \frac{k_z^2}{k^2} \kappa^2 (\omega^2 - k_z^2 v_{A,z}^2) - 4\Omega^2 \frac{k_z^4 v_{A,z}^2}{k^2} = 0, \tag{8.46}$$

where $k^2 = k_z^2 + k_R^2$, and the Alfvén speed in the z-direction is defined as $v_{A,z} = (B_z^2/4\pi\rho)^{1/2}$. For Keplerian rotation, $\kappa = \Omega$. This equation can be solved easily for ω, which yields

$$\omega_0^2 = k_{z0}^2 + \frac{k_{z0}^2}{2 k_0^2} - 2\sqrt{\frac{k_{z0}^4}{k_0^2} + \frac{k_{z0}^4}{16 k_0^4}}. \tag{8.47}$$

---

[11] These follow simply by putting $v_R \to v_R + \delta v_R$ (and similarly for the other variables) in equations (6.9), (8.30), and (8.32), and then using these same equations to eliminate terms that depend solely on the unperturbed quantities.

Here, $\omega_0 \equiv \omega/\Omega$, and k_0 and k_{z0} are, respectively, k and k_z expressed in units of $\Omega/v_{A,z}$. Note that ω^2 reaches its minimum value of $-\frac{9}{16}\Omega^2$ when $k^2 = k_z^2 = \frac{15}{16}(\Omega/v_{A,z})^2$. For $k_R = 0$, the modes become stable when $k_z^2 > 3(\Omega/v_{A,z})^2$, and in the long wavelength limit $\omega^2 \simeq -3(v_{A,z}k_z)^2$.

To appreciate the physical meaning of this instability, let us examine the fastest growing mode, which occurs when $k^2 = k_z^2 = \frac{15}{16}(\Omega/v_{A,z})^2$ and $\omega^2 = -(9/16)\Omega^2$. Solving the linearized equations (8.39)–(8.45) with these values, we get

$$\delta v_R = \delta v_\phi, \tag{8.48}$$

$$\delta B_R = -\delta B_\phi, \tag{8.49}$$

$$\delta B_R = i\frac{4}{3}\frac{k_z B_z}{\Omega}\delta v_R, \tag{8.50}$$

$$\delta B_\phi = -i\frac{5}{4}\frac{4\pi\rho\Omega}{k_z B_z}\delta v_\phi, \tag{8.51}$$

$$\frac{|\delta B_R|^2}{8\pi} = \frac{5}{3}\frac{\rho|\delta v_R|^2}{2}. \tag{8.52}$$

With these solutions, we can now return once more to equations (8.39)–(8.45) and see how the instability grows. The perturbation δv_R, which is generated by δv_ϕ through the Coriolis force term $2\Omega\,\delta v_\phi$ in equation (8.40), induces the perturbation δB_R through equation (8.43). The shearing in the disk, affecting the term $(R\,d\Omega/dR)\delta B_R$ in equation (8.45), leads to the perturbation δB_ϕ through the term δB_R. This in turn enhances δv_ϕ through the right-hand side of equation (8.42). Thus, a positive feedback loop is established. Some of the other terms in the linearized equations act to stabilize the perturbation, but they are overwhelmed by the positive feedback in the unstable modes. However, for modes with a large wavenumber, the term $ik_z B_z\delta v_\phi$ in equation (8.45) and the term $ik_z B_z\delta B_R/4\pi\rho$ in equation (8.40) will overwhelm the positive feedback and render the mode stable.

For the unstable modes, equation (8.52) says that the turbulent kinetic energy density is approximately equal to the turbulent magnetic field energy density. Apparently, the final saturated state of the system approaches equipartition. Numerical simulations[12] confirm this basic result, and go further in demonstrating that the ratio of final turbulent energy densities is only weakly dependent on the initial and subsequent physical conditions. We shall therefore find it convenient to parametrize the dynamo-generated energies in a rotational flow according to

$$\frac{\langle\delta B^2\rangle}{8\pi} = C_0\frac{1}{2}\langle\rho\,\delta v^2\rangle, \tag{8.53}$$

where the constant C_0 has a value between 1 and 10, depending on the vertical profile of the Keplerian structure. Note, however, that this formulation does not yet tell us what the actual magnetic energy density is—only that it is comparable to that of the turbulent gas. The final piece of the puzzle that will permit this evaluation will come from the solution of the energy equations themselves (see equations 8.54–8.56).

---

[12]See Brandenburg et al. (1995), Hawley, Gammie, and Balbus (1995), and Stone et al. (1996).

To see which magnetic field component (or components) dominates, assume that the turbulent field generated by the dynamo constitutes the total field in this system. The equations governing the evolution of the magnetic energy density are then

$$\frac{1}{2}\frac{\partial B_\phi^2}{\partial t} = R\frac{d\Omega}{dR}B_\phi B_R + B_\phi \vec{\nabla} \times (\delta\mathbf{v} \times \mathbf{B})_\phi + \frac{\eta c^2}{4\pi}B_\phi |\nabla^2\mathbf{B}|_\phi, \tag{8.54}$$

$$\frac{1}{2}\frac{\partial B_R^2}{\partial t} = B_R \vec{\nabla} \times (\delta\mathbf{v} \times \mathbf{B})_R + \frac{\eta c^2}{4\pi}B_R |\nabla^2\mathbf{B}|_R, \tag{8.55}$$

$$\frac{1}{2}\frac{\partial B_z^2}{\partial t} = B_z \vec{\nabla} \times (\delta\mathbf{v} \times \mathbf{B})_z + \frac{\eta c^2}{4\pi}B_z |\nabla^2\mathbf{B}|_z, \tag{8.56}$$

where, as usual, η is the resistivity of the plasma. These equations all follow from the chain rule of differentiation and the application of equation (8.32), in which $\mathbf{J}$ is replaced with the curl of $\mathbf{B}$ from Ampère's law.

We saw earlier in equation (8.49) that in the linear regime, the amplitudes of the azimuthal and radial components of the magnetic field are equal when the perturbation is small. However, the final turbulent state is affected by the nonlinear character of the magnetohydrodynamic equations. As the amplitudes of the perturbation increase, nonlinear effects become more important. Due to shearing in the Keplerian flow, the average value of $RB_\phi B_R \, d\Omega/dR$ is positive. The energy equations (8.54)–(8.56) show that this term contributes to a growing anisotropy of the turbulent magnetic field, in the sense that more and more magnetic field energy is channeled into the azimuthal direction. For a Keplerian flow with $R \, d\Omega/dR = -\frac{3}{2}\Omega$, the growth rate due to the shearing of this structure is larger than that associated with any other dynamo process. So, in the final state, the azimuthal component of the magnetic field dominates the field energy density.

In summary, then, we have the following situation with regard to a radiatively inefficient, magnetized disk. As the gas flows inward and spirals into an approximately Keplerian structure at a distance from the black hole corresponding to the circularization radius, a linear instability first stretches the magnetic field lines carried by the gas and produces a radial component of $\mathbf{B}$. The magnetic field generated during this step is approximately in equipartition with the turbulent kinetic energy and counts for a small fraction of its final intensity, but it nonetheless provides the seed for the next step. Second, the shearing in the Keplerian flow stretches the radial magnetic field in the azimuthal direction, increasing the magnetic field energy density. The energy comes from the rotation of the gas and, during this process, a significant fraction of angular momentum may be transported outward due to torques arising from the Maxwell stress $B_R B_\phi/4\pi$. Finally, some of the magnetic field energy is converted into kinetic energy, which is eventually converted into thermal energy by viscous dissipation through the Lorentz force term $\mathbf{B} \cdot \vec{\nabla} \times (\delta\mathbf{v} \times \mathbf{B})$. In addition, the magnetic field energy may be dissipated through Ohmic resistivity.

The final step is to solve the energy equations (8.54)–(8.56) and find the magnetic energy density in the turbulent plasma. From dimensional analysis, we infer that

$$\frac{\langle \rho\delta v^2\rangle^{1/2}}{\langle \rho\rangle^{1/2}\zeta} \propto R\frac{d\Omega}{dR}, \tag{8.57}$$

where ζ is the scale height of the flow and we use $\langle \rho \delta v^2 \rangle^{1/2}/\langle \rho \rangle^{1/2}$ to represent the turbulent velocity. Thus, from equations (8.53) and (8.56) we obtain

$$\langle B^2 \rangle^{1/2} \propto \langle \rho \rangle^{1/2} \zeta R \frac{d\Omega}{dR}, \tag{8.58}$$

which is one of the main results we have been seeking, since it provides us with a measure of the MRI-induced magnetic field throughout the (Keplerian) accreting region.[13] For a Keplerian flow, $\Omega = (GM/R^3)^{1/2}$, and therefore $R \, d\Omega/dR = -\frac{3}{2}\Omega$.

At the beginning of this section, we suggested that a magnetic field in radiatively inefficient accretion flows can provide a viscosity to drives the matter inward. Now that we have an estimate of this field (equation 8.58), we are in a position to begin our derivation of the kinematic viscosity ν, analogous to the prescription we had in the thin-disk case (equation 7.9).

One approach is to invert the sense of equation (7.8) and use it to write a prescription for ν in terms of the torque τ_{out},

$$\nu = \frac{2}{3} \frac{\tau_{\text{out}}}{2\pi \Sigma R^2 \Omega}. \tag{8.59}$$

But τ_{out} is the stress (units of force per unit area) times the radius, integrated over a cylinder at radius R, with surface area $\approx 2\pi R H$. Thus, if we define the quantity $W_{R\phi}$ to be the total vertically integrated stress, equation (8.59) transforms into

$$\nu = \frac{2}{3} \frac{W_{R\phi}}{\Sigma \Omega}. \tag{8.60}$$

In general, $W_{R\phi}$ is the sum of the Maxwell and Reynolds stresses. It is a pressure, or a force per unit area, which is generally not isotropic—hence its tensor structure, which allows one to write all three force components acting on a surface element with any given unit normal vector. For example, in the case of the electromagnetic field, with electric and magnetic components E_i and B_i, the Maxwell stress tensor may be written as[14]

$$T_{ij} = \frac{1}{4\pi} E_i E_j + B_i B_j - \frac{1}{2}(E^2 + B^2)\delta_{ij}, \tag{8.61}$$

where δ_{ij} is the Kronecker delta, and the indices (i, j) here run over the three cylindrical coordinates (R, ϕ, z). The parcel $T_{\alpha\beta}$ of 16 quantities (in four-dimensional spacetime) gives the force per unit area applied to any surface by the electromagnetic field.

With the dynamo in full force, $\mathbf{B}$ saturates at near its equipartition value (see above), and T_{ij} easily dominates over the Reynolds stress.[15] Thus, we may write

$$W_{R\phi} \approx \beta_v \int dz \frac{\langle B^2 \rangle}{8\pi} \tag{8.62}$$

---

[13] The validity of this relationship has been confirmed by the numerical simulations of Brandenburg et al. (1995). This work includes an analysis of the magnetohydrodynamic dynamo at two different radii, one of which is 5 times smaller than the other. More recent calculations by Hawley (2000) have added additional support for this result.

[14] See, e.g., Melia (2001a).

[15] See, e.g., Balbus, Gammie, and Hawley (1994).

(the average inside the integral being taken over time). Note that even though this approximation is valid simply on the basis that the Reynolds stress is relatively small, its validity is enhanced by the fact that the turbulent velocity (which accounts for the kinetic, or Reynolds, stress) and $\mathbf{B}$ are generated by the same process, so both should be scalable by $\mathbf{B}$ (see equation 8.53). Numerical simulations show that the proportionality constant β_v changes very slowly with R and typically falls in the range ~ 0.1–0.2.

Now, to evaluate $W_{R\phi}$ in equation (8.62), we need to know the vertical extent over which the stress is acting. Though our derivation of the scale height in equation (7.43) was predicated on the assumption that the disk is thin, for the purpose of estimating $W_{R\phi}$, we will take the liberty of adopting ζ as a reasonable representation of the disk thickness here as well—an approximation that is certainly reasonable even in cases where $\zeta \sim R$.

Thus, from equation (7.43), we see that

$$(\zeta \Omega)^2 \approx \frac{P}{\rho}, \tag{8.63}$$

and we may therefore write equation (8.58) as

$$\frac{B^2}{8\pi} = \beta_P P, \tag{8.64}$$

where β_P is another proportionality constant, whose value (indicated from the numerical simulations) is ~ 0.03. Thus,

$$W_{R\phi} \approx \beta_v \int dz \beta_P P, \tag{8.65}$$

or

$$W_{R\phi} \approx \beta_v \beta_P c_s^2 \Sigma \tag{8.66}$$

(for simplicity, ignoring here any possible vertical gradient in temperature.) Evidently,

$$v = \frac{2}{3} \beta_v \beta_P \frac{c_s^2}{\Omega}. \tag{8.67}$$

But from equation (7.43), we have that

$$\zeta = \frac{c_s}{\Omega} \tag{8.68}$$

and, therefore,

$$v = \frac{2}{3} \beta_v \beta_P c_s \zeta$$
$$\approx \frac{1}{3} \beta_v \beta_P c_s H. \tag{8.69}$$

It might at first seem surprising that this expression for v reduces to the same formulation as our α-viscosity prescription in section 7.1. However, the physics behind v does not change from application to application; only the details vary, depending on the physical state of the system. Ultimately, the units of v are velocity times distance, and the only natural scales one may use to characterize the viscosity at any given R are the local sound speed and the vertical size of the disk. The specifics of how the transfer of momentum is effected are all absorbed into the proportionality constant multiplying c_s and H, which we do expect to have a value reflecting the mechanism giving rise to the torque.

In the α-disk prescription, the viscosity is said to arise from the exchange of small parcels of plasma flowing back and forth across a given radius. These carry with them their specific angular momenta, which, however, vary with radius. The net effect is a torque by one ring on another, since the exchanged angular momentum densities do not cancel exactly. To characterize the unknown velocity of these plasma eddies and the unknown length scale over which they flow, we introduced the (unknown) constant α, expected to be less than one. Magnetized disks subject to the MRI derive, instead, their viscous dissipation from the action of the magnetic field on the underlying plasma. Still, because of the similar formulation between the two, we can see a direct connection with the α-disk prescription by making the elegant and simple identification

$$\alpha \equiv \tfrac{1}{3}\beta_v\beta_P, \qquad (8.70)$$

with which we recover equation (7.9), and which now characterizes the *anomalous* viscosity of our magnetized disk.

Now that we understand how important the MRI-induced viscosity can be, however, we must incorporate the viscous dissipation as a heating term in the energy equation (8.31) and as a source or sink of momentum in equation (8.30).

Equation (6.9) is always valid, and under these circumstances, equation (8.32) does not change either. However, viscosity alters the momentum of the orbiting gas, so equation (8.30) must be modified as follows:

$$\rho\frac{\partial \mathbf{v}}{\partial t} + \rho(\mathbf{v}\cdot\vec\nabla)\mathbf{v} = -\vec\nabla\left(P + \frac{B^2}{8\pi} + Q\right) - \rho\vec\nabla\Phi + \frac{1}{4\pi}(\mathbf{B}\cdot\vec\nabla)\mathbf{B}$$
$$+ 2\hat{\mathbf{e}}^j\nabla^i(S_{ij}\rho\nu), \qquad (8.71)$$

where we have used a vector identity to split the second term on the right-hand side of equation (8.30); P is now understood to represent the nonmagnetic pressure, $\rho R_g T/\mu + P_{\text{rad}}$, and the unit vector $\hat{\mathbf{e}}^j$ is attached to the j-coordinate. In addition,

$$S_{ij} \equiv \frac{1}{2}\left(\frac{\partial v_i}{\partial x^j} + \frac{\partial v_j}{\partial x^i} - \frac{2}{3}\delta_{ij}\frac{\partial v_k}{\partial x^k}\right). \qquad (8.72)$$

When the flow is Keplerian, one can easily show that $S_{R\phi} = -\tfrac{3}{4}\Omega$.

The energy equation (8.31) must similarly be modified with the addition of a term on the right-hand side to represent the dissipation caused by ν. The heating rate (energy per unit volume, per unit time) is given as[16]

$$\Gamma \equiv 2\nu\rho\mathbf{S}^2, \qquad (8.73)$$

---

[16]For a more detailed derivation of this equation, and others involving a kinematic viscosity, see Landau and Lifshitz (1987).

so that

$$\frac{d}{dt}(\rho u) + \rho u(\vec{\nabla} \cdot \mathbf{v}) = -(P + Q)(\vec{\nabla} \cdot \mathbf{v}) + \eta \mathbf{J}^2 + \Gamma. \tag{8.74}$$

This now completes the set of equations we need in order to solve for the structure of radiatively inefficient disks.

But given the level of complexity inherent in any realistic modeling of these objects, one is compelled to solve the equations we have developed in this section using methods of discretization, evolving them on a spatial grid with sufficient resolution to bring out the essential features of this problem, e.g., the wavelengths associated with the fastest growing modes in the MRI. For example, recent three-dimensional magnetohydrodynamic simulations have been carried out to model the accretion disk in Sagittarius A*, the supermassive black hole at the galactic center (see section 12.2).

These calculations[17] make use of the pseudo-Newtonian gravitational potential[18] to mimic the dynamically important marginally stable circular orbit of the full Schwarzschild metric at $r = 3r_S$ (see figure 3.2). In this prescription,

$$\Phi = -\frac{GM}{r - r_S}, \tag{8.75}$$

with which the specific angular momentum of a circular orbit is

$$l_{\text{circ}} = \frac{(GMr)^{1/2} r}{r - r_S}. \tag{8.76}$$

In addition, these simulations ignore radiation transport and energy losses, which are, by assumption, dynamically unimportant in the radiatively inefficient system we have here. And the angular momentum is transported by Maxwell and Reynolds stresses arising from magnetic field and velocity correlations in the MRI-induced turbulence.

The general features emerging from these calculations include three principal flow components (see figure 8.2): (1) a hot, rotationally supported disk extending down to the marginally stable orbit; (2) an extended, low-density coronal backflow enveloping the disk; and (3) a distinctive, jet-like outflow perpendicular to the disk in the vicinity of the black hole. Near the equator, thermal pressure dominates magnetic (i.e., the MRI-generated magnetic field is somewhat below equipartition), whereas the surrounding region is strongly magnetized.

Although this simulation lacks a formal treatment of the energetics required for a detailed calculation of the disk's spectrum, one can nonetheless estimate the synchrotron emissivity as a function of position within the disk. The result of this

---

[17]See Hawley and Balbus (2002). Their simulations use $128 \times 32 \times 128$ grid zones in cylindrical coordinates (R, ϕ, z). The radial grid extends from $R = 1.5r_S$ to $R = 220r_S$. There are 36 equally spaced zones inside $R = 15r_S$, and 92 zones increasing logarithmically in size outside this radius. In the z-direction, there are 50 zones equally spaced between $-10r_S$ and $10r_S$, with the remainder of the zones logarithmically stretched to the z-boundaries at $\pm 60r_S$. The azimuthal domain is uniformly gridded over $\pi/2$ in angle. The radial and vertical boundary conditions are simple zero-gradient outflow conditions, with which no flow is permitted into the computational domain. The ϕ-boundary is periodic.

[18]This formalism was introduced by Paczyński and Wiita (1980).

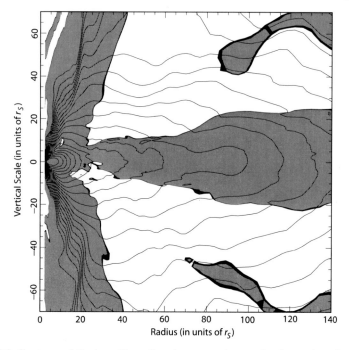

Figure 8.2 Contours of the logarithm of total pressure (magnetic plus gas), at the end of a three-dimensional magnetohydrodynamic simulation of accretion within $\sim 100\, r_S$ of the black hole. Cylindrical radius is shown in the horizontal direction (in units of r_S) and the z-coordinate is along the vertical axis. The shaded regions overlaid on top of the contours reveal where gas pressure exceeds magnetic. Note that gas pressure dominates in the disk, the hot inner torus, and along the funnel wall where matter is being expelled. However, magnetic pressure dominates in the bulk of the coronal envelope atop the disk. (Image from Hawley and Balbus 2002)

analysis, summarized in figure 8.3, shows that the highest peak frequencies are $\sim 2.5 \times 10^{11}$ Hz, produced in the very inner portion of the disk. Very importantly, the gas temperature here varies between 4×10^{10} and 10^{11} K, much higher than the temperature ($\sim 10^5$–10^6 K) one gets in thin-disk accretion. The radiative inefficiency of these structures means that the infalling plasma retains most of the dissipated gravitational energy, and advects it through the event horizon. Just outside the black hole, the internal energy of the gas is so high that the temperature reaches billions of degrees.

We will revisit the physics of these disks in chapter 12, where we consider the behavior of weakly accreting AGNs. We will learn that the spectral and polarimetric guidance provided by the observations has greatly influenced how we view the environment within $\sim 10 r_S$ of these objects, which appear to be fed by a hot, tenuous plasma settling into a Keplerian rotation; this gas subsequently winds its way inward under the action of an anomalous viscosity from the MRI-induced turbulence. The radio and mm spectral components in these sources are produced in this region, as is the variable (synchrotron) IR and (inverse Compton) X-ray flux, when present.

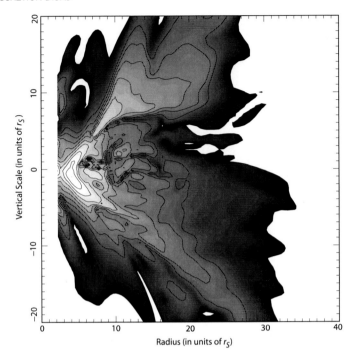

Figure 8.3 Contour map and grayscale of peak frequencies for synchrotron emission at the end of the simulation shown in figure 8.2. The horizontal axis shows the (cylindrical) radius (in units of r_S), and the vertical one gives the z-coordinate. Note that the spatial scale here is smaller than in the previous figure. The contours are equally spaced in log frequency, with a peak at 2.5×10^{11} Hz. The highest frequencies emerge from the inner edge of the torus, and hence from the smallest spatial scale. (Image from Hawley and Balbus 2002)

SUGGESTED READING

The theory of thick disks has been under development since the early 1980s. Recent numerical simulations have confirmed many of the results obtained previously using mostly analytical arguments. See Jaroszyński, Abramowicz, and Paczyński (1980), Abramowicz, Calvani, and Nobili (1980), Blaes (1987), and Igumenshchev, Chen, and Abramowicz (1996).

The importance of magnetic fields on the dynamics of infalling plasma is starting to receive more and more attention with the growing realization that compact objects often accrete at rates well below those expected on the basis of pure hydrodynamics alone. The dissipation of magnetic-field energy heats the gas and supports it against gravitational collapse. See Shvartsman (1971), Kowalenko and Melia (1999), Hawley (2000), Hawley and Balbus (2002), and Igumenshchev and Narayan (2002).

Chapter Nine

Pulsing Sources

We now begin to apply what we have learned from the high-energy surveys, and the development of theoretical tools in previous chapters, to study individual classes of object—sometimes focusing on the specific archetypal sources themselves as representative members of their group. The various categories we will define are based primarily on observational criteria, though often at least some theoretical insight is needed to help distinguish between otherwise confusing similarities in spectra, variability, and spatial distribution. Our first analysis will focus on the subject of variable high-energy sources.

As we stated at the very beginning of this book, rapid source variability (on timescales as short as milliseconds in some cases) is one of the features that distinguishes high-energy astrophysics from many other branches of astronomy. In a way, we are "preprogrammed" to react attentively when we see such quick evolution in our environment. Our ability to sense change is essential for survival. For example, visual motion provides us with a source of information that can serve several functions, including (1) establishing the three-dimensional structure of an otherwise two-dimensional visual scene, (2) guiding our balance and postural control for interaction with other objects around us, and (3) estimating distances and time to collision with obstructions. On a more fundamental level, perceiving change in a predator's vision of the land facilitates the capture of prey identified with that motion, most famously demonstrated through the incredible visual acuity of an eagle. We too, as members of the animal kingdom, react keenly to any stimulus generated by change in an otherwise "bland" external medium.

Highly variable sources, particularly the transient ones, therefore easily command our attention. And such was the case, on those long winter nights back in 1967, when Jocelyn Bell and colleagues were sitting transfixed by radio pulses they were receiving on a newly commissioned radio telescope. Though they would not realize the complete meaning of their results until several years later, the graduate student Bell and her thesis adviser, Antony Hewish, had just made what many still consider to be one of the most significant discoveries in radio astronomy. But radio pulsars, as they are now known, are not just radio sources; they also radiate in X-rays and γ-rays, and constitute an important class of high-energy object. It is fitting, therefore, to begin our classification of the sources we introduced in chapters 1 and 2 with these highly variable entities.

9.1 RADIO PULSARS

In the mid-1960s, Antony Hewish at Cambridge University realized that the then newly discovered technique of interplanetary scintillation could be used to identify quasars (themselves discovered only a few years earlier) and designed a large radio telescope to do so. Interplanetary scintillation is the apparent fluctuation in the intensity of a radio source due to the diffraction of radio waves as they transmit through the turbulent solar wind in interplanetary space. Given a known fluctuation size in the diffracting medium, smaller (in projection) sources scintillate more rapidly than the bigger ones as the column density changes along our line of sight; this happens because a greater fraction of the emitting region is occulted in the smaller objects.

But the most interesting result they obtained with the new telescope was actually something totally unexpected—in late 1967, they started to detect pulsations from an unknown source at right ascension 19 h 19 min with a periodic separation of about 1.3 s. A second source, with period 1.2 s, was seen soon afterward near right ascension 11 h 33 min, followed by another two (at 08 h 34 min and 09 h 50 min) in quick succession.[1] Within a year, the radio pulsar census had grown to an amazing class of 27. For designing the telescope and facilitating this discovery, Hewish was awarded the Nobel prize in physics in 1974.

The link between these new pulsating radio sources and fast spinning neutron stars was realized very quickly. The proposal had already been made several years earlier that a neutron star with a magnetic field of 10^{10} G might exist at the heart of the Crab nebula (shown as an X-ray structure in figure 9.1).[2] And by 1968, the idea that the rapid rotation of such a highly magnetized neutron star could be the source of energy in the Crab had already caught on.[3] Right from the beginning, the radio emission from these sources was interpreted as being due to synchrotron radiation (equation 5.117) by particles accelerated in the pulsar magnetosphere along curved magnetic field lines (see section 4.2).

Since then, more than 1700 radio pulsars have been discovered (see figure 9.2), including two of the best-known objects in all of astronomy: the fast 33 ms-pulsar in the Crab Nebula,[4] and the 89 ms-pulsar in the Vela supernova remnant.[5] Both of these (now well-studied) pulsars reside within supernova remnants, which at the time of their discovery was a powerful indicator that neutron stars are born in core-collapse supernovae from massive main sequence stars.

The distribution of roughly 400 pulsars, known within 20 years of Bell and Hewish's discovery, is shown in galactic coordinates in figure 9.3. The size of the circles in this diagram represents a pulsar's average flux density. Massive stars burn through the main sequence quickly compared to the age of the Galaxy, so neutron stars produced in supernova explosions tend to begin their existence close to star-forming regions; as we shall see, most of them therefore lie along the galactic plane since they are most easily detectable when they're young and have not yet had sufficient time to wander off toward large galactic latitudes.

---

[1] A firsthand account of these events has been published by Jocelyn Bell in the *Annals of the New York Academy of Science* (Bell 1977). The scientific results were published by Hewish et al. (1968).

[2] This was based on the pioneering work of Hoyle, Narlikar, and Wheeler (1964).

[3] Pacini (1967), then a postdoctoral fellow at Cornell, had actually published this idea a few months before the discovery of Bell and Hewish. Subsequent work by Gold (1968) introduced the concept of a rotation-powered pulsar that loses its rotational energy via the emission of electromagnetic radiation by the rotating magnetic dipole, and the acceleration of relativistic particles.

[4] See Staelin and Reifenstein (1968).

[5] See Large et al. (1968)

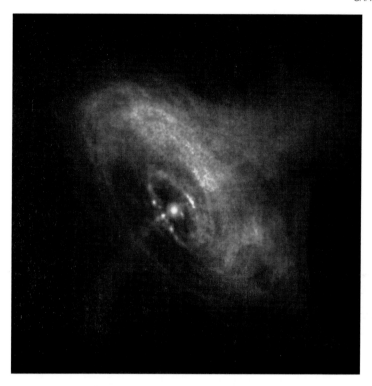

Figure 9.1 Both the Crab Nebula and Crab pulsar are clearly visible in this *Chandra* image, which shows tilted rings of high-energy particles ejected outward over a light-year from the rapidly spinning neutron star. The inner ring has a diameter of about a tenth of a light-year. Jet-like structures extend outward from the central object perpendicular to the rings. The Crab Nebula radiates via synchrotron emission; its bell-shaped appearance is probably due to the interaction of this large, magnetized bubble with nearby clouds of gas and dust. (Image courtesy of NASA/CXC/SAO)

Figure 9.4 illustrates the projection onto the galactic plane of the spatial distribution of pulsars known by 2005. This sample is, of course, much bigger than that shown in figure 9.3, but the main point here is that even with the vastly expanded collection of objects in this class, most of them lie within a few kiloparsecs of the Sun. Therefore, only a small fraction of the total galactic population has thus far been found.

The observed periods of pulsars, whose distribution is illustrated in figure 9.5, range over three and a half orders of magnitude, from approximately 1.5 ms to 5 s. The majority of them fall between 0.2 and 2 s. Note, however, that most of the short-period objects are members of binary systems (indicated as shaded regions in this figure), so additional physics is involved in shaping this portion of the distribution. As we shall see later in this section, these ms-pulsars probably contain old neutron stars reactivated by the transfer of mass and angular momentum from a companion star. In general, there is no strong dependence of duty cycle on period, but complex waveforms with two or more peaks tend to be detected more often at the longer periods.

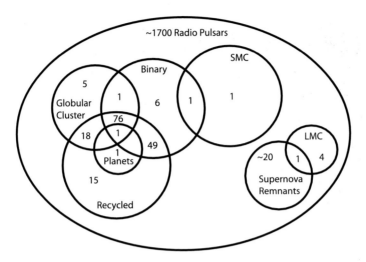

Figure 9.2 The known pulsar population now exceeds 1700, including 80 binary and millisecond pulsars associated with the disk of our Galaxy, and 103 in 24 of the galactic globular clusters. This Venn diagram shows the number, and their respective locations, of the various types of pulsars now known. LMC and SMC denote, respectively, the large and small Magellanic Clouds. (Image from Lorimer 2005)

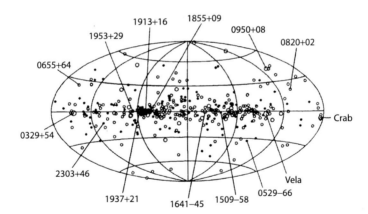

Figure 9.3 The location of approximately 400 radio pulsars shown in galactic coordinates. Two of the more interesting ones, Crab and Vela, are individually labeled. The size of the circles gives a rough indication of each pulsar's average 400-MHz flux density ($<10\,$mJy, 10–$100\,$mJy, 100–$1000\,$mJy, and $>1000\,$mJy). Pulsar surveys tend to be nonuniform across the sky, due to various contaminating factors that affect the system noise in the detector. As such, pulsars with average 400-MHz flux densities above $20\,$mJy represent a relatively complete flux-limited sample over the whole sky; those with flux densities as small as $1\,$mJy probably do not. (Image from Taylor and Stinebring 1986)

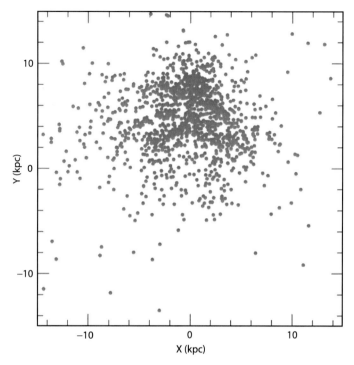

Figure 9.4 The known radio pulsars projected onto the galactic plane. The galactic center is at (0, 0) and the Sun is located at (0, 8.5) kpc. (Image from Lorimer 2005)

Pulsars joined the high-energy fraternity in 1969, when X-ray (1.5–10 keV) pulsations were discovered in the Crab.[6] Soon afterward, the first satellite dedicated exclusively to γ-ray astronomy (known as the Small Astronomy Satellite 2, or SAS-2) confirmed the existence of γ-ray pulsations from the Crab as well.[7] SAS-2 also discovered γ-pulsations from the Vela Pulsar at about the same time,[8] which turned out to be the brightest γ-ray source in the sky. Here already was a major surprise—the Vela γ-ray lightcurve (see the upper right-hand panel in figure 9.7) is characterized by two relatively sharp peaks, separated by 0.4 in phase, though not phase-aligned with the radio and optical pulses. This was also the era in which several unidentified γ-ray sources were detected, including *Geminga* (which we will discuss shortly), a faint source in the Gemini region. Its name actually derives from the Milanese dialect,[9] in which *gh'é minga* means "it is not there," or "it does not exist."

---

[6]These were first reported by Fritz et al. (1969) and then confirmed by Bradt et al. (1969) three months later.

[7]See Kniffen et al. (1974).

[8]See Thompson et al. (1975).

[9]See Bignami and Caraveo (1996) for a detailed history of Geminga, from its discovery to its final identification.

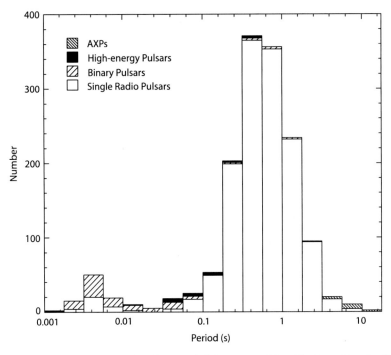

Figure 9.5 Most known radio pulsars have periods between ∼0.2 and 2 s, but with a significant tail in the distribution extending down to ≈ 0.0015 s. Most of the short-period pulsars (say, below 0.1 s) are members of gravitationally bound binary systems (indicated by shading in the histogram), probably containing old neutron stars reactivated by the accretion of mass and angular momentum from an evolving companion star. Several pulsars have now been detected via their X-ray and γ-ray emission (called "high-energy" pulsars in this diagram), and several others constitute a relatively new class of object—the anomalous X-ray pulsars (or AXPs), which we discuss later in this section. (Image from Manchester et al. 2005; see also Taylor, Manchester, and Lyne 1993)

Interestingly, although pulsars were first discovered as radio sources, a confirmation that they possess strong magnetic fields came only several years later, with a remarkable spectral observation of Hercules X-1 with the *Uhuru* satellite (see section 1.5).[10] (This object is a different type of pulsar, accreting in a binary system, which we will consider with the rest of its class in section 9.2.) An excess spectral component between ∼40 and 60 keV was interpreted as resonant electron cyclotron emission in the hot polar plasma of the rotating neutron star (see equation 5.61). The corresponding magnetic field strength, inferred from equation (5.60), is ∼5 × 10^{12} G. This high-energy observation yielded the first direct measurement of a pulsar's magnetic field, and confirmed the expectation that such sources must be highly magnetized, rapidly spinning neutron stars.

---

[10]See Tananbaum et al. (1972).

We can learn a great deal about the physics of pulsar emission from measurements of its period and period derivative. A typical neutron star has a radius $R_{ns} \approx 10$ km and a mass $M_{ns} \approx 1.4 M_\odot$. Its moment of inertia is therefore

$$I_{ns} = \tfrac{2}{5} M_{ns} R_{ns}^2 \approx 10^{45} \text{ g cm}^2. \tag{9.1}$$

Taking the Crab pulsar as an example, for which the spin period is $P_{spin} = 33.403$ ms, the rotational energy is

$$E_{rot} = \frac{1}{2} I_{ns} \left(\frac{2\pi}{P_{spin}} \right)^2 \approx 2 \times 10^{49} \text{ ergs}. \tag{9.2}$$

The Crab pulsar also has a measured period derivative $\dot{P}_{spin} = 4.2 \times 10^{-13}$ s s^{-1}, implying an energy loss rate

$$\frac{d E_{rot}}{dt} \equiv \dot{E}_{rot} = -4\pi^2 I \frac{\dot{P}_{spin}}{P_{spin}^3} \approx 4.5 \times 10^{38} \text{ ergs s}^{-1}. \tag{9.3}$$

But a rotating magnetic dipole radiates energy at a rate

$$\dot{E}_{dip} = -\frac{2\ddot{m}^2}{3c^3}. \tag{9.4}$$

This is a generalization of the power in equation (5.17), though with the electric dipole moment replaced with the corresponding magnetic dipole moment **m**. According to equation (7.80), **m** may be written in terms of B_0, the magnetic field at the polar cap, as

$$m = \tfrac{1}{2} B_0(R) R^3, \tag{9.5}$$

and with $\ddot{m} \approx m(2\pi/P_{spin})^2$ (in terms of the period P_{spin}), we therefore have

$$\dot{E}_{dip} = -\frac{8\pi^4}{3c^3} \frac{B_0^2 R^6}{P_{spin}^4}. \tag{9.6}$$

(Of course, $\ddot{m}$ depends on the inclination angle of the magnetic dipole relative to the spin axis, but we will ignore this factor of order unity for this discussion.)

Thus, equating $\dot{E}_{rot}$ with $\dot{E}_{dip}$, we conclude that

$$B_0 = 1.3 \times 10^{19} (P_{spin} \dot{P}_{spin})^{1/2} \text{ G}, \tag{9.7}$$

which for the Crab implies that $B_0 = 7.6 \times 10^{12}$ G. In addition, equation (9.7) shows that $\dot{P}_{spin} \propto P_{spin}^{-1}$, as long as B_0 is more or less constant. Thus, the quantity

$$\tau_{spin} \equiv \int_{t_0}^{t} d P_{spin} \dot{P}_{spin}^{-1} = \frac{1}{2} P_{spin} \dot{P}_{spin}^{-1} \tag{9.8}$$

represents the so-called *spin-down* age of the pulsar, from the time t_0 of its birth, to the current time t. This characteristic time is a measure of how long it takes the pulsar to lose its rotational energy via magnetic-dipole braking. For the Crab, $\tau_{spin} \approx 1258$ yr. Given that Chinese astronomers observed the supernova explosion that produced the Crab pulsar in A.D. 1054, its actual age today is 953 yr, about 25% shorter than its spin-down age. Thus, although these two numbers don't match exactly, τ_{spin} is nonetheless still a reasonable estimate of the pulsar's age, and is often the only measure of time that we have for these objects since none of the others were observed historically.

In light of equation (9.6), the observed spin-modulated power of pulsars is actually a puzzle, since it accounts for only a tiny fraction of the expected emission. In the radio and optical bands, the observed luminosity is only about 10^{-7}–10^{-5} of $\dot{E}_{dip}$; it is roughly 10^{-4}–10^{-3} of $\dot{E}_{dip}$ in X-rays, and the fraction increases to about 10^{-2}–10^{-1} in γ-rays. This discrepancy is usually taken as indirect evidence that a significant fraction of the pulsar's rotational energy is carried off by a pulsar wind, constituting a mixture of relativistic particles and electromagnetic fields, which often produces a pulsar-wind nebula (PWN, also called a plerion or synchrotron nebula) radiating at radio, optical, and X-ray wavelengths (see figure 9.1).

All in all, one could say that we have a good idea of the pulsar's magnetic field, its magnetic-dipole braking rate $\dot{E}_{dip}$, and its spin-down age. But details are still lacking on the specific physical processes acting in the pulsar's magnetosphere that produce its broadband spectrum, from radio to γ-rays. As far as the high-energy emission is concerned, several ideas have been proposed, and each has its own merits. Future observations, e.g., with GLAST (see plate 4), will help to differentiate between the presently viable scenarios and further refine our understanding of these objects.

So to set the stage for future measurements, let us consider the possible sources of high-energy emission from these objects, which include both thermal and non-thermal processes. In at least one case (Geminga), both are clearly evident in the spectrum (see figure 9.6). Photospheric emission from the hot surface of a cooling neutron star produces a modified blackbody spectrum, extending from the optical through the soft X-ray range. Charged particles accelerated in the magnetosphere (see section 4.2) attain a power-law distribution, and their emission is correspondingly a power law, also beginning in the optical band, but in this case stretching all the way into the γ-ray domain. Some of these charged particles may bombard the star's polar caps and heat them sufficiently to radiate a soft thermal X-ray spectrum, which may be distinguished from the cooling component by the significantly smaller area. And on a larger scale, the PWN radiates some of the energy injected into the expanding plasma via synchrotron processes. In binary systems, the relativistic pulsar winds may collide with the companion star, or its mass outflux, and thereby also convert some of the directed (kinetic) energy into thermal energy that is eventually radiated.

Figure 9.7 shows the pulse profiles of the six spin-powered pulsars (including the Crab and Vela pulsars, and the enigmatic Geminga) detected by EGRET (see section 1.5). Their differential spectra are shown collectively in figure 9.8. These objects show no variability in their overall intensity, spectral behavior, or the shape of their pulse profile. Three of the γ-ray pulsars, Vela, Geminga, and PSR B1706–44, show a definite turnover in their photon spectra at energies ~1 GeV. Often, such a fall-off in flux suggests an upper limit to the particle acceleration, and is thus an indirect measure of the physical conditions (e.g., the magnetic field intensity) giving rise to such an effect.

The γ-ray pulse profiles of the EGRET pulsars are more complex than their radio counterparts. For example, the leading peak of the Vela pulse profile is very

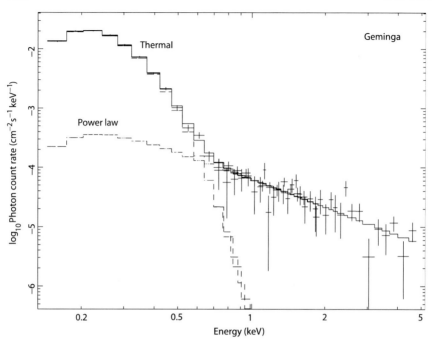

Figure 9.6 The X-ray spectrum of Geminga consists of a thermal component arising from the neutron-star surface, and a power-law contribution dominating toward higher energies (see also figure 9.8). The soft X-ray data are from ROSAT, whereas the harder emission was observed with ASCA. (Image from Halpern and Wang 1997)

sharp, requiring a very narrow beaming of the high-energy photons. Because of this, and the fact that the γ-ray pulses and radio pulses are generally out of phase, we interpret the high-energy emission from pulsars as originating in a different location (or locations) than the radio flux. The complexity in the pulse profile might also be an indication that the high-energy emission region is extended, perhaps throughout much of the magnetosphere.

Models for the high-energy emission in pulsars generally fall into one of two main camps: the *polar-cap* models, in which the emission zone is close to the polar cap, and the *outer-gap* models, in which the acceleration zone is close to the pulsar's light cylinder. The light cylinder is an imaginary surface with radius

$$R_L \equiv \frac{c}{P_{\text{spin}}/2\pi}, \tag{9.9}$$

defined to be where the azimuthal velocity of the co-rotating magnetic field reaches lightspeed. Beyond this region, the field lines break away from the magnetosphere and become traveling waves.

An important motivation for hypothesizing a source of γ-rays so distant from the stellar surface is that within the intense magnetic field close to the pulsar, γ-rays may interact with virtual photons in the magnetosphere and "materialize," i.e., they may produce an electron–positron pair, written as $\gamma + B \rightarrow e^+ + e^-$. This is

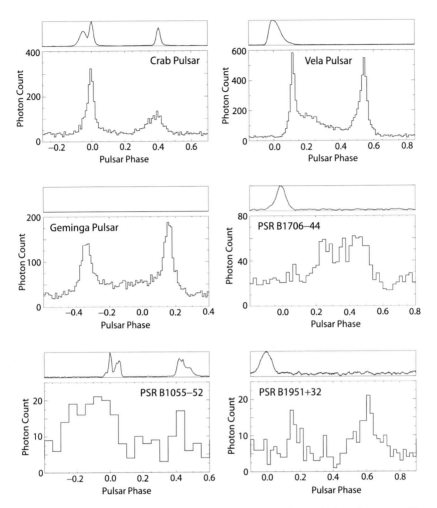

Figure 9.7 Of the many hundreds of known spin-powered pulsars, six have been unambiguously detected by EGRET. They show no strong evidence of variability in overall intensity, spectrum, or pulse profile. The γ-ray pulse profiles are generally more complex than their radio counterparts. Four of the γ-ray pulse profiles have two distinct peaks separated by 0.4–0.5 in pulsar phase. However, with the exception of the Crab pulsar, the γ-ray pulse profiles are not correlated with the radio profiles (shown here with arbitrary units on the vertical axis). Note also that the leading peak of the Vela γ-ray pulse profile has a FWHM of less than 0.03 in rotation phase, implying a very narrow beaming of high-energy photons. (Image courtesy of J. M. Fierro)

particularly relevant to very bright sources, such as the Crab and Vela pulsars, in which the efficiency for getting γ-rays out of the polar cap is small due to the large optical depth arising from this interaction. But since the dipole magnetic field drops off as r^{-3} (see equation 7.80), higher γ-ray emissivities are possible well away from the polar cap. In either case, the dominant photon emission mechanisms are

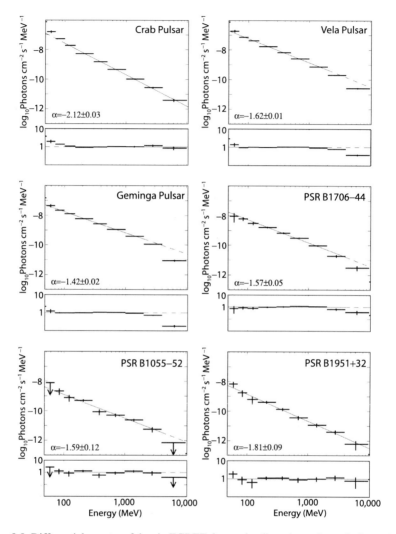

Figure 9.8 Differential spectra of the six EGRET-detected radio pulsars shown in figure 9.7. With the exception of the Crab pulsar, the spectra are very hard, with measured spectral indices ~ -1.4 to -1.8. Vela, Geminga, and PSR B1706–44 show a definite turnover in their photon spectra around ~ 1 GeV. (Image courtesy of J. M. Fierro)

synchrotron and curvature radiation and inverse Compton scattering of soft thermal X-ray photons emitted by the hot stellar surface. As we have already noted, future observations, e.g., with GLAST, will include measurements of the relative phases between peaks of the pulse profiles at different energies, and thereby distinguish between the polar-cap and outer-gap scenarios.

What is already clear, however, is that age plays a key role in determining a pulsar's high-energy profile. The Crab pulsar, archetypal member of the young

($\lesssim$5000 yr) class, produces a spectrum dominated by charged particles accelerated along curved field lines in its magnetosphere. It is a bright source in X-rays and γ-rays, as well as in radio and optical-UV. Young Crab-like pulsars are strong nonthermal X-ray emitters, with a luminosity $L_x \sim 10^{34}$–10^{36} ergs s^{-1} in the ROSAT energy range. X-ray images of the Crab remnant (see figure 9.1) also reveal a strong particle outflow, creating an inner nebula with a torus ($r \approx 0.38$ pc), an inner ring ($r \approx 0.14$ pc), and a jet and counterjet.

When pulsars reach a spin-down age $\sim 10^4$–10^5 yr, they develop characteristics like those of Vela. Though the latter is the only one detected in the optical band thus far, about ten members of this old pulsar class have been detected in X-rays, and four of them are also γ-ray emitters. Unlike the young sources, Vela-like pulsars have misaligned pulse profiles at different energies. They also have a soft—primarily thermal—X-ray spectrum with a temperature $\sim 10^6$ K, whereas the Crab's spectrum is nonthermal (as we have seen).

But there are also similarities between the young and old pulsars. For example, the Vela PWN resembles the inner region of the Crab nebula (see figure 9.1): it has an inner torus, a ring, and two jets. Very importantly, both the Crab and Vela nebulae have a symmetry axis, interpreted as the projection of the pulsar's angular momentum vector in the plane of the sky, co-aligned with the direction of proper motion. In these two objects, the neutron star must have received its *natal kick* along the rotation axis of the progenitor star. This may be a common occurrence in all pulsars.

Whereas the Vela-like pulsars are old enough for their magnetospheric emission to have subsided relative to the thermal stellar surface radiation, the middle-aged pulsars, such as Geminga, PSR B0656+14, and PSR B1055−52, display both non-thermal and thermal emission in their X-ray spectra (see figure 9.6). For these sources, the UV through soft X-ray spectral components are thermal, whereas the power-law component dominates toward very high energies.

However, by the time a pulsar has reached the spin-down age of $\sim 10^6$ yr, even its surface emission has subsided below easily detectable levels, since its temperature is then too low for the star to produce any significant X-ray luminosity. These old pulsars are visible as X-ray sources only at small distances from Earth. ROSAT and ASCA have detected the faint X-ray glow from only three such objects, PSR B1929+10, PSR B0950+08, and PSR B0823+26. Their spin-down ages all lie within the range 2–30 Myr, and they are located within $\sim$0.2–0.4 kpc of the Sun. These objects are so faint that if we interpret their X-ray luminosity as arising from the polar caps, the emission regions must be very small, each no bigger than about 100 m^2. Correspondingly, no γ-ray emission has ever been detected from them. A possible explanation for this is that the γ-radiation would normally be produced via inverse Compton scattering, and since the seed photons produced via thermal emission at the stellar surface are much fewer in number, the upscattered radiation is itself therefore necessarily of much lower intensity than one would observe in a young pulsar.

Before we leave this class of pulsars and move on to consider their closely related brethren—the X-ray pulsars—we ought to at least mention several other more recently uncovered objects that appear to have much in common with the radio sources we have been discussing here. In recent years, several apparently young neutron stars have been discovered with strong pulsations in X-rays (known as *anomalous X-ray pulsars*, or AXPs), and some others producing flashes of γ-rays, labeled *soft γ-ray repeaters* (SGRs). These objects are in some ways similar, in that their periods ($\sim$5–12 s) are much longer than those of typical radio pulsars. However, they do differ significantly in their γ-ray characteristics. Whereas an SGR γ-ray outburst can release as much as $\sim 10^{42}$–10^{44} ergs of energy per event, no γ-ray activity has ever been seen in AXPs.

Only a handful of AXPs are known, but they are regular targets of many X-ray observatories because of their elusive nature (see also figure 9.10 below). Though it is not clear whether they are powered by magnetic-dipole braking, one can nonetheless estimate their magnetic field strength according to equation (9.7) and, given a period of $\sim$5–12 s and a period evolution $\dot{P}_{spin} \approx (0.05 - 4) \times 10^{-11}$ s s^{-1}, one can estimate a spin-down age $\tau_{spin} \approx 3$–100 kyr, with a corresponding field $B \sim 10^{14}$–10^{16} G. If this estimate is correct, AXPs are clearly not typical radio pulsars. Instead, they appear to be members of a relatively new class of neutron stars known as *magnetars*—objects with superstrong magnetic fields.[11]

The label "magnetars" is even more apt for SGRs, of which four are now known. With periods of $\sim$5–8 s and a period derivative $\dot{P}_{spin} \sim 10^{-10}$ s s^{-1}, these sources appear to be young ($\sim$1–10 kyr) neutron stars dwelling within supernova remnants. They are not only very powerful transient sources of γ-rays, but also very bright quiescent X-ray sources with luminosities $L_x \sim 10^{34}$–10^{36} ergs s^{-1}.

Recent observations[12] of the AXP known as 1E 1048.1–5937, have revealed the presence of an expanding hydrogen shell centered on the pulsar, with a diameter of $\sim$30 pc (assuming a source distance of 2.7 kpc) and a velocity of $\sim$7.5 km s^{-1}. Interpreting this structure as a bubble blown out by the wind of a massive star, one infers a mass of 30–40$M_\odot$ for the source. Yet no such star is now visible in the AXP field of view. A natural conclusion is that the progenitor of 1E 1048.1–5937 must have been the culprit producing this shell, suggesting that magnetars originate from more massive exploding stars than do regular radio pulsars. In ways that we do not yet fully understand, this may explain why the birthrate of magnetars is much smaller than that of radio pulsars (since the mass function drops with increasing stellar mass), and perhaps why they are so strongly magnetized.

Finally, at the other end of the magnetic strength scale, we find the ms-pulsars—neutron stars spinning with a period of only several milliseconds. Distinguished from the majority of pulsars by their very short periods and small period derivatives, they have spin-down ages (equation 9.8) of 10^9–10^{10} yr and relatively weak magnetic fields $\sim 10^8$–10^{10} G (equation 9.7).

A clue into how these anomalies of nature may have been produced is provided by the fact that over three-fourths of the known ms-pulsars in the galactic disk are members of binaries with a compact companion star, whereas only one percent or so of ordinary pulsars may be found coupled in this fashion. In a contact binary (see

---

[11]The original idea for the existence of such objects was proposed by Thompson and Duncan (1995).
[12]See Gaensler et al. (2005).

section 6.3), mass is transferred from the main sequence companion to its compact partner, a process that transfers not only matter, but also angular momentum, as we noted in our discussion of Roche lobe overflow. Though these neutron stars are not manifested as X-ray pulsars now (see below), they apparently underwent a phase of accretion in the past that spun them up.[13] Since the common view is that these pulsars would have turned off by now because of the loss of rotational energy over their long life, they are often called "recycled" pulsars.

Over half of the roughly 100 known recycled radio pulsars reside in the galactic plane. The rest are in globular clusters, which facilitate the process of recycling. In cases where the ms-pulsar is solitary, it is believed that its erstwhile companion must have evaporated completely or was tidally disrupted after the ms-pulsar formed.

Thanks to prolonged observational campaigns with high-energy satellites, we now understand radio pulsars much better than we did in the past. Several hundred (including the X-ray pulsars below) of the almost two thousand such objects known have been detected with X-ray and/or γ-ray instruments. With their strong gravitational and magnetic fields, these neutron stars provide a rare opportunity for us to study the behavior of matter and radiation under truly extreme physical conditions not attainable on Earth. They are unique laboratories where nuclear physics, general relativity, and relativistic electrodynamics all play a role in shaping the electromagnetic signals they send us.

9.2 X-RAY PULSARS

Accretion-powered X-ray pulsars were among the first sources observed in X-ray astronomy, and continue to be some of the most widely studied objects in (or near) the Galaxy. The current census lists about 100 such sources, 28 of which are located in the Magellanic Clouds. Three of them have been identified in other galaxies, M31 and M33.[14]

A sample of this class is shown in figure 9.9, along with closely related binary sources that produce X-ray bursts, which we will study in section 11.1. Thanks to their characteristic timing signatures, these objects were recognized from the beginning as rotating, magnetized neutron stars, though clearly distinct from radio pulsars. Some of the more prominent members of this group are identified by name in figure 9.9, including Hercules X-1 (or Her X-1), which we introduced in the previous section as the first neutron-star source yielding compelling evidence for the presence of an intense magnetic field.

The term *X-ray pulsar* was originally used to designate objects powered by accretion of matter from a companion star in an interacting binary, as we discussed briefly in section 7.3.2. The intent was to distinguish between such sources and the broader class of radio pulsars, which, as we have seen, tend to be isolated and powered

---

[13]Many authors have written on this subject. For a sampling, see Bisnovatyi-Kogan and Komberg (1974) and Alpar et al. (1982).

[14]See, e.g., Dubus et al. (1999).

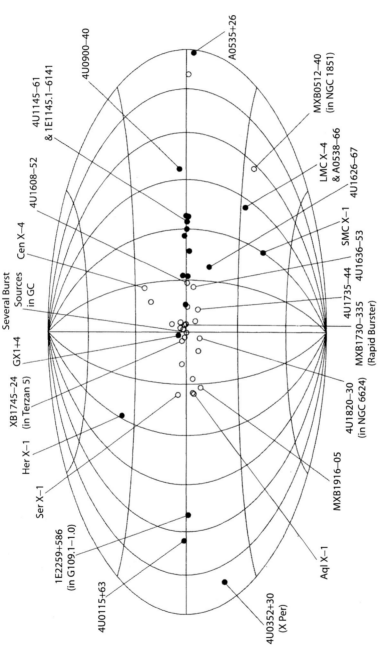

Figure 9.9 Diagram showing the positions of 21 X-ray pulsars and 27 X-ray burst sources (see also figure 11.1) in galactic coordinates. Some of the more prominent members of each class are identified by name. The former are rotating neutron stars accreting matter onto their polar caps, whereas the latter are neutron stars burning hydrogen and helium accumulated across their entire surface. Note that X-ray pulsars tend to be distributed primarily along the galactic equator, whereas X-ray burst sources are concentrated toward the center of the Galaxy. (Image from Joss and Rappaport 1984)

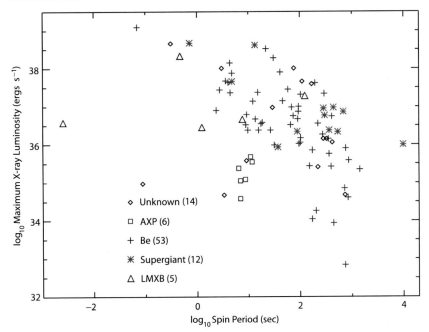

Figure 9.10 Accretion-powered X-ray pulsars were among the first sources observed in X-rays. These objects are powered by accretion of matter, funneled onto their magnetic polar caps (see figure 7.10), from a companion star in a close, interacting binary. Of the 100 or so such objects now known, 28 are located in the Magellanic Clouds. The observed spin periods range from 0.0025 s to about 3 h, though most lie between ∼1 and 1000 s. Their subdivision into separate classes is based on the spectral type classification of the mass donor companion star. Most X-ray pulsars have massive companions, either OB supergiants or Be stars. Only a few of the neutron stars in low mass X-ray binaries (LMXBs) have exhibited measurable pulsations. The anomalous X-ray pulsars we considered in section 9.1 are characterized by soft X-ray spectra, clearly different from those of the other pulsars. (Image from Mereghetti 2001)

by the loss of rotational energy. But our discussion in the previous section clearly demonstrated that the sensitivity of modern instruments permits us to see X-ray pulsations in radio pulsars as well. So the old terminology is not well suited to current observational trends. Instead, it is more appropriate to distinguish the two classes of objects by referring to them as "accretion-powered" and "rotation-powered" pulsars.

The observed spin periods of these sources tend to fall in the range 2.5 ms to about 3 h, though most of them are between ∼1 and 1000 s. These are plotted in figure 9.10, along with their maximum observed X-ray luminosity. Observational selection effects probably affect the discovery rate for periods less than 1 s and greater than 1000 s. Nevertheless, the pulsar distribution is consistent with a constant number per logarithmic period interval, undoubtedly reflecting the evolution of a typical member of this class.

X-ray (or more specifically, accretion-powered) pulsars generally fall into two dominant subclasses based on the companion's spectral type. Most of them have massive binary partners, either OB supergiants or Be stars. The more numerous

X-ray sources in low-mass binary systems (LMXBs)—those in which the companion is typically smaller than the Sun (see figure 9.13)—display many interesting high-energy phenomena (including X-ray bursts, as we shall see shortly), but generally lack periodic coherent signals indicative of a neutron-star spin. As such, the LMXB membership of accretion-powered pulsars is greatly underrepresented (see figure 9.10). One possible explanation for this phenomenon is that neutron stars in LMXBs are weakly magnetized and therefore accrete over the whole stellar surface, mitigating any possible pulsed emission associated with rotation. After all, as we saw in the previous section, LMXBs are believed to be the progenitors of ms-pulsars, whose surface magnetic field is $\sim 10^8$–10^{10} G. Equation (7.88) indicates that at this intensity, the magnetic field would be far too weak to channel the accreting plasma onto the polar caps of the neutron star.

A sample of pulse profiles is shown in figure 9.11. The pulses have large duty cycles and a modulation factor of $\sim$20–90%. The fact that $\sim$50% duty cycles

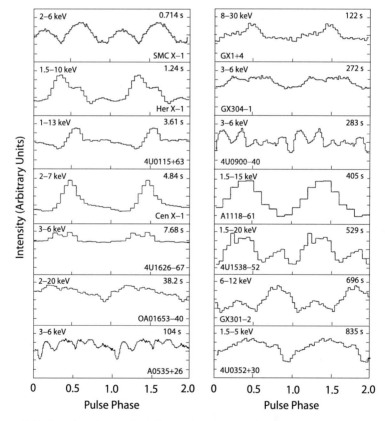

Figure 9.11 Sample pulse profiles for 14 of the known binary X-ray pulsars. In each case, the data are folded modulo the inferred pulse period and plotted against pulse phase for two complete cycles. The profiles are characterized by large duty cycles of over 50% (compared with only $\sim$3% in radio pulsars), modulation factors of $\sim$20–90%, and no obvious trend in pulse morphology as a function of pulse period. (Image from Joss and Rappaport 1984)

are common in accretion-powered pulsars, whereas this number drops to ∼3% in the case of radio pulsars, means that the emission is significantly less beamed in the former than in the latter. It is widely believed that the pulsations result from the anisotropic geometry associated with polar-cap accretion, as indicated in figure 7.10. A misalignment of the dipole and spin axes produces a periodically variable aspect of the emission region.

In the case of LMXBs, the accretion is due to Roche lobe overflow. This leads to the formation of a disk and subsequent channeling along the magnetic field lines onto the polar cap regions. On the other hand, when the companion is a massive star, not filling its Roche lobe, accretion results from the gravitational capture (section 6.2) of its wind.

Compression of the plasma falling onto the polar caps converts gravitational energy into heat, which is then radiated away by a combination of bremsstrahlung and synchrotron processes. The X-rays must transfer through the magnetized, optically thick medium in the accretion column, which some simulations suggest is unstable, possibly leading to the phenomenon of cooling via the evaporation of "photon bubbles" (see section 7.3.2).

The spectra of X-ray pulsars (figure 9.12) have no sharp features, emblematic of the blending we believe is taking place via a superposition of emission components at different heights within the funnel. Most of the power is emitted in the range ∼2–20 keV, with a rapid falloff in flux above ∼20 keV, reminiscent of the bremsstrahlung shoulder we discussed in section 5.2. We therefore interpret this to mean that the maximum temperature attained by the shocked gas once it reaches the stellar surface is of this order.

In section 10.1, we will attempt to prove that certain X-ray binaries actually contain a black-hole accretor rather than a neutron star. The most common argument made to lay the foundation for such an assessment is based on a measurement of the compact object's mass. As we shall see, there exists a straightforward method of "weighing" the X-ray source once we constrain the binary parameters, including the orbital period, the projected semi-major axis of the compact star, the velocity of the companion and, if present, the duration of the X-ray eclipse, which provides us with the companion's size.

Figure 9.13 shows the result of applying this technique to the case of 4U1626−67, an X-ray pulsar in a highly compact binary system, whose companion is a ∼0.1 $M_\odot$ late-type dwarf. With a binary separation of only ∼1 light-second, this entire system would easily fit within the Sun! As we shall see, this binary model also applies to typical galactic bulge X-ray sources, particularly the X-ray bursters (section 11.1).

9.3 CATACLYSMIC VARIABLES

Now imagine taking an accretion-powered pulsar and replacing the neutron star with a white dwarf. To be sure, there are important differences between the two, including (possibly) the star's mass, radius, and magnetic field strength. But binaries with an accreting, magnetic white dwarf do exist and are known as magnetic Cataclysmic

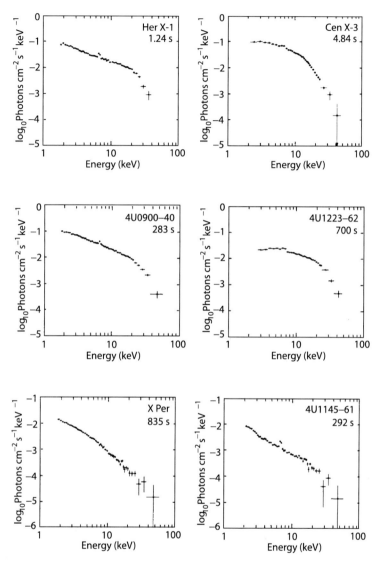

Figure 9.12 Sample X-ray spectra, averaged over pulse phase, for six representative X-ray pulsars. The broadband emission is typically devoid of prominent features, aside from iron K-shell emission near 7 keV. Notice also the rapid falloff in flux above ~20 keV. (Image from White et al. 1983)

Variables (mCVs); we introduced these systems in section 7.3.1 to discuss the physics of accretion columns.

We learned earlier that the magnetic field in mCVs funnels the accretion flow near the star, producing an accretion shock above the polar cap, and pulsed optical and X-ray emission. Figure 9.14 shows the modulated X-ray and UV lightcurves for two

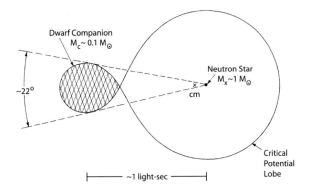

Figure 9.13 X-ray pulsars typically have massive companion stars, with $M_c \sim 20\ M_\odot$ and radii $R_c \sim 20\ R_\odot$. However, at least one X-ray pulsar, 4U1626–67, is in a highly compact binary system, as shown here. The center of mass is indicated by the x-mark. The companion (KZ TrA) is a late-type dwarf ($\sim 0.1\ M_\odot$) whose light results primarily from reprocessed X-radiation. The size of the orbit indicated in this figure is based on a measurement of the orbital period and the lack of variable Doppler delays (which limits the binary's inclination relative to the line of sight). This model also applies to many of the galactic bulge X-ray sources, such as the X-ray bursters (see section 11.1). (Image from Joss and Rappaport 1984)

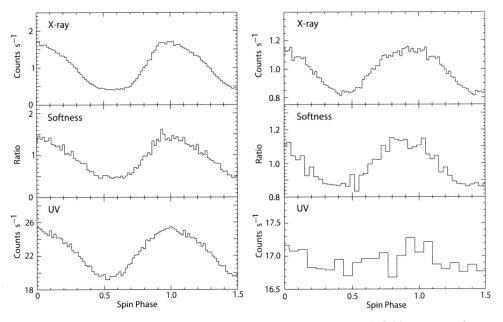

Figure 9.14 In an intermediate polar (IP), the white dwarf has a magnetic field strong enough to influence the accretion flow, but not strong enough to completely disrupt the accretion disk and to synchronize the spin of the white dwarf to the binary period. AO Psc (left panels) and V1223 Sgr (right panels) exhibit sinusoidal X-ray and UV modulations. The former correspond to the 0.2- to 12-keV band observed with XMM-*Newton*, whereas the latter represent the 2050- to 2450-Å. The middle panels show the $(0.2 - 4)/(6 - 12)$-keV softness ratio. (Image from Evans and Hellier 2005)

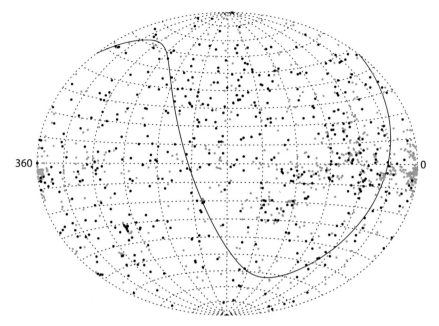

Figure 9.15 Distribution of the ∼500 known Cataclysmic Variables in galactic coordinates. The galactic center wraps around at the right/left edge of the plot, and the solid line indicates declination $\delta = 0$. Cataclysmic Variables with measured orbital periods are shown as black dots; those without a known period are shown as gray dots. (Image from Gänsicke 2005)

such systems, based on recent XMM-*Newton* observations. Although the modulation is energy-dependent, the pulsational period is independent of wavelength, indicating that all of these components are produced within the same region. As noted in section 7.3.1, the overall spectrum from these sources is due to a combination of bremsstrahlung and cyclotron emission within the funnel and in the stellar surface immediately adjacent to the polar caps.

However, not all Cataclysmic Variables (CVs) are magnetic, though at least one-third of white dwarfs in binaries do have a measurable magnetic field, compared with only ∼2 % of those that are isolated.[15] Binary membership clearly has something to do with a white dwarf's field strength, possibly through the action of repeated nova eruptions, which may uncover submerged field lines below the stellar surface. The distribution of known CVs (shown in figure 9.15) is rather isotropic. These sources are relatively faint, so we tend to see predominantly those near Earth; many of these objects are no more than a few hundred parsecs away.

Though the rate of discovery of CVs was at first rather slow, already some 27 systems with measured orbital periods were known by the late 1970s. It was noted even then that the sample showed a strong bias toward short-period binaries and a glaring deficiency of CVs in the orbital period range ∼1.5–3.25 h. But the ROSAT survey would soon produce a large increase in the number of known mCVs, permitting a

---

[15]See, e.g., Angel, Borra, and Landstreet (1981), and Ritter (1984).

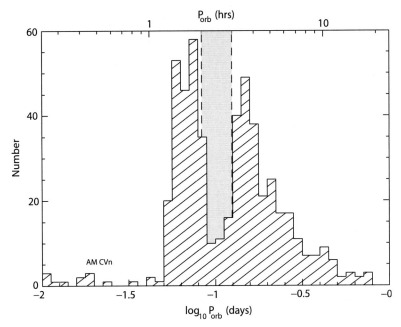

Figure 9.16 The period distribution of the Cataclysmic Variables shown in figure 9.14 is characterized by a sharp cutoff at ~80 min, a deficit of systems in the 2- to 3-h gap (shown here as a gray vertical swath), and a steady drop-off in numbers toward longer orbital periods. Systems with periods shorter than 80 min (El Psc, V485 Cen, and the AM CVn stars) have unusual donor companions. (Image from Gänsicke 2005)

better statistical analysis of the period distribution.[16] A histogram created from a more recent catalog is shown in figure 9.16, where a rather sharp cutoff in the number of sources emerges at ~80 min, in addition to the relative paucity of objects in (what is now known as) the 2- to 3-h gap. The fact that systems crowd toward smaller periods suggests that the orbital period distribution of CVs must be closely related to their evolution, i.e., to the rate at which they lose orbital angular momentum.

The mCVs divide naturally into two subclasses: the DQ Herculis (or DQ Her) stars (named after the archetypal binary DQ Herculis)[17] and the AM Herculis (or AM Her) stars (for which the archetype, not surprisingly, is AM Herculis).[18] Of the roughly 100 mCVs known, about two-thirds are members of the AM Herculis class, shown as a function of period in figure 9.17. DQ Her binaries show evidence for the presence of an accretion disk, and have a white dwarf spin period P_{spin} much smaller than their orbital period P_{orb}, suggesting that the white dwarf has a small magnetic field and is not strongly coupled to the binary. Indirect estimates of their field strengths fall in the range ~5–30 MG.[19] AM Her binaries, on the other hand,

[16] See Voges et al. (1999).
[17] See, e.g., Warner (1985).
[18] See, e.g., Liebert and Stockman (1985).
[19] See Wickramasinghe and Ferrario (2000) for a review of these objects.

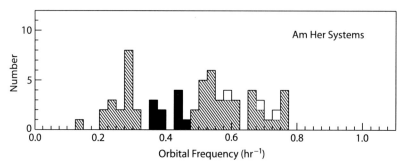

Figure 9.17 A subset of the CVs contain a magnetized accretor and are known either as AM Her binaries (with the white dwarf spin synchronized to the binary period) or DQ Her binaries (in which the white dwarf is asynchronous). About two-thirds of the roughly 100 magnetic CVs (or mCVs) fall into the first category, and are characterized by dipolar magnetic fields $\sim$7–200 MG; the DQ Her systems have weaker magnetic fields, typically in the range 5–30 MG. This figure shows the distribution of AM Her binaries as a function of (synchronized) period. The filled portions of the histogram correspond to systems with confirmed orbital periods, and the black portions highlight objects in the canonical 2- to 3-h period gap (see figure 9.16). (Adapted from Webbink and Wickramasinghe 2002)

contain a white dwarf synchronized to the binary ($P_{\rm spin} = P_{\rm orb}$), no accretion disk, and larger white dwarf magnetic fields ($\sim$7–200 MG).[20]

The known DQ Her systems typically have orbital periods $P_{\rm orb} \gtrsim 3$ h. Their degenerate dwarf is believed to have been spun up to a short rotation period $P_{\rm spin}$ by the accretion torque, reminding us of the manner in which the ms-pulsars are believed to have been produced (see section 9.1). The observed modulation of their lightcurve at $P_{\rm spin}$ is due partly to emission directly from the degenerate dwarf and partly to reprocessing of the pulsed X-ray emission.

In contrast, the known AM Her systems typically have $P_{\rm orb} \lesssim 3$ h (see figure 9.17). The magnetic dipole moment m_1 of the degenerate dwarf (see equation 7.80) is sufficiently strong to render R_m larger than the size of the primary's Roche lobe, so no steady disk can exist inside the inner Lagrangian point L_1. This m_1 is also large enough to couple the degenerate dwarf to its companion and force synchronization between $P_{\rm spin}$ and $P_{\rm orb}$, in spite of the accretion torque. The optical light from these systems is strongly polarized at the level of $\gtrsim$10%.

While the above properties adequately characterize the known DQ Her and AM Her binaries, theory indicates that systems can exist that have some properties of each. Therefore, a more appropriate definition of these subclasses is that DQ Her binaries are mCVs with *asynchronously* rotating degenerate dwarfs, while AM Her systems are those with *synchronously* rotating degenerate dwarfs (see figure 9.18).

It is believed that DQ Hers evolve into AM Hers. According to this hypothesis, DQ Her and AM Her systems are intrinsically similar but are observed at different evolutionary epochs. This picture is attractive for several reasons. First, as a close

---

[20]Magnetic fields are measured from Zeeman and cyclotron spectroscopy, providing estimates either of the effective dipolar photospheric field or of magnetic fields at accretion spots on the white dwarf surface.

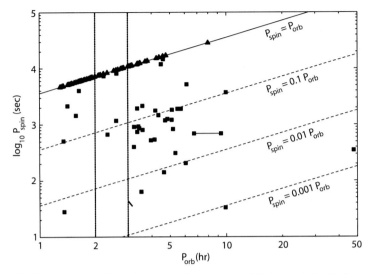

Figure 9.18 The spin period P_{spin} (in seconds) versus the orbital period P_{orb} (in hours) of the mCVs. AM Her binaries (also known as polars) are indicated by triangles, and DQ Her binaries (also known as intermediate polars) are represented by squares. Synchronization occurs when the magnetic coupling between the two stars in the binary overwhelms the accretion torque and slows down the white dwarf. The termination line on this plot corresponds to a situation in which the two periods are equal, and all the DQ Her binaries fall below this line because they have a spin period shorter than their binary period. The four "nearly synchronous" DQ Her binaries (with $0.9 > P_{spin}/P_{orb} > 0.7$) are V697 Sco, HS0922+1333, RX J0425+42, and V381 Vel; the six "rapid rotators" (with $P_{spin}/P_{orb} < 0.01$) are WZ Sge, AE Aqr, GK Per, V533 Her, DQ Her, and XY Ari; the six conventional DQ Her binaries within or below the period gap are DD Cir, HT Cam, V795 Her, RX J1039.7–0507, V1025 Cen, and EX Hya. The remaining 23 DQ Her binaries above the period gap are V709 Cas, 1RXSJ154814.5–452845, 2236+0052, V405 Aur, YY Dra, PQ Gem, V1223 Sgr, AO Psc, HZ Pup, UU Col, 1RXSJ062518.2+733433, FO Aqr, V2400 Oph, BG CMi, TX Col, 1WGA1958.2+3232, TV Col, AP Cru, V1062 Tau, LS Peg, RR Cha, RX J0944.5+0357, and V1425 Aql. (Image from Norton et al. 2004)

binary evolves, its orbital period P_{orb} decreases (to a minimum period P_{orb}^{min}, which for CVs is apparently ~80 min; see figure 9.16). Thus, the known DQ Her systems with periods $P_{orb} \gtrsim 3$ h will eventually have periods $P_{orb} \lesssim 3$ h, like those of the known AM Her binaries. Second, the Roche lobe of the degenerate dwarf shrinks as P_{orb} decreases, and therefore the size of the disk also decreases. At the same time, the Alfvén (or magnetic) radius R_m (equation 7.88) for disk accretion increases because the mass transfer rate decreases. Eventually, these two radii cross, and the disk must disappear. Third, magnetic coupling (of almost any kind) between the magnetic degenerate dwarf and the (magnetic) secondary increases rapidly as the binary separation a (equation 6.77) and the mass transfer rate $\dot{M}$ both decrease with decreasing P_{orb}, so that synchronization becomes more likely.

Let us now continue our discussion from section 6.3 and try to understand, at least qualitatively, how close binary evolution may proceed. For binary periods

$P_{orb} \gtrsim 10\,h$, secondaries with a mass $\lesssim 1.2\,M_\odot$ must be significantly evolved in order to fill their Roche lobe (Equation 6.98). It is therefore believed that such binaries are driven by nuclear evolution of the secondary.[21] Assuming conservative mass transfer (in which $\dot{M}_1 = -\dot{M}_2$), we infer that $\dot{P}_{orb} > 0$ in these systems since $q < 1$ (where $q \equiv M_2/M_1$). The binaries AE Aqr (with $P_{orb} = 9.9\,h$) and GK Per (with $P_{orb} = 46\,h$) are thought to be evolving due to this mechanism.

Binaries with orbital periods $P_{orb} \lesssim 10\,h$ contain a secondary star whose nuclear evolution proceeds too slowly to drive mass transfer at the rates observed.[22] These binaries are therefore believed to evolve due to the loss of orbital angular momentum,

$$\dot{L} = \dot{L}_1 + \dot{L}_2 + \dot{L}_{orb}, \tag{9.10}$$

where $\dot{L}_1$ and $\dot{L}_2$ are the rates of change of the spin angular momenta of the degenerate dwarf and the secondary, respectively, and $\dot{L}_{orb}$ is the rate of change of the orbital angular momentum of the binary.

Gravitational radiation drives the evolution of the binary by extracting orbital angular momentum.[23] Simulations shows that this mechanism gives a minimum orbital period of about 1.2 h, corresponding to the transition of the secondary[24] from the main sequence to the degenerate sequence, and therefore successfully accounts for the observed cutoff seen at $P_{orb} \approx 1.3\,h$ (see figure 9.16). However, evolution due to gravitational wave emission gives mass transfer rates as large as those observed only for $P_{orb} \lesssim 3\,h$.

Consequently, other mechanisms that could drive binary evolution effectively at intermediate binary periods ($3\,h \lesssim P_{orb} \lesssim 10\,h$) have been investigated. One possibility receiving a great deal of attention over the years is magnetic braking,[25] in which the strongly magnetized secondary star couples to its stellar wind, losing angular momentum in its efflux ($\dot{L}_{MB} = \dot{L}_2$). Since the secondary is largely convective, tidal coupling between it and the binary system is very effective.[26] This keeps the secondary synchronized, so that the angular momentum is lost from the binary system itself ($\dot{L}_{MB} = \dot{L}_{orb}$). For large but conceivable values of the companion's magnetic field ($B_2 \sim 300\text{--}1000\,G$) and stellar wind ($\dot{M}_{wind} \sim 10^{-10}M_\odot\,yr^{-1}$), magnetic braking can produce mass transfer rates as large as those observed even well beyond a period of 3 h.

The gap in the orbital period distribution shown in figures 9.16 and 9.17 may result from the fact that the systems have a very low mass transfer rate between periods of 2 and 3 h. This situation would be realized if the binary's evolution were driven by a torque $\dot{L}_{add}$ stronger than that due to gravity, as long as its influence were to cease suddenly at $P_{orb} \approx 3\,h$. In this picture, a gap occurs because the the secondary star becomes bloated when driven out of thermal equilibrium, i.e., when

---

[21] See, e.g., Webbink, Rappaport, and Savonije (1983).

[22] This is consistent with observational evidence that these secondary stars are not significantly evolved (Patterson 1984).

[23] See Kraft, Mathews, and Greenstein (1962).

[24] Main sequence stars get smaller as they lose mass, but degenerate stars get bigger. Thus, as the secondary star becomes degenerate, it remains in contact with the Roche lobe even though P_{orb} and the binary separation a are both increasing after the turn-around at minimum period.

[25] Early investigators of this idea were Verbunt and Zwaan (1981) and Taam (1983).

[26] Some estimates place the tidal timescale under these circumstances at only $\sim 100\,y$; see Zahn (1977).

$\tau_{\rm evol} < \tau_{\rm KH}$, where $\tau_{\rm evol} \equiv |L|/|\dot{L}|(\approx \tau_{\rm add})$ is the evolutionary timescale, and $\tau_{\rm KH}$ is the Kelvin-Helmholz thermal timescale (the characteristic time for the star to lose its internal energy). Then, when the additional torque ceases ($\dot{L}_{\rm add} \to 0$), the secondary star shrinks back to its main sequence radius and comes out of contact, ending mass transfer. The system continues to evolve to shorter binary periods and smaller separations, but now on the much longer timescale $\tau_{\rm grav}(\gg \tau_{\rm add})$. The secondary comes back into contact when the radius of its Roche lobe equals the main sequence radius of the star, and mass transfer is re-established, but at the much lower rate given by evolution due solely to the emission of gravitational waves.

This scenario seems to be viable because the secondary becomes fully convective at $P_b \approx 3$ h.[27] That means that magnetic braking may be disrupted at this point, since the magnetic field of the secondary loses its radiative anchor and, being buoyant, is expelled from the star. In this case, $\dot{L}_{\rm add} = \dot{L}_{\rm MB}$.[28]

In this picture, whether or not a gap arises, whether or not a disk is permitted to exist (subject to the magnitude of the Alfvén radius in equation 7.88), and whether or not the accreting white dwarf synchronizes to the binary period (i.e., $P_{\rm spin} \to P_{\rm orb}$), all depend on the strength of the primary's magnetic field and how it evolves with the binary's evolution. In terms of the magnetic moment

$$m_1 = \tfrac{1}{2} B_0 R_{\rm wd}^3, \qquad (9.11)$$

where B_0 is the magnetic field at the polar cap, and $R_{\rm wd}$ is the white dwarf's radius (see equation 7.80), four regimes have been identified for DQ Her/AM Her evolution: (1) When $m_1 \lesssim 10^{31}$ G cm^3, the magnetic field of the degenerate dwarf is unable to funnel the accretion flow; these systems are not DQ Her binaries and may show little or no evidence of a magnetic field. This situation is shown schematically in the top panel of figure 9.19. (2) Systems with 10^{31} G cm$^3 \lesssim m_1 \lesssim 10^{33}$ G cm^3 are always DQ Her binaries. (3) Systems with 10^{33} G cm$^3 \lesssim m_1 \lesssim 10^{35}$ G cm^3 are DQ Her binaries that evolve into AM Her binaries, assuming m_1 is constant throughout their evolution (an assumption that may not be valid). Situations (2) and (3) are illustrated in the middle panel of figure 9.19. (4) Systems with 10^{35} G cm$^3 \lesssim m_1$ are always AM Her binaries. This is an extreme situation in which the white dwarf's magnetic field not only funnels the accreting plasma onto its polar caps, but at the same time completely disrupts the accretion disk so that matter crossing the inner Lagrangian point at L_1 is immediately captured by the dipole field of the primary.

In the context of DQ Her to AM Her evolution, m_1 probably increases as the binary evolves because the DQ Her and AM Her systems are segregated in period, with the DQ Hers tending to lie longward of the gap, while AM Hers dominate toward smaller periods. But in recent years, some mCVs have started to crop up *within* the gap itself, calling into question the view of a simple evolution through it. Indeed, a quick inspection of figures 9.16 and 9.17 shows that the gap is actually less pronounced for the AM Her binaries than for CVs as a whole. However, an interesting idea has been

---

[27] See, e.g., Robinson et al. (1981).
[28] This suggestion was first proposed by Rappaport, Verbunt, and Joss (1983).

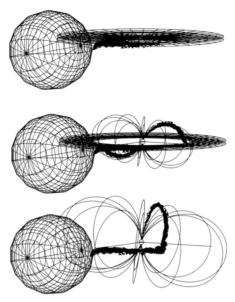

Figure 9.19 There are two main classes of magnetic Cataclysmic Variables (mCVs): the AM Her systems (also known as polars) and the DQ Her binaries (also known as intermediate polars). In the former, the magnetic field of the white dwarf is strong enough to lock the whole system into synchronous rotation. It also prevents the formation of an accretion disk and funnels the inflowing plasma directly onto the white dwarf's polar caps (as seen in the bottom diagram). In intermediate polars, the magnetic field is weaker, but is still capable of disrupting the inner edge of the disk. From there, the captured gas flows along the field lines toward the white dwarf's polar caps (middle diagram). In nonmagnetic CVs, on the other hand, a well-defined disk forms and the accretion is mediated primarily through the equatorial plane of the white dwarf (top diagram). (Image from Wu et al. 2003)

proposed to not only account for these unusual mCVs appearing where the system should be dormant, but also to indirectly affirm the viability of the magnetic braking picture of DQ Her to AM Her evolution. AM Her binaries with a highly magnetized white dwarf can trap the secondary's wind and inhibit the loss of angular momentum from the system.[29] For very strong magnetic moments ($m_1 \gtrsim 4 \times 10^{34}$ G cm^{-3}), magnetic braking is attenuated altogether and mass transfer is driven primarily by gravitational radiation. In that case, the secondary never comes out of contact with its Roche lobe, and the system continues to evolve continuously through the period gap.

The fact that the strong magnetic field of the degenerate dwarf dramatically alters the optical, UV, and X-ray appearance of mCVs has been known for a long time. That it can also alter the evolution of the binary itself has come to light only in more recent times. Spin-up and spin-down of the magnetic degenerate

[29] See Webbink and Wickramasinghe (2002).

dwarf temporarily speeds up and slows down the evolution of DQ Her binaries.[30] Cooperative magnetic braking speeds up the evolution of AM Her binaries when the degenerate dwarf has a particularly strong ($B_1 \gtrsim 5 \times 10^7$ G) magnetic field. Under some circumstances, synchronization of the magnetic degenerate dwarf may even inject angular momentum into the binary, driving the system apart and producing a *synchronization-induced period gap*, yielding an entirely new way of producing ultrashort period binaries.

SUGGESTED READING

Read about the discovery of pulsars from both a personal account and a scientific perspective in Bell (1977) and Hewish et al. (1968).

A recent review on radio pulsar statistics, including their galactic distribution and birth rate, their period evolution, and the phenomenon of isolated recycled pulsars, may be found in Lorimer (2005).

For a more historical perspective on radio pulsars, see also Taylor and Stinebring (1986) and Phinney and Kulkarni (1994).

Arguably the least understood category of young neutron stars is the growing class of anomalous X-ray pulsars. A review of recent observational progress on these objects may be found in Kaspi (2007).

A review of X-ray pulsars and other X-ray binary sources by the world's leading experts may be found in Lewin et al. (1997).

Evidence continues to grow that some neutron stars possess very strong magnetic fields (the so-called magnetars) above the quantum critical strength at which the cyclotron energy equals the electron rest mass. For a comprehensive review of the physical processes associated with such objects, see Harding and Lai (2006).

Magnetic Cataclysmic Variables are reviewed in Wickramasinghe and Ferrario (2000) and Wu et al. (2003).

And for a more in-depth examination of the binary evolution in Cataclysmic Variables, see Lamb and Melia (1987).

---

[30]See Ritter (1984) and Lamb and Melia (1987).

Chapter Ten

Black Holes in Binaries

After the unexpected discovery of radio pulsars in 1967, the identification of neutron stars and cataclysmic variables proceeded rather quickly, in part because their spectra, luminosity, and variability compared favorably with established (or developing) theory. By comparison, the gradual uncloaking of black holes has been much more difficult because the distinction between black-hole sources versus other types of compact objects is based primarily on what is *not* there, rather than what can be measured directly. For example, the absence of a boundary layer (section 7.2.3) might be an indication that the accretor does not have a hard surface; it could also be due to the fact that the disk is disrupted before it reaches the star. Therefore, relying directly on spectral information has not been a very trustworthy method of ascertaining whether a compact high-energy source is in fact a black hole. Instead, the process of black-hole identification has involved several observations, at different wavelengths, and careful sleuthing that incorporates different types of analysis, including the secondary's motion within the binary.

After decades of discovery, we now know that X-ray binaries generally come in two varieties: the high-mass (HMXBs) sources, in which the companion star is an O-B supergiant donor, and the low-mass (LMXBs) systems that we have already surveyed in previous chapters. The LMXBs typically have short orbital periods ($\lesssim 10\,\mathrm{h}$) and K-M donor stars. An important observational difference between these two categories is that the optical flux in LMXBs is triggered by reprocessing of the X-rays irradiating the accretion disk around the compact object, whereas it is dominated by the hot supergiant star itself in HMXBs. This distinction has an overriding influence on how one obtains critical system parameters, such as the binary separation and orbital period.

As fate would have it, the first strong evidence for the existence of black holes came from X-ray and optical observations of the X-ray binary Cygnus X-1, containing the 9th-magnitude supergiant star HD 226868.[1] An artist's view of the Cygnus X-1 binary system is shown in plate 14, highlighting the blue supergiant nature of the companion star and the hidden X-ray source accreting via an accretion disk. As we shall see shortly, however, the current crop of black-hole systems transcends binary type and includes representatives not only from LMXBs and HMXBs, but possibly also from what some are now calling "intermediate"-mass X-ray binaries.

Today, a total of about 20 binary systems are known to contain a compact object too massive to be a neutron star or a degenerate dwarf of any kind, with a mass $M_{bh} > 3\,M_\odot$. In addition, there are some 20 others for which the black-hole identification is

---

[1] See Webster and Murdin (1972) and Bolton (1972).

not as strong, though spectral clues suggest that at least some of them are probably black holes too. But before we move too far ahead of ourselves and begin our discussion of black-hole binaries in earnest, let us first take a moment to examine why certain X-ray sources are perceived to contain black holes, and why we believe that this identification is not likely to change with any new set of observations in the future.

10.1 EARLY DISCOVERY OF BLACK HOLES

Early observations of the HMXB Cygnus X-1 showed that the supergiant HD 226868 moves with a velocity $\approx$75 km s^{-1}, in a 5.6-day orbit about an unseen companion. Through careful analysis, one can show that for this particular binary, the mass of the compact object must therefore be at least 4 $M_\odot$. After we consider how this mass is measured, we shall see why this limit is of such critical importance.

The orbital motion of the two stars in this system is the most reliable way of "weighing" them. Let us put ourselves in the co-rotating frame, and identify the center-of-mass (CM) of the binary, which allows us to write

$$a = a_C + a_X \tag{10.1}$$

for the binary separation. Here a_C and a_X are, respectively, the distance from the CM to the companion and the distance from the CM to the X-ray source. By definition of the CM,

$$M_C a_C = M_X a_X. \tag{10.2}$$

(We will not yet use the label M_{bh} in order to keep this derivation as general as possible. Only later, when we realize that the inferred value of M_X is too high for it to be a neutron star or white dwarf, will we use the more specific designation M_{bh}.)

Doppler shift measurements of the companion's lightcurve yield the projected orbital speed

$$v_C = \frac{2\pi}{P_{orb}} a_C \sin i, \tag{10.3}$$

where i is the inclination angle between the normal to the orbital plane and the line of sight, and P_{orb} is the binary's period. Since periodic variations in the companion's measured flux give both v_C and P_{orb}, this equation therefore provides a measure of $a_C \sin i$.

However, we can also use X-ray measurements to determine the light-travel time across the binary's diameter, which gives us the quantity $(a_C/c) \sin i$. From Kepler's law (equation 6.77), we know that

$$\frac{G(M_C + M_X)}{a^3} = \left(\frac{2\pi}{P_{orb}}\right)^2, \tag{10.4}$$

and a simple manipulation of equations (10.1) and (10.2) yields

$$a = \frac{M_C + M_X}{M_X} a_C. \tag{10.5}$$

Thus, we may now combine equations (10.3), (10.4), and (10.5), to arrive at one of the most celebrated equations in X-ray astronomy,

$$f(M_C, M_X, i) \equiv \frac{(M_X \sin i)^3}{(M_C + M_X)^2} = \frac{v_C^3 P_{\text{orb}}}{2\pi G}. \tag{10.6}$$

The left-hand side is known as the *mass function*, which is evidently measurable directly from knowledge of the quantities P_{orb} and v_C.

In the specific case of Cygnus X-1 (see also section 10.2), this analysis results in the value

$$a_C \sin i = (5.82 \pm 0.08) \times 10^6 \text{ km}. \tag{10.7}$$

Since the eccentricity ($\lesssim 0.02$) in this source is known to be very small (again, from the lightcurve of HD 226868), the binary's orbit is essentially circular, and

$$f = (0.252 \pm 0.01) \, M_\odot. \tag{10.8}$$

According to equation (10.6), the only remaining unknown in our estimation of M_X is therefore the mass of the companion, M_C. Unfortunately, this is often where our tracking produces precious few clues, but such was not the case for Cygnus X-1, and this is the reason why it became the first binary for which we could justify a black-hole identification for its unseen star.

If the optical companion were a relatively normal OB supergiant, its mass would be as high as $\gtrsim 20 \, M_\odot$. The maximum value for M_X follows from equation (10.6) setting $\sin i = 1$, which in this case would give $M_X \sim 6 \, M_\odot$. However, HD 226868 is likely to be undermassive for its spectral type as a result of mass transfer and binary evolution (recalling our discussion in section 9.3), possibly by as much as a factor 2–3. On the other hand, the distance to Cygnus X-1 (based on a survey of many stars in its vicinity) is about 2.5 kpc. Thus, we know that $M_C \gtrsim 8.5 \, M_\odot$ in order for it to produce the luminosity we measure at Earth. A plausible lower limit to M_C is therefore $\sim 10 \, M_\odot$, and combined with an upper limit to the inclination of 60°, based on the absence of X-ray eclipses, we infer a very reliable lower limit,

$$M_{\text{bh}} = M_X \gtrsim 4 \, M_\odot, \tag{10.9}$$

for the mass of the unseen, compact object.

The significance of this result rests on the fact that neutron stars cannot support themselves against gravitational collapse for arbitrarily large masses. We already know that white dwarfs are restricted by the Chandrasekhar limit ($\approx 1.44 \, M_\odot$), above which it costs less energy for the electron to fuse with a proton and form a neutron (via inverse beta decay, $e^- + p \rightarrow n + \nu_e$) than to keep climbing the Fermi ladder. Above this mass, the dying star collapses into a configuration dominated by a sea of neutrons, whose degeneracy sets a size of about 10 km for a one-solar-mass object.

Thus, as far as we know, a neutron star is the most compact astronomical object attainable by nature, a black hole of similar mass being the sole exception. But whereas all the matter fed into the latter (classically) finds its way to a singularity at the origin, the neutrons in the former must be supported against collapse by a pressure dictated by the equation of state (EoS), which is uncertain in the high-density regime due to poorly constrained many-body interactions. If general relativity is correct, however, even this lack of precision in the EoS may be bypassed with an argument based on causality.[2]

For a given pressure P and density ρ in the neutron star, the sound speed is $c_s \equiv (\partial P/\partial \rho)^{1/2}$. The sound speed increases as the star collapses, eventually approaching the speed of light c. When the star is so compact that c_s would need to exceed c in order for P to prevent total collapse to a singularity, different parts of the star could no longer communicate causally with each other, and hydrodynamic support against an irreversible infall would become inevitable. The original calculations showed that the maximum mass a neutron star can have before this catastrophe occurs is $\sim 3.2\, M_\odot$. More recent simulations have updated this number somewhat,[3] but everyone now appears to have settled on the consensus value of $\sim 3\, M_\odot$ for the maximum neutron-star mass, which we will hereafter call M_{ns}^{max}. We will see shortly (figure 10.2) how dramatically observations have confirmed this limit.

With a lower limit to its mass of $\sim 4\, M_\odot$, the compact object in Cygnus X-1, therefore, instantly became a compelling black-hole candidate, regardless of whatever spectral or variability traits it may possess. But it was not alone for very long. Within only a few years of gaining notoriety, Cygnus X-1 was joined as a black-hole source by a new X-ray object, A0620-00, discovered by the satellite *Ariel* V. The latter, however, is not an HMXB but, rather, an X-ray nova associated with an LMXB undergoing dramatic episodes of enhanced mass transfer triggered by viscous–thermal instabilities in the disk (see section 7.4). We will revist X-ray novae in section 10.3, but for now it suffices for us to know that the X-ray flux during such an event can increase from nondetection to about 50 times that of the Crab.[4]

Overwhelmed by the intense optical light from the irradiated disk, the companion remains undetected during outburst. But the X-rays switch off after a few months, and eventually the companion's spectrum reemerges as the dominant source of visible light from the LMXB, allowing us to measure its orbital velocity and period. In A0620-00, the companion is a mid-K star moving at 457 km s^{-1} with a period of 7.8 h. The mass function for this binary system is therefore $f = 3.2 \pm 0.2\, M_\odot$, the largest ever measured up to that point. Conservative estimates place the companion's mass at $\sim 0.25\, M_\odot$, and since there are no X-ray eclipses from this system, we infer that the inclination angle is $i < 85°$. The mass of the compact star in A0620–00 is therefore $M_{bh} = M_X \gtrsim 3.2\, M_\odot$, which is again greater than M_{ns}^{max}, making this object the second compelling case for a binary system harboring a black hole.

Having said this, notice that both of these black hole "candidates" have measured masses close to M_{ns}^{max}. This troubled many who were eager to prove once and for all that stellar-sized black holes do indeed exist in nature. But a little over a decade later, the Japanese satellite *Ginga* discovered a new X-ray transient in outburst, named GS 2023+338 (also known as V404 Cyg). Even before its binary parameters could be used to estimate the compact object's mass, already the fact that V404 Cyg's luminosity saturated at $L_x \sim 10^{39}$ ergs s^{-1} meant that we were probably witnessing

[2] See Rhoades and Ruffini (1974).
[3] See, e.g., Kalogera and Baym (1996).
[4] One Crab = 2.43×10^{-9} ergs cm^{-2} s^{-1} keV^{-1} (=1 mJy) (averaged over 2–11 keV) for a Crab-like spectrum with photon index 2.08.

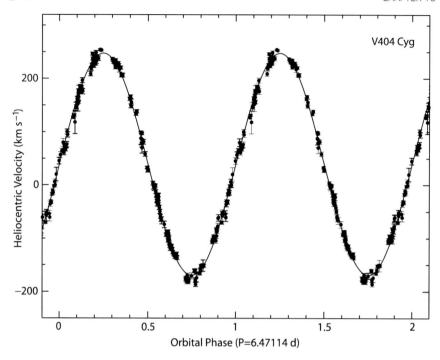

Figure 10.1 Radial velocity curve of the K0 companion in the transient low-mass X-ray binary V404 Cyg during quiescence. (Image from Casares, Charles, and Naylor 1992)

the maximum possible X-ray power radiated at the Eddington limit.[5] According to equation (1.19), this means the accretor has a mass of $7-8 M_\odot$, well above the neutron-star limit.

In quiescence, spectroscopic analysis of this new system revealed a K0 star moving with a velocity of 211 km s^{-1} in a 6.5-day orbit (see figure 10.1), for which the mass function is therefore $f = 6.3 \pm 0.3 M_\odot$. Thus, independently of the value of M_C and i, this translates into a lower mass limit $M_{bh} = M_X \sim 6 M_\odot$ for the unseen star, well above those of Cygnus X-1 and A0620-00, and well above the neutron-star limit. Since then, many other black-hole binaries have also been revealed through dynamical studies of X-ray transients during quiescence. Like V404 Cyg, seven of them have mass functions in excess of $5 M_\odot$, and therefore constitute our best solar-sized black-hole discoveries to date.

The current catalog of 20 black-hole binaries, confirmed on the basis of dynamical arguments, is given in Table 10.1, ordered in terms of descending orbital period.[6] The vast majority of these sources are transients, identified after the systems were discovered as X-ray novae. Only three of them are persistent sources: Cygnus X-1,

---

[5] See Życki, Done, and Smith (1999).

[6] The masses in this table were compiled by Orosz (2003) and Charles and Coe (2006). Note that the masses quoted here represent the most likely values, once the companion type and binary inclination are taken into account. In some cases, these masses are much higher than their calculated minimum values.

Table 10.1 Confirmed Black-Hole Binaries.

Name	P_{orb} [h]	$f[M_\odot]$	Donor Spect. Type	Class	$M_{bh}[M_\odot]$
GRS 1915 + 105	804.0	9.5 ± 3.0	K/M III	L/Trans	14 ± 4
V404 Cyg	155.3	6.09 ± 0.04	K0 IV	L/Trans	12 ± 2
Cygnus X-1	134.4	0.244 ± 0.005	09.7 Iab	H/Persist	10 ± 3
LMC X-1	93.80	0.14 ± 0.05	07 III	H/Persist	>4
XTE J1819–254	67.60	3.13 ± 0.13	B9 III	I/Trans	7.1 ± 0.3
GRO J1655–40	62.90	2.73 ± 0.09	F3/5 IV	I/Trans	6.3 ± 0.3
BW Cir	61.10	5.74 ± 0.29	G5 IV	L/Trans	>7.8
GX 339–4	42.10	5.8 ± 0.5	—	L/Trans	
LMC X-3	40.90	2.3 ± 0.3	B3 V	H/Persist	7.6 ± 1.3
XTE J1550–564	37.00	6.86 ± 0.71	G8/K8 IV	L/Trans	9.6 ± 1.2
4U 1543–475	26.80	0.25 ± 0.01	A2 V	I/Trans	9.4 ± 1.0
H1705–250	12.50	4.86 ± 0.13	K3/7 V	L/Trans	6 ± 2
GS 1124–684	10.40	3.01 ± 0.15	K3/5 V	L/Trans	7.0 ± 0.6
XTE J1859 + 226	9.200	7.4 ± 1.1	—	L/Trans	
GS2000 + 250	8.300	5.01 ± 0.12	K3/7 V	L/Trans	7.5 ± 0.3
A0620–003	7.800	2.72 ± 0.06	K4 V	L/Trans	11 ± 2
XTE J1650–500	7.700	2.73 ± 0.56	K4 V	L/Trans	
GRS 1009–45	6.800	3.17 ± 0.12	K7/M0 V	L/Trans	5.2 ± 0.6
GRO J0422 + 32	5.100	1.19 ± 0.02	M2 V	L/Trans	4 ± 1
XTE J1118 + 480	4.100	6.3 ± 0.2	K5/M0 V	L/Trans	6.8 ± 0.4

LMC X-1, and LMC X-3. Several of these binaries contain an A-F companion star, and apparently constitute a new category—the intermediate-mass X-ray binaries (IMXBs)—which may be the evolutionary predecessors of LMXBs.

Additional support for a black-hole designation of these tabulated sources is provided by observations that focus on their spectral and timing characteristics. We have already encountered an example of this, when we used the saturated X-ray luminosity of V404 Cyg to argue that the mass of the compact object should be about 7–8 $M_\odot$. Now that we have dynamical evidence that its mass is probably 12 ± 2 $M_\odot$, the spectrally derived value is a reasonable confirmation of this more direct and accurate determination.

Other features in the source spectrum or X-ray lightcurve that may be used to support a black-hole designation include, first, the absence of a boundary layer (see section 7.2.3). As we mentioned earlier, if the boundary layer is present, then clearly the source has a hard surface and is most likely a neutron star. If not, then the star may lack a hard boundary, in which case matter must be flowing through an event horizon. The absence of pulsations (see section 7.3.2) or X-ray bursts (see section 11.1) may also indicate a black hole. Any of these features would be associated with phenomena on the stellar surface itself, or within a magnetosphere anchored to the surface. Again, if none of these are seen, it is at least possible that the star lacks a hard boundary. Finally, matter flowing through the event horizon advects a fraction of the dissipated gravitational energy with it. Thus, black holes in binaries are expected, on average, to be dimmer than their neutron-star counterparts.

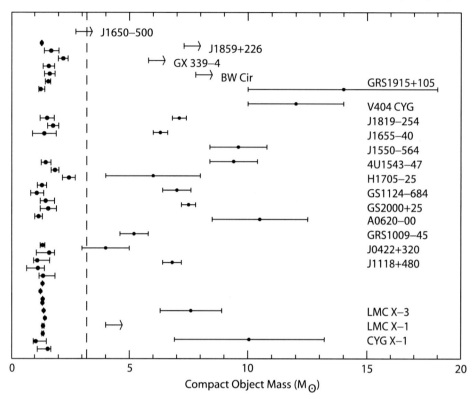

Figure 10.2 Measured masses of compact objects in X-ray binaries. The objects on the left of the dividing (dashed) line at 3.2 $M_\odot$ include radio pulsars, high-mass X-ray binaries, and low-mass X-ray binaries. These compact objects are all believed to be neutron stars. To the right of the dividing line, the equation of state cannot support the compact objects against collapse. These X-ray sources are therefore believed to be black holes. (Image from Casares 2001)

Observationally, black-hole systems are seen to be ∼100 times fainter than quiescent neutron-star binaries with the same orbital period, providing prima facie evidence that an event horizon is responsible for this major difference.[7]

Table 10.1 is displayed graphically in figure 10.2, along with neutron-star HMXBs and LMXBs, and several radio pulsars. The originally-derived maximum neutron-star mass ($M_{ns}^{max} \approx 3.2\,M_\odot$) is represented as a dashed vertical line. The division between objects that are known to be neutron stars and those whose mass places them in the black-hole category is rather remarkable. Most of the neutron stars cluster around the Chandrasekhar limit (∼1.44 $M_\odot$), suggesting that they have accreted very little since their formation. Some in the HMXB and LMXB categories are more massive, but none exceed the M_{ns}^{max} limit—a powerful confirmation of the causality argument that produced this value. On the other hand, there is no obvious clustering of black holes with mass, presumably because there is no equivalent to

[7] See, e.g., Menou et al. (1999).

a Chandrasekhar limit for black holes. They can form with any mass, and remain black holes as they continue to grow.

Lest we leave this introductory discussion with a false impression, we should emphasize that the collection of stellar-sized black holes in Table 10.1 is, of course, but a tiny sample of what is believed to be the complete galactic population. Dynamical studies of X-ray transients reveal that roughly $\frac{3}{4}$ contain a black hole. And an extrapolation of the number of these sources uncovered since 1975 with an outburst duty cycle of $\sim$10–100 years, suggests that there must be a dormant population of at least 1000 or so such systems.[8]

From a theoretical standpoint, we expect a number that is probably much bigger than this. Stellar evolution models predict a population closer to 10^9 stellar-mass black holes in the Galaxy.[9] But though the catalog we have to work with is just a small subset of the entire distribution, two dominant categories of black-hole binaries have nonetheless already emerged from the list. In the next two sections, we consider the physical principles underlying the key observational characteristics exhibited by these two classes—the steady HMXBs and the transient LMXBs.

10.2 THE ARCHETYPAL HMXB CYGNUS X-1

Since its discovery in 1964,[10] Cygnus X-1 has been closely monitored by many high-energy instruments, from X-rays to γ-rays. Though many other black-hole systems are now known, its role as an archetype for this class has not diminished, and it is still unquestionably the most recognizable member of the HMXB subgroup—i.e., systems that accrete at a much higher rate than their more numerous LMXB counterparts (see section 10.3).

Cygnus X-1's spectrum has been known since the early 1970s to be a blend of thermal and nonthermal components (see figure 10.3). Such black-hole binaries undergo transitions in which one or more of these spectral features dominates the high-energy emission.[11] In general (and in oversimplified) terms, the thermal contribution to the spectrum may be modeled as a multitemperature blackbody (see section 7.2.2, particularly figure 7.4), originating from the inner region of the accretion disk, whereas the nonthermal component is adequately modeled as a power law originating from the particle distribution in equation (5.112), where the spectral index x is typically $\sim$2. The power law extends to much higher energies than the thermal component, often tailing off exponentially at $\sim$100 keV.

As shown in figure 10.3, the typical black-hole binary spectrum sometimes exhibits a broadened Fe K_α emission line (at roughly 6.7 keV). We shall see in chapter 12 that this spectral feature's asymmetric shape, seen more famously in AGNs, is due to an emission region spread across many gravitational redshifts.

---

[8] See, e.g., Romani (1998).
[9] See Brown and Bethe (1994).
[10] See Bowyer et al. (1965).
[11] See, e.g., McClintock and Remillard (2006).

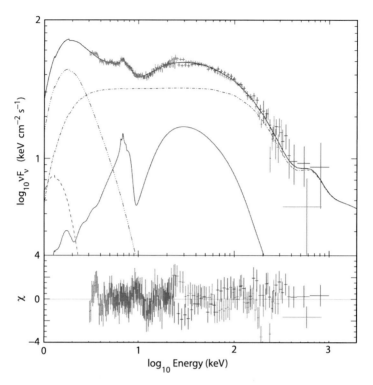

Figure 10.3 In June 7-11, 2003, Cygnus X-1 was observed by all three instruments aboard the INTEGRAL satellite: JEM-X, IBIS/ISGRI, and SPI. The joint spectrum is best fitted with a hybrid thermal/nonthermal Comptonization model, with a power-law injection of relativistic electrons, a reflection component (solid), the disk thermal emission (dashed), the Comptonized emission (dot-dashed), and an additional warm Comptonization soft component (3-dot-dashed). At high energy, the spectrum shows a rapid drop-off around 100 keV, but with continued nonthermal emission. The lower panel shows the χ^2 values of the fit. (Image from Malzac et al. 2006)

Nowadays, observations such as this are providing a powerful probe of the spacetime close to the black hole's event horizon.

It is also evident in figure 10.3 that the spectrum includes a bump at $\sim$10–30 keV. This feature is interpreted as the fraction of power-law photons reflected by the relatively cool material orbiting in the equatorial plane, which becomes visible when the inclination of the binary allows us to view the accretion disk largely face-on. (We shall return to the subtle—though critically important—second bump at $\sim$1 MeV shortly)

Cygnus X-1 is most often found in the so-called low hard state (LHS), defined by a relatively low flux in soft X-rays ($\sim$1 keV) and a high flux in hard X-rays ($\sim$100 keV). Figure 10.4 is a compilation of four sets of observations made by several high-energy instruments from 1991 to 1996. The LHS was seen by *Ginga* and OSSE in 1991, and then again by *Beppo*-SAX in 1998. The high-energy power-law

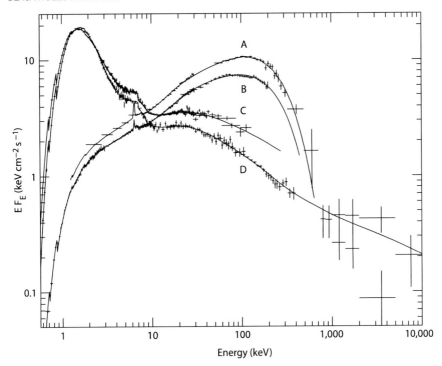

Figure 10.4 Example broadband spectra of Cygnus X-1 from pointed observations. (A) Hard state: *Ginga*-OSSE spectrum from 6 June 1991; (B) hard state: *Beppo*-SAX spectrum from 3–4 May 1998; (C) soft state: ASCA-RXTE spectrum from 30–31 May 1996; and (D) soft state: combined *Beppo*-SAX spectrum from 22 June 1996 and that from *Compton*-GRO OSSE and COMPTEL observations on 14–25 June 1996. The solid curves represent the best-fit Comptonization models (thermal in the hard state, and hybrid, thermal–nonthermal, in the other states). (From Zdziarski et al. 2002)

component in this state has a spectral index in the range ∼1.4–2.2. Occasionally, Cygnus X-1 switches to the high soft state (HSS), which was detected by ASCA and RXTE in 1996, and then again by *Beppo*-SAX that same year.[12] The high-energy power law in this state is much softer (∼2.4), and the bolometric luminosity is dominated by the thermal component, which peaks at a few keV (see also figure 10.3). Sometimes, this source is detected in an intermediate state (IMS), which, as the name suggests, is a transitory configuration exhibiting a relatively soft hard X-ray spectrum (with spectral index ∼2.1–2.3) and a moderately strong soft thermal component; the IMS appears when the system is switching from LHS to HSS, or back again.

Simultaneous observations of Cygnus X-1 (and other sources) in the radio and X-ray bands have revealed that during the LHS, strong radio emission arises from

---

[12]The concept of high-energy states and transitions between them originated when Tananbaum, Gursky, Kellogg, Giacconi, and Jones (1972) observed a global spectral change in Cygnus X-1, in which the soft X-ray flux (2–6 keV) decreased by a factor of 4, while the hard X-ray flux (10–20 keV) increased by a factor of 2. Similar X-ray transitions were subsequently seen in A0620-00 (Coe, Engel, and Quenby 1976) and in many other sources as well.

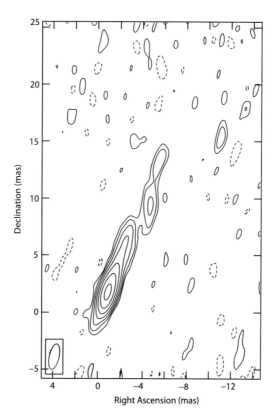

Figure 10.5 VLBA and phased VLA image of Cygnus X-1 at 8.4 GHz. The lowest contour is 0.157 mJy beam^{-1} and each of the other contours represent a factor of 2 increase. The jet-like feature extends out about 15 mas from the core, with an opening angle of $<2°$. At an assumed distance of 2 kpc, this size corresponds to $\sim$30 A.U. (From Stirling et al. 2001)

a jet-like feature extending out from an otherwise steady core (figure 10.5). In contrast, the jet is rather weak during a HSS, though the core emission is always present. We know that the radio emission must be associated with the compact object in this binary, because the radio lightcurve displays a sinusoidal modulation with a minimum at superior conjunction of the X-ray source.

The fact that the radio jet is present in one state and not in the other is an indication that transitions from the LHS to the HSS, and vice versa, must be associated with significant changes in the geometry of the accretion flow. In the HSS, the geometrically thin, optically thick, accretion disk (section 7.2) must extend down to the last stable orbit (see section 3.4.1), since the spectrum is dominated by the thermal disk component, peaking at a few keV. The hard X-ray emission is probably produced in a nonthermal corona above the equatorial plane.[13]

[13]It doesn't surprise us to learn that Cygnus X-1 has many characteristics in common with certain AGNs. For example, the presence of a nonthermal corona, possibly linked to magnetic flares bursting out of the accretion disk, reminds us of similar hot structures in Seyfert galaxies (Nayakshin and Melia 1997). See also Gierliński et al. (1999).

It is believed that a transition from the HSS to the LHS is associated with the truncation of the thin disk at several hundred Schwarzschild radii. At the same time, a geometrically thick, optically thin, hot disk forms between the truncation radius and the event horizon (see sections 7.4 and 8.1). It is in this region that the high-energy radiation is produced via thermal Comptonization.[14] A reversal back to the HSS results when the cold disk refills its inner region, either by penetrating into the hot corona or by reclaiming the hot gas as it cools and collapses back down toward the plane.

Since the radio jet is tightly correlated with the LHS, the jet and the X-ray emitting region in this state appear to be physically connected, leading to the now widely accepted view that the corona (or hot, thick disk) of the LHS constitutes the base of the jet.[15]

To develop a physical model for this phenomenon, we begin by noting that under certain circumstances, geometrically thin, optically thick accretion disks become unstable in their inner regions, and evolve into hot, optically thin configurations in which the electron temperature may reach or exceed the electron's rest mass (see section 7.4).[16] When this happens, the creation of electron–positron pairs can significantly modify the properties of the plasma. An important element in our understanding of these structures is therefore whether there exists a critical accretion rate above which the rate of pair creation (from either $\gamma\gamma \to e^+e^-$ or $\gamma e^- \to e^-e^+e^-$) unavoidably exceeds the rate of pair annihilation ($e^+e^- \to \gamma\gamma$).

Fortunately, Cygnus X-1 is not the only object for which this kind of analysis can bear fruit; it shares several important characteristics with other sources, including the *Einstein*-discovered object 1E 1740.7–2942, the most luminous hard X-ray/soft γ-ray emitter in the vicinity of the galactic center (see figure 12.9). Though there is no dynamical evidence for a black hole in this system, it is generally accepted that its spectral characteristics make it a likely member of the black-hole binary class.[17] Its unusual nature is underscored by the detection, in 1990, of a broad feature between $\sim$250 and $\sim$500 keV in its γ-ray spectrum, with detectable emission up to $\sim$700 keV. The peak of this bump at $\sim$500 keV (with a FWHM $\sim$240 keV) suggests that this effect may be attributed to e^+–e^- annihilation, possibly correlated with the presence of high-energy photons with energy $> m_e c^2$ in the vicinity of the compact star.

At millimeter wavelengths, 1E 1740.7–2942 has a radio structure composed of a double-sided jet emanating from a compact and variable core, reminiscent of the 8.4 GHz jet in Cygnus X-1 (figure 10.5), though much more powerful and extended.[18] At least, 80% of this radio jet is composed of electron–positron

---

[14]The earliest model for such a hot, inner disk structure was proposed by Shapiro et al. (1976). Subsequently, Misra and Melia (1993, 1995) developed a physical picture for how the hot, bloated plasma behaves, including the ejection of an electron–position pair plasma to form a radio jet on larger scales.

[15]See Misra and Melia (1993), Fender et al. (1999), Merloni and Fabian (2002), and Markoff, Nowak, and Wilms (2005).

[16]This concept was first introduced by Lightman and Eardley (1974) and Pringle (1976).

[17]See Bouchet et al. (1991) and Sunyaev et al. (1991).

[18]See Mirabel et al. (1992).

pairs.[19] Thus, together with the apparent annihilation line seen from this source, the predominant e^+–e^- composition of the jet suggests that the central compact object is disgorging an intense flux of high-velocity electron–positron pairs into the surrounding medium.

We don't exactly know yet how these electron–positron pairs are produced, but an important clue seems to be provided by the occasional appearance of a measurable flux of γ-rays at ~ 1 MeV, in both Cygnus X-1 and 1E 1740.7–2942. The combined JEM-X, IBIS/ISGRI, and SPI spectrum shown in figure 10.3, as well as spectrum D in figure 10.4, are instances of this phenomenon. This bump is produced by bremsstrahlung-self-Compton processes within the inner, hot corona.[20]

But the ~ 1-MeV radiation is so intense within the funnel-like depression in the gravitationally condensed inner region of the disk (see, e.g., figure 8.1) that many (actually, most) of these γ-rays materialize into $e^+ e^-$ pairs (via $\gamma\gamma \to e^+ e^-$) within only 5–10 r_S of the black hole (where $r_S \equiv 2GM/c^2 \approx 30\, M_{bh}/10\, M_\odot$ km). These leptons are subsequently accelerated outward (to velocities $\beta \equiv v/c \sim 0.4$–$0.6$) by the intense radiation field.[21] Simulations show that roughly 90% of these leptons annihilate before they reach a vertical height $z \sim 30\, r_S$ above the disk, by which time the particle density n_e within this nascent "jet" is too low to sustain a substantial annihilation rate, and the remaining pairs continue to propagate into the medium surrounding the black-hole binary. The highly variable broad annihilation line seen in 1E 1740.7–2942 by *GRANAT*[22] presumably originates at the base of the $e^+ e^-$ outflow.

The picture we have just painted for the high-energy activity in Cygnus X-1 and black-hole candidates such as 1E 1740.7–2942 is rather comprehensive, and involves several processes all acting in concert to produce the observed broadband spectrum of these sources. There are many other issues we have not touched on; some are areas of active current research, and our understanding of these objects will no doubt continue to evolve with time.

For example, the appearance of an asymmetric Fe K_α line offers us the concrete possibility of mapping the spacetime near the black hole in Cygnus X-1, but it would be helpful to have a better sense of its temporal variability. As far as the jet is concerned, what we have described here is supported by VLA observations, which in the case of 1E 1740.7–2942 reveal a spectral index $\alpha = 0.81 \pm 0.1$ (assuming a flux density $S_\nu \propto \nu^\alpha$) in its symmetrically aligned jets. The jet radiation is therefore due to synchrotron emission (see section 5.5). However, based on what we know about the e^+–e^- pair creation and acceleration near the black hole, it is unlikely that these leptons receive a relativistic boost on their way out into the jet. Some other mechanism must continuously refresh their energy. This may be

---

[19] See Chen et. al. (1994).

[20] See also Liang and Dermer (1988) and Misra and Melia (1995).

[21] This behavior, more commonly seen in superluminal AGN jets, is due to the transfer of energy from the radiation field to the particles via Compton scattering. We will have a more thorough discussion of this process in section 12.3

[22] See Bouchet et al. (1991) and Sunyaev et al. (1991).

electromagnetic or it may be acceleration by strong shocks within the outflow (see section 4.3).

Cygnus X-1 and its cohort are among the most interesting and exciting objects in the sky, and we will learn a great deal more about them as we improve the threshold of our detectors, and study their spectral variability with greater precision.

10.3 X-RAY NOVAE

There are about 200 cataloged X-ray binaries in the Galaxy,[23] each containing a neutron star or black hole accreting from a companion. The great majority of these are persistent sources, though 30 or so are transient, known only because of their nova-like outbursts, which occur anywhere from once every several years, to perhaps once per century or longer.

As we noted in Table 10.1, almost all the black-hole binaries are X-ray novae, characterized by episodic outbursts at X-ray, optical, and radio frequencies. Conventional wisdom has it that the outburst is caused by a sudden dramatic increase in mass accretion rate through the disk. In these systems, the mass transfer from the donor star is not sufficient to sustain a continuous viscous flow onto the compact object, and matter fills the outer region of the disk. At some point, a critical condition is met, triggering an outburst. The nature of this critical condition is still subject to some debate, though it probably has something to do with the strong temperature dependence of the viscosity, to which we shall return shortly. The X-ray flux rises on a timescale of several days, and subsequently declines over a period of weeks to months. During outburst, the X-ray flux may be several million times larger than its value during source quiescence.

A well-known member of this group is X-ray nova GRS 1124–684, detected by the Franco-Soviet *GRANAT* satellite and by the Japanese X-ray observatory *Ginga*.[24] Several images taken at different energies during the flare are shown in figure 10.6, demonstrating, in particular, the emergence of an electron–positron annihilation line in the 430- to 530-keV band. Our discussion in the previous section has prepared us to interpret this as the result of numerous energetic photon–photon interactions within a hot centralized cauldron activated during the burst. Some observational support for this conclusion is offered by the observed line variability, on a timescale of <7 h, which sets an upper limit of only 7×10^{14} cm (roughly 47 A.U.) to the size of the emitting region. Thus, aside from the fact that GRS 1124–684 is an LMXB, whereas Cygnus X-1 is an HMXB and the binary nature of 1E 1740.7–2942 is unknown, all three of these sources share an important characteristic: their inner disk region is sufficiently hot to produce energetic photons that materialize and emerge as an abundant flux of electron–positron pairs.

---

[23] See, e.g., van Paradijs (1995).

[24] These sources are labeled according to the satellite that first discovered them. In the case of Nova Muscae, it is also variously called GRS 1124–684 and GS 1124–684, because both *GRANAT* and *Ginga* discovered it at about the same time. See Goldwurm et al. (1992) and Kitamoto et al. (1992).

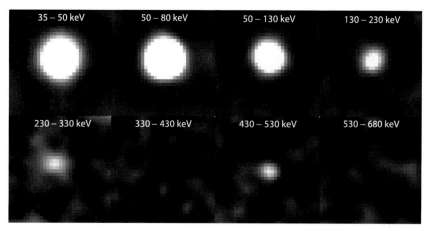

Figure 10.6 Images of X-ray Nova Muscae (GRS 1124-684) taken by SIGMA during a flare in 1991. Notice that the source fades in the 330- to 430-keV band and then re-emerges in the 430- to 530-keV band where the positron annihilation radiation is centered. X-ray novae are a class of transient sources thought to contain an accreting black hole. Some X-ray novae can become the brightest objects in the X-ray sky, with fluxes exceeding 1 Crab or more. The origin of high-energy (>50 keV) radiation from X-ray novae is one of the most important open problems in high-energy astrophysics. Of particular interest is the nature of the variable 511-keV feature observed during flares such as this. (From Goldwurm et al. 1992)

After the X-ray outburst in GRS 1124-684 subsided, the counterpart was identified with a star of spectral type $\sim$K3/5, orbiting about the compact object with a period of 10.4 h. The mass function (equation 10.6) was inferred to be $3.01 \pm 0.15 \, M_\odot$ (see Table 10.1), yielding a minimum black-hole mass for this system of $3.75 \pm 0.43 \, M_\odot$. This is larger than the maximum possible neutron-star mass, M_{ns}^{max}, so like Cygnus X-1, Nova Muscae is a strongly confirmed black-hole binary.

We can learn a great deal about the accretion process in X-ray novae but esti-mating their average luminosity. Many of them are not detected as X-ray sources in quiescence, but we can take the total energy emitted during an outburst and divide it by the interval from the last event (in cases where at least two events have been seen). Such estimates are only qualitative, but they provide us with useful infor-mation regarding, e.g., whether or not the companion is filling its Roche lobe (see section 6.3). Interestingly, when they can be measured in this way, the time-averaged luminosities are all in the range 10^{35}–10^{36} ergs s^{-1}. So with the conversion factor given in equation (1.5), this translates into a mass accretion rate $\sim$$10^{15}$ g s^{-1}. By comparison, many of the persistent LMXBs, such as the X-ray pulsars we considered in section 9.2, accrete at the much higher rate $\sim$$10^{17}$–$10^{18}$ g s^{-1}. The possibility therefore exists that somewhere between 10^{15} and 10^{17} g s^{-1} there lies a critical accretion rate (thought by many to be $\sim$$10^{16}$ g s^{-1}) below which the mass transfer becomes unstable.

Oddly enough, however, optical observations of these transient LMXBs in qui-escence reveal that mass transfer from the secondary is continuing even between outbursts. The optical spectrum clearly reveals the presence of hydrogen Balmer emission lines,[25] from which we infer an optical disk luminosity $\sim 10^{32}$ ergs s^{-1}. The fact that mass transfer continues even between flaring events means that the secondary star almost certainly fills its Roche lobe. Yet the transferred matter is not finding its way directly toward the black hole. The optical emission may come from either the outer accretion disk heated by viscous dissipation, or from a hot spot where the stream collides with the outer edge of the disk (see figure 7.5).

With this geometry in mind, we can use the Roche lobe parameters of section 6.3 to estimate the required mass-transfer rate onto the outer disk in order to sustain the observed optical disk luminosity. The value corresponding to systems such as A0620-00 and GRO J1655–40 is $\sim 6 \times 10^{15}$ g s^{-1}, only slightly below the putative critical value for stable disk accretion. Remarkably, this mass transfer rate, inferred from the binary's optical emission properties, is the same order of magnitude as the long-term average we estimated earlier. For this reason, we conclude that a large fraction of the transferred matter is stored in the accretion disk during the quiescent interval. And since the potential energy of the accumulated mass can account for the total energy released in the next outburst, this picture supports the idea of a disk instability as the trigger that initiates the nova event.

But there is still some uncertainty regarding exactly how the trigger functions. There is even a competing model that does not depend on a disk process at all, though observations do not seem to favor it at the moment. The disk instability model[26] relies on a very steep temperature dependence of the opacity in a partially ionized accretion medium. Though we may not yet recognize it here, this process is very similar to the instability we discussed in section 7.4, based on the strong temperature-dependence of the cooling coefficient in figure 7.11. Changes in opacity (and emissivity) strongly affect the cooling mechanism's dependence on temperature and lead to transitions between the stable thermal states of the system, which in this figure correspond to regions of positive gradient in the displayed functions.

An accretion disk is stable both in a cool, neutral state (corresponding to temperatures below $\sim 10^5$ K in figure 7.11), and in a hot, fully ionized state, i.e., above $\sim 10^8$ K. In these regions, an increased gravitational dissipation rate is countered by an increased emissivity, maintaining overall thermal balance. However, when the mass flow rate onto the outer part of the disk is within a range where an increased dissipation rate leads to an increased temperature, but a decreased cooling efficiency (essentially between $\sim 10^5$ and $\sim 10^8$ K), then the structure undergoes a thermal limit cycle. In quiescence, the accretion disk is in the cool state. As matter accumulates at the outer edge, both the surface density and temperature increase, and when the latter exceeds $\sim 10^5$ K, the thermal instability sets in, driving the disk to the very hot plateau. This is presumably accompanied by an enhanced accretion rate and a nova outburst, which persists until the surface density drops and the disk transitions once again back down to its cool state.

[25] See, e.g., Filippenko et al. (1995) and McClintock et al. (1995).

[26] This topic has had a long history, mainly because it was first developed for novae in white-dwarf systems. Some highly relevant papers include Meyer and Meyer-Hofmeister (1981), Mineshige and Wheeler (1989), Cannizzo (1993), and Lasota, Hameury, and Huré (1995).

We use the term "presumably" here because the resulting luminosity variations depend critically on what is assumed for the viscosity parameter α appearing in the prescription of kinematic viscosity (equation 7.9). To obtain a dwarf-nova type of outburst, the viscosity parameter must be assumed to be larger in the hot state than in the cold configuration, which is still not known quantitatively. But given the right choice of variation in α, such models can explain the observed properties of X-ray novae quite well, including the rise time, the decay rate, and the recurrence frequency of outburst.

An alternative scenario holds that matter is not being stored in a disk at all. Rather, the subphotospheric layers of the companion star are heated by the penetration— and subsequent deposition—of hard X-rays (with energy $>7\,\mathrm{keV}$) produced by the black hole. These layers slowly expand and ultimately move the atmosphere into an unstable regime, leading to a sudden mass transfer. This in turn builds up an the accretion disk and causes a transient, enhanced mass infall rate onto the compact object.

The difficulty with this model is that the observed X-ray flux in quiescence appears to be insufficient to induce the mass transfer instability. Together with the fact that we know from the optical spectrum of these objects that mass is being transferred across the inner Lagrangian point, even between outbursts, this dearth of hard X-rays in the quiescent state suggests quite compellingly that we must be dealing with a thermal instability in the disk, as discussed above.

10.4 QPOS IN BLACK-HOLE BINARIES

From a physics perspective, one of the principal reasons for studying X-ray binaries is the unique opportunity they provide for examining the behavior of matter and radiation under the influence of strong gravity. General relativity has been tested and confirmed extensively in weak fields (i.e., those for which $2GM/c^2r \ll 1$), but not yet for the dynamics of particles subject to superstrong gravitational accelerations near a compact object, where the gravitational binding energy is comparable to the constituents' rest mass. As we have seen earlier in this book, particularly in chapter 3, general relativity makes several extreme predictions for these regions, including the existence of an event horizon, a marginally stable orbit (within which no other stable orbits exist), strong inertial-frame dragging, and a gravitationally induced precession at rates comparable to the orbital motion itself (see below). These influences are present in all X-ray binaries, including those in which the compact object is a neutron star. The reason we have delayed broaching this subject until now, however, is that general relativistic effects manifest themselves most prominently in cases where the accretor is a black hole, and these considerations are the focus in the remainder of this chapter.

Given the relatively small size of the compact object in these binaries (even by Earth standards!), the characteristic time over which the various relativistic effects exert their influence is measured in fractions of a second. According to equation (1.23), we can expect a variability time scale of $\sim 0.1\,\mathrm{ms}$ at $\sim 15\,\mathrm{km}$ and $\sim 2\,\mathrm{ms}$ at $100\,\mathrm{km}$ for a $1.4\,M_\odot$ neutron star, and $\sim 1\,\mathrm{ms}$ at $3r_S$ ($\sim 100\,\mathrm{km}$)

for a $10\,M_{\odot}$ black hole. As we learned in chapter 1, detecting such rapid variability was the principal motivation for the design of the *Rossi* X-ray Timing Explorer (RXTE), which discovered rapid changes in these binaries with spectacular consistency.

All of the material within the compact object's Roche lobe is bound, which means that the forces of attraction invariably lead to harmonic motions. Whether these are cyclic in nature, reflecting the underlying Keplerian rotation, or pulsational, having to do, e.g., with vertical restoring forces in the accretion disk (see figure 7.3), we expect dynamical effects close to the black hole (ignoring neutron stars for now) to be associated with a characteristic frequency. More often than not, we see several frequencies at once. Therefore, it is useful to study the timing behavior of black-hole binaries with Fourier analysis, which provides a *power density* $P(\nu)$, quantifying the relative importance (or power) of emission at frequency ν. In the case of black-hole binaries, the observed frequencies at which measurable power is produced tend to fall in the mHz to kHz range.

The so-called *power spectrum* is a distribution, usually expressed graphically, demonstrating the variability components together as a function of frequency. If a variability component is aperiodic, we expect it to cover many frequency resolution elements. Broad structures in a power spectrum therefore do not represent anything of specific interest to harmonic behavior, and are called *noise*. Often the noise appears as a power law, $P(\nu) \propto \nu^{-\mu}$, where the power-law index μ is typically between 0 and 2. Narrow features, on the other hand, are where most of the interesting physics is encoded; these are called quasi-periodic oscillations (QPOs), since they are not generally sufficiently pure to be represented by a single frequency (in which case they would clearly be referred to as "periodic" instead).

Low-frequency QPOs, in the 0.1- to 30-Hz range, have now been detected in about 14 black-hole binaries, and their frequencies and amplitudes tend to be correlated with the spectral parameters for both the thermal and power-law components. This type of QPO can vary in frequency on timescales as short as one minute, though they generally remain stable and persistent for much longer. In one particular example—the longest-period source in Table 10.1, GRS 1915+105—a 2.0- to 4.5-Hz QPO has been seen in every single RXTE observation conducted over a 6-month period.[27] This level of stability suggests that low-frequency QPOs are probably associated with the flow of matter in the accretion disk. Note, however, that their frequencies are much lower than the Keplerian frequencies near the marginally stable orbit (see discussion following figure 3.2).

High-frequency QPOs, in the range 40- to 450-Hz, occur less frequently, though by now at least five black-hole binaries have revealed them. These oscillations are transient and subtle, but they attract more interest than their low-frequency counterparts because they are consistent with the behavior of matter near the marginally stable orbit of a $\sim 10\,M_{\odot}$ black hole. Figure 10.7 shows the power density spectra of two black-hole binaries with very obvious high-frequency QPOs. A close inspection of these frequencies reveals that they seem to occur in a 3:2 ratio, which implies a

---

[27] See Muno et al. (2001).

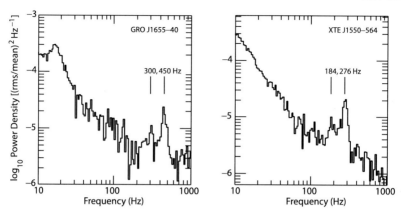

Figure 10.7 High-frequency quasi-periodic oscillations observed in two black-hole binary systems, for the energy range 13–30 keV. (From Remillard and McClintock 2006)

coupling due to some resonance condition, possibly in concert with one or more of the strong gravitational effects we hope to study.[28]

The correct mechanism generating these QPOs in black-hole systems has not yet been identified. Several proposals have received attention in the case of low-frequency QPOs, all focusing on the fact that these oscillations are strongest at photon energies above 6 keV, where the power-law component dominates over the thermal disk emission. These models include global disk oscillations, radial oscillations associated, e.g., with spiral shocks, and oscillations in some transition region separating the cool disk from a hotter Comptonization zone. An alternative to these simple hydrodynamic processes invokes the presence of spiral waves in a magnetized disk, with a transfer of energy out to the radius where material co-rotates with the spiral wave.[29]

The fact that the high-frequency QPOs appear to be closely associated with dynamical phenomena near the disk's marginally stable orbit, offers us great hope of someday identifying the actual physical mechanism responsible for producing them. And though we won't have a definitive answer by the conclusion of our discussion here, we can at least begin to analyze what types of frequency arise from general relativistic effects in this region.

To do so, we must first generalize the orbit equations (3.89), (3.90), and (3.91) to the case where the black-hole spin parameter a (Equation 3.101) is nonzero. The procedure is the same as before, beginning with the geodesic equation (3.78), followed by an evaluation of the Christoffel symbols for the Kerr metric (equation 3.97). Understandably, this derivation is longer than in the Schwarzschild case, since the metric coefficients (equation 3.104) are more complicated.[30]

---

[28] A suggestion along these lines, involving orbiting blobs of accreting matter, was made by Abramowicz and Kluźniak (2001). General relativistic coordinate frequencies had also been invoked earlier by Stella, Vietri, and Morsink (1999) to explain fast QPOs.

[29] See Tagger and Pellat (1999).

[30] This task was first undertaken by Carter (1968) and Bardeen, Press and Teukolsky (1972).

The generalization of the X^3 equation (3.89) is

$$\Sigma \frac{d\phi}{d\tau} = -\left(a\frac{\tilde{E}}{c} - \frac{\tilde{l}_\phi}{\sin^2\theta}\right) + \frac{a}{\Delta}\left[\frac{\tilde{E}}{c}(r^2 + a^2) - \tilde{l}_\phi a\right], \tag{10.10}$$

where the functions Σ and Δ are defined in section 3.4.2, and $\tilde{E}$ and $\tilde{l}_\phi$ are, respectively, the specific energy and angular momentum of the particle parallel to the symmetry axis. For motion in the equatorial plane ($\theta = \pi/2$), one can easily verify that in the limit $a \to 0$, this expression reduces to the previous form of the ϕ-equation for a Schwarzschild metric with $\tilde{l} = \tilde{l}_\phi$.

Equation (3.90) for X^0 becomes

$$\Sigma \frac{dT}{d\tau} = -\frac{a}{c}\left(a\frac{\tilde{E}}{c}\sin^2\theta - \tilde{l}_\phi\right) + \frac{r^2 + a^2}{\Delta c}\left[\frac{\tilde{E}}{c}(r^2 + a^2) - \tilde{l}_\phi a\right], \tag{10.11}$$

which will easily reduce to its Schwarzschild form with a simple manipulation when $a \to 0$. The X^1 (or r) equation is a little more involved because it contains several angular momentum components separately. Defining the quantity

$$\tilde{q} \equiv \tilde{l}_\theta^2 + \cos^2\theta\left[a^2\left(1 - \frac{\tilde{E}^2}{c^2}\right) + \frac{\tilde{l}_\phi^2}{\sin^2\theta}\right], \tag{10.12}$$

where $\tilde{l}_\theta$ is the component of specific angular momentum associated with the poloidal coordinate θ, we have

$$\Sigma \frac{dr}{d\tau} = \pm\left\{\left[\frac{\tilde{E}}{c}(r^2 + a^2) - \tilde{l}_\phi a\right]^2 - \Delta\left[c^2 r^2 + \left(\tilde{l}_\phi - a\frac{\tilde{E}}{c}\right)^2 + \tilde{q}\right]\right\}^{1/2}. \tag{10.13}$$

In these (and all subsequent) expressions, the upper sign refers to prograde orbits (i.e., co-rotating with $\tilde{l}_\phi > 0$), while the lower sign refers to retrograde motion (counterrotating with $\tilde{l}_\phi < 0$).

Finally, while the X^2-equation may be ignored when $a = 0$ (since we can rotate the coordinate system arbitrarily in that case), the poloidal motion is an essential component of the overall dynamics in the Kerr metric, because here the gravitational acceleration depends on both r and θ:

$$\Sigma \frac{d\theta}{d\tau} = \pm\left\{\tilde{q} - \cos^2\theta\left[a^2\left(c^2 - \frac{\tilde{E}^2}{c^2}\right) + \frac{\tilde{l}_\phi^2}{\sin^2\theta}\right]\right\}^{1/2}. \tag{10.14}$$

Note that $\tilde{q} = 0$ is a necessary and sufficient condition for the motion initially in the equatorial plane to remain in that plane for all time. Any orbit that crosses the equatorial plane has $\tilde{q} > 0$.

Fortunately, the scope of our analysis will permit us to use a reduced form of these (otherwise) complicated equations. Our primary interest lies with the structure of disks near the event horizon of black holes, and we may therefore restrict our attention to circular orbits in the equatorial plane. Thus, in equation (10.13), both

$dr/d\tau$ and $d^2r/d\tau^2$ may be set to zero—two conditions that lead to a simultaneous solution for $\tilde{E}$ and $\tilde{l}_\phi$:

$$\tilde{E} = \frac{c^2\left[r^{3/2} - r_S\,r^{1/2} \pm a(r_S/2)^{1/2}\right]^2}{r^{3/2}\left[r^{3/2} - \frac{3}{2}r_S\,r^{1/2} \pm 2a(r_S/2)^{1/2}\right]} \tag{10.15}$$

and

$$\tilde{l}_\phi = \frac{\pm c(r_S/2)^{1/2}\left[r^2 \mp 2a(r_S/2)^{1/2}r^{1/2} + a^2\right]}{r^{3/4}\left[r^{3/2} - \frac{3}{2}r_S\,r^{1/2} \pm 2a(r_S/2)^{1/2}\right]^{1/2}}. \tag{10.16}$$

More importantly, the first general relativistic frequency we encounter arises from a generalization of equation (7.11) to Keplerian motion in a Kerr metric. By definition,

$$\Omega_K(r) = \frac{d\phi}{dT}. \tag{10.17}$$

From equations (10.10) and (10.11) with $\sin\theta = 1$, and using the solutions for $\tilde{E}$ and $\tilde{l}_\phi$ in equations (10.15) and (10.16), we find that

$$\Omega_K(r) = \pm\frac{c\,(r_S/2)^{1/2}}{r^{3/2} \pm a\,(r_S/2)^{1/2}}. \tag{10.18}$$

The Keplerian frequency ν_K is simply $\Omega_K/2\pi$. One can easily verify that $\nu_K \approx 184$ Hz at a radius of 100 km from a 10 $M_\odot$ black hole. The frequencies in figure 10.7 are clearly in this range, so the evidence strongly suggests that general relativistic effects near the marginally stable orbit are responsible for the high-frequency QPOs.[31]

If particle orbits were indeed perfect circles, then ν_K would be the only frequency we could associate with harmonic motion in the circum-black-hole environment. But in reality, plasma in the disk may deviate away from its Keplerian path in two independent directions, each of which is associated with a restoring force, producing its own (additional) harmonic oscillation. Material in the disk may execute small excursions in radius (driven by equation 10.13), with a periodicity specified by the so-called *epicyclic* frequency ν_r, and may also venture off the equatorial plane (modulated by equation 10.14), giving rise to a third frequency that we will label ν_θ.

The procedure for finding ν_r is rather straightforward. Forming an equation for dr/dT from equations (10.11) and (10.13), one relinquishes the requirement that υ_r be strictly zero, and instead includes a perturbation in radial velocity $\delta\upsilon_r$. At the same time, a nonzero υ_r implies a corresponding perturbation $\delta\upsilon_\phi$ in the azimuthal direction, subject to the equation of motion produced from a merger of equations (10.10) and (10.11). Then, assuming these perturbed velocities to be much smaller than υ_ϕ, the dynamical equations are linearized (meaning that only terms first order in $\delta\upsilon_r$ and $\delta\upsilon_\phi$ are retained). The resulting equations are harmonic and their solution is

---

[31]The astute reader will notice that this argument could have been made on the basis of Newtonian gravity alone. However, the point here is that if we are indeed dealing with phenomena this close to the black hole, then the relativistic corrections appearing in equation (10.18) cannot be ignored.

characterized by the frequency

$$\nu_r(r) = \nu_K(r) \left[1 - 6 \left(\frac{r_S}{2r} \right) - 3a^2 \left(\frac{r_S}{2r} \right)^2 \pm 8a \left(\frac{r_S}{2r} \right)^{3/2} \right]^{1/2}. \qquad (10.19)$$

Physically, this motion is simply a slight oscillation in radius, executing cycles outward and inward on a given orbit. Notice that for the Schwarzschild metric (with $a = 0$), ν_r vanishes at $r = 3r_S$, a result that is entirely consistent with the fact that no restoring force exists to bring a particle back to the marginally stable orbit once it has crossed to smaller radii (see figure 3.2).

A similar procedure for perturbative motion in the vertical direction,[32] analyzed with the use of equations (10.10), (10.11), and (10.14), produces an analogous harmonic oscillation with its own distinctive frequency

$$\nu_\theta(r) = \nu_K(r) \left[1 \mp 4a \left(\frac{r_S}{2r} \right)^{3/2} + 3a^2 \left(\frac{r_S}{2r} \right)^2 \right]^{1/2}. \qquad (10.20)$$

All three of these frequencies are interesting and important in analyzing QPOs from black-hole binaries because, in general, $\nu_r \neq \nu_K$ and $\nu_\theta \neq \nu_K$. Thus, (slightly) eccentric orbits "waltz" back and forth at the *periastron precession* frequency $\nu_{per} = \nu_K - \nu_r$, and orbits tilted relative to the equatorial plane of a spinning black hole "wobble" at the *nodal precession* frequency $\nu_{nod} = \nu_K - \nu_\theta$. Remembering that no stable orbits are possible inside of the marginally stable orbit, which changes with black-hole spin a, we see right away that a positive identification of observed QPOs with any of these key frequencies provides us with a straightforward way of "measuring" the spin of the central object, or at least delimiting its possible range of values.[33]

The general idea that observed QPOs may be identified with any of these general relativistic frequencies is applied often in the primary literature, invoking orbiting clumps, vortices, and many other types of disturbance on top of a smooth, featureless disk. The difficulty in making a compelling case for any particular model, though, is that black-hole binaries are very complex systems, with many interlinked phenomena. If a magnetic field is present, say anchored in the disk, the complications multiply severalfold.

This is the state in which we find ourselves with regard to research on black holes in binaries. Using straightforward dynamical arguments, we can now be certain of a black-hole identification in at least 20 systems. Their spectra comprise several components, each of which reflects the physics in different regions of the circum-black-hole environment. These binaries undergo dramatic transitions in their physical state, evidenced by corresponding shifts in the radiation pattern they produce. One of the most promising avenues of future research, with a direct bearing

---

[32] See, e.g., Kato (1990).

[33] It is important to note that this type of analysis can also be carried out for supermassive black holes, even though they lack a binary companion. Accretion from the interstellar medium produces an accretion disk even in these objects, since infalling matter always carries with it a finite specific angular momentum. For a specific application of these ideas to the black hole at the galactic center, see Melia et al. (2001).

on our understanding of general relativity, is the pursuit of better timing information associated with rapid variability just outside the black hole's event horizon. Understanding how and why distinctive frequencies arise in the power density spectrum will undoubtedly provide us with a much deeper understanding of the spacetime surrounding these very interesting and exotic astrophysical objects.

SUGGESTED READING

Black-hole masses are often inferred from stellar motions within the binaries they occupy. See Orosz (2003) and Charles and Coe (2006).

The X-ray properties of black-hole binaries are reviewed in Remillard and McClintock (2006).

The formation, physical properties, and demographics of black holes are discussed in Kaper et al. (2001).

Some black holes in binaries undergo transitions in which several high-energy features alternately dominate the spectrum. See McClintock and Remillard (2006).

The concept of high-energy states and transitions between them originated with early observations of Cygnus X-1 and A0620-00. See Tananbaum, Gursky, Kellogg, Giacconi, and Jones (1972) and Coe, Engel, and Quenby (1976).

Black holes in binaries have many similarities with their much larger brethren in the nuclei of galaxies (see chapter 12). It is therefore not surprising that some of them should look like smaller versions of quasars. The term "microquasar" has been coined for stellar-mass black holes that look like active galactic nuclei in projection. See, e.g., Mirabel et al. (1992) and Mirabel (2001).

The great majority of black holes in binaries are detected primarily through their nova-like outbursts, which occur with a frequency of anywhere from once every few years up to once per century or longer. They have therefore been termed "X-ray novae." For a comprehensive review of their multiwavelength emission and dynamical properties, see Tanaka and Shibazaki (1996) and Liang (1998).

Nova outbursts are often modeled with a disk instability. Read more about this type of scenario in Meyer and Meyer-Hofmeister (1981), Mineshige and Wheeler (1989), Cannizzo (1993), and Lasota, Hameury, and Huré (1995).

The study of quasi-periodic oscillations (QPOs) in black-hole binaries can probe the spacetime near the event horizon in these systems. Read a comprehensive review on this phenomenon in van der Klis (2000).

Chapter Eleven

Bursting Stars

Though myriad transient phenomena dot the heavens, two particular types of explosive events stand out in our purview of high-energy astrophysics. One of these categories, thus far detected only from within our own Galaxy, is the sudden release of powerful bursts of X-rays from otherwise docile neutron stars. We need only remind ourselves of the "doomsday" marshmallow analogy from chapter 1 to realize the prodigious quantities of energy available in such compact environments to propel these systems to catastrophic outcomes.

The second category of energetic transient events is defined principally by the γ-rays they put out. These gamma-ray bursts (GRBs) are now known to occur in distant galaxies. One such recently detected event produced the brightest known explosion in the cosmos—second only to the Big Bang itself.

X-ray and gamma-ray bursts are not the only highly dynamical phenomena in astronomy, but they are certainly among the most interesting, since they have much to teach us about stellar evolution and large-scale structure, not to mention fundamental physics, including nuclear physics at very high densities, and the propagation of relativistic plasma through the interstellar medium.

But though they share an explosive identity, these two classes of events also differ in important ways. For example, X-ray bursts barely do damage to the underlying star, which therefore not only survives the event, but also repeats it—every few hours. In contrast, gamma-ray bursts completely commit the underlying object to the event—in a spectacular display of cosmic fireworks not easily seen in any other astrophysical context.

11.1 X-RAY BURST SOURCES

X-ray bursts were discovered independently by several groups in 1975, and were later studied intensively with the Small Astronomy Satellite 3 (*SAS*-3). Follow-up monitoring of the sources producing these events has been coordinated by every other X-ray and γ-ray mission flown since that time.[1] Of the 200 or so X-ray binaries we introduced in section 10.3, about 100 are LMXBs, and, of these, over half are accreting neutron stars, most of which are now known to produce X-ray bursts. The most recent population study was carried out with RXTE, spanning more than nine years of observation. The burst sample from that period includes over 1000

---

[1]The original discovery was made by Belian, Conner, and Evans (1976) and Grindlay et al. (1976). The first comprehensive study of this phenomenon was carried out by Lewin and co-workers using *SAS*-3 (Lewin and Joss 1981).

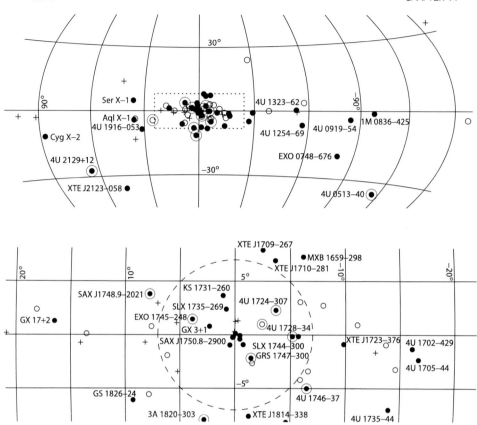

Figure 11.1 Sky distribution of bursters: +, bursters not observed by RXTE; •, those detected by RXTE; and ○, those observed but not detected by RXTE. Sources within globular clusters are additionally indicated by a larger concentric circle. The galactic center region (dashed rectangle in the upper panel) is magnified further in the lower panel. The four (unlabeled) sources closest to the galactic origin are (clockwise from lower left) SAX J1747.0−2853, GRS 1741.9−2853, KS 1741−293, and 2E 1742.9−2929. The dashed circle shows the approximate perimeter of the galactic bulge. Among the bursts observed by RXTE, some show H-ignition of mixed H/He fuel, some are initiated by He-ignition in pure He, and others are driven by He-ignition in mixed H/He fuel. See also figure 9.9. (From Galloway et al. 2006)

events, with an array of burst properties representing a wide range of theoretically predicted burst profiles. In the sky distribution of bursters observed by RXTE (shown in figure 11.1), the majority are concentrated toward the galactic center, suggesting a population of old binaries. At least 10 have been found in globular clusters.

The mean quiescent (i.e., interburst) source luminosity L_0 of the RXTE sample is $0.3 - 2 \times 10^{37}$ ergs s^{-1}, roughly one-tenth of an Eddington limit (equation 1.19). When an X-ray burst goes off, it typically has a rise time $\lesssim 1$ s, lasts 3–1000 s, and recurs on a timescale $t_r \sim 10^3$–10^6 s. The bursts themselves have luminosities $L_b \sim 10^{39}$ ergs s^{-1} and total energies $E_b \sim 10^{39}$–10^{40} ergs. Therefore, it appears that neutron stars radiate at or near $L_{\rm edd}$ during the events themselves. Of these bursting sources, 35 exhibit photospheric expansion, which, as we shall see, is consistent with the view that the neutron star's atmosphere is blown off when its luminosity approaches this limit.

There are actually two kinds of X-ray burst. The type we have identified here is called Type I, believed to be the result of thermonuclear flashes in the freshly accreted material on the surface of the neutron star. A second, less common, type of X-ray burst (labeled Type II) is shorter and erratic, and shows no evidence of thermonuclear burning and subsequent cooling in the stellar atmosphere. These are instead due to accretion instabilities in the surrounding disk, which lead to a temporary enhancement of the accretion rate onto the star, along the lines we investigated earlier with white-dwarf novae and (black-hole) X-ray novae, though not as energetic.

The primary evidence for the thermonuclear interpretation of Type I bursts comes from a comparison between the time-integrated persistent and burst fluxes. The ratio of the former to the latter, usually called α, may be used as a measure of the relative efficiency of the two processes responsible for producing them. This ratio may be evaluated as $E_b/t_r L_0$, and is observed to have a value ~ 20–300. A quick inspection of equations (1.1) and (1.2) reveals that these values are comparable to those predicted, assuming the burst energy is liberated in nuclear burning, whereas the persistent luminosity is due to the dissipation of gravitational energy. Figure 11.2 shows the quiescent spectrum of a prominent X-ray burst source, A1742-294, located within $\sim 1°$ of the galactic center. The emission accounting for this spectrum is thermal, and is generated by the dissipation of gravitational energy as the accreted matter settles into the surface of the neutron star.

The thermonuclear flash model has been very successful in reproducing the basic features of the X-ray burst phenomenon, including the short rise times, the recurrence timescales, the luminosities, energies released during the events, the spectral softening seen during the decaying portion of the burst, and (as we have just seen) the ratio of time-averaged burst luminosity to the persistent accretion power.[2] Figure 11.3 shows the X-ray lightcurve of a particularly strong burst from the aforementioned A1742-294, demonstrating many of the characteristics we have just described. What is particularly striking about this event is the clarity with which we can see the delayed onset of X-ray emission at the highest energies, and the gradual shift of emitted power toward lower energies during the burst decay, signaling the initial rapid heating of the burning layers and their subsequent cooling as the nuclear fuel is spent.

X-ray bursts are caused by the unstable burning of freshly accreted H/He on the surface of the neutron star, accumulated over a period of a few hours to form a layer ~ 10 m thick. The accreted matter contains very small amounts of heavier elements, since these binary systems are very old—given their distribution around the galactic center—and belong to Population II. In a few cases, matter is accreted from a He white dwarf companion, so the burning layer in those cases has very little, if any, hydrogen.

---

[2] Some notable papers that helped develop the thermonuclear flash model include those by Joss (1977), Lamb and Lamb (1978), Taam (1980), and Ayasli and Joss (1982).

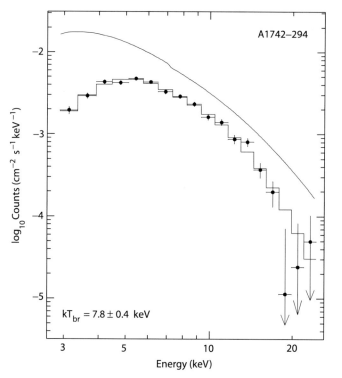

Figure 11.2 The X-ray burster A1742-294 is the brightest persistent X-ray source among those located within ∼1° of the galactic center (see figure 11.1), accounting for about $\frac{1}{3}$ of the 2- to 20-keV total flux from this region. This figure shows the typical persistent spectrum of this source measured with the ART-P telescope aboard the GRANAT observatory on 9 September 1990. Dots indicate the pulse-height spectrum, in counts $cm^{-2}\,s^{-1}\,keV^{-1}$. The histogram shows the best-fit optically thin bremsstrahlung emission, with the temperature indicated in the lower left corner. The model photon spectrum is indicated by the solid curve. The difference between the latter and the histogram is accounted for by absorption along the line of sight and the energy-dependent response matrix of the detector. (Image from Lutovinov et al. 2001)

The nuclear fuel is compressed and heated hydrostatically, and both the density and temperature at the bottom of the accreted layer increase until eventually the hydrogen starts to burn steadily into helium. If the temperature in the neutron-star envelope is sufficiently low to begin with, the hydrogen burns into helium in a thin shell via electron capture or the pycnonuclear (p-p) chain.[3] At higher densities, the helium burns steadily into carbon in another thin shell via the pycnonuclear triple-α reaction. Finally, at still higher densities, a sequence of electron capture reactions occurs, transforming carbon into heavy elements.

---

[3]The term thermonuclear refers to processes that depend primarily on the temperature, where the particle speed is essential, e.g., to overcome a Coulomb barrier. In low-temperature environments, the main dependence is often on the particle density, and these processes are called pycnonuclear.

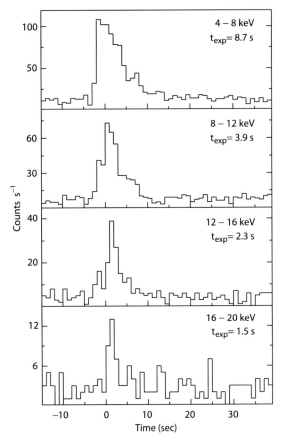

Figure 11.3 The strongest (and arguably most interesting) burst from A1742-294 was detected on 18 October 1990, during which the peak X-ray 3- to 20-keV flux was ∼1.5 Crab. This figure shows the measured lightcurves in various energy bands during this event. Notice that the peak flux in the hard energy bands is reached later than those of the soft energy bands, i.e., the burst rise time (∼5 s) is considerably longer in the 12- to 16-keV band than that (∼1–2 s) in the 4- to 8-keV band. The e-folding exponential decay time of the burst is denoted t_{exp}. The total burst duration in the hardest energy band is severalfold shorter than in the softest band. Evidently, the source's spectrum softened appreciably by the end of the burst. (Image from Lutovinov et al. 2001)

At high temperatures in the neutron-star envelope, H burns into He via the hot carbon–nitrogen–oxygen (CNO) process, and He continues to burn further via the triple-α reaction. Explosions eventually occur because these processes are thermally unstable.[4] As the temperature increases, the rate of collisions between the various nuclear species goes up, enhancing the burning rate, which in turn releases more energy, raising the temperature, and feeding a runaway cycle. These situations lead to various combinations of H and He flashes that we see as X-ray bursts.

---

[4] See, e.g., Hansen and Van Horn (1975).

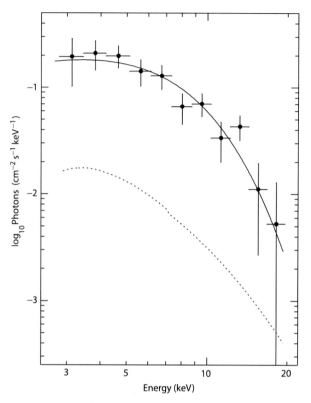

Figure 11.4 Photon spectrum of A1742-294 averaged over the entire 18 October 1990 burst. The solid curve represents the best-fit blackbody spectrum, compared with source's persistent spectrum, indicated by the dotted line. The blackbody spectrum during the burst has a characteristic temperature $kT_{bb} \approx 2.15 \pm 0.19$ keV, and is associated with a spherical emission surface with radius $R_{bb} \approx 6.4 \pm 1.4$ km. The total 3- to 20-keV luminosity is $L_{bb} \approx 1.2 \times 10^{38}$ ergs s^{-1}. (Image from Lutovinov et al. 2001)

The impact of runaway nuclear burning on the surface temperature of the neutron star is demonstrated dramatically in figure 11.4, which shows the photon spectrum of A1742-294 averaged over the powerful burst shown in figure 11.3. The burst spectrum is essentially a blackbody, indicating that the radiation was emitted within an optically thick medium before escaping. Interestingly, under the assumption that the emitting surface is spherical, one derives a photospheric radius of only ~6.4 ± 1.4 km. That is small for a neutron star, suggesting that only a portion of its surface was burning and/or radiating at any given time during the event. Below, we will consider several recent developments that shed more light on how the nuclear explosion takes place and develops across the surface.

Bursts for which the peak flux reaches levels comparable to the Eddington limit frequently exhibit a characteristic spectral evolution during the first few seconds, with an increase in blackbody radius and a corresponding dip in temperature, though the overall flux remains approximately constant. This is thought to result from the temporary expansion of the X-ray emitting photosphere once the critical luminosity is reached. Since the photons must now propagate through an increasing volume, their energy dilutes as they diffuse outward through the expanding medium. The effective temperature decreases as the burst flux in excess of the Eddington value is converted into kinetic and gravitational potential energy of the expanding envelope.

In recent years, thanks in part to RXTE, our understanding of X-ray bursts has deepened with a better appreciation of the role played by the neutron star's magnetic field. It has long been thought that a strong magnetic field stabilizes the nuclear burning in the freshly accreted material by funneling the hydrogen and helium onto its polar caps, where the heightened density and temperature cause the nuclear fuel to burn more rapidly and thus avoid the pileup that inevitably produces a thermonuclear explosion.

However, RXTE has discovered that many X-ray bursts exhibit large-amplitude brightness oscillations.[5] The natural interpretation for this is that the nuclear burning is not uniform over the stellar surface and that the oscillations are due to rotation of the neutron star. It appears that ignition of the thermonuclear flash occurs at a single point and then propagates around the star. In one particular burster, 4U 1636–536, enhanced nuclear burning apparently takes place at two nearly antipodal spots on the stellar surface. The surprising implication is that the weak external dipole magnetic field of this compact object succeeds in funneling the infalling plasma preferentially onto the magnetic poles, and at the same time, prevents the accumulating matter from spreading out over the surface. Contrary to earlier expectations, this does not stabilize the nuclear burning against thermonuclear runaway. Instead, the flash (or flashes) in this object occur at the magnetic poles.

Of course, this now raises the very important question of how the nuclear burning actually propagates across the stellar surface, a subject of considerable interest to computational astrophysicists who model these processes with high-resolution grids and a state-of-the-art nuclear reaction network and equation of state. The developing wisdom in this field has it that nuclear burning can propagate on a neutron star in three different ways.[6] In a deflagration wave (i.e., an ordinary flame), the burning front is convectively or turbulently unstable and propagates at a velocity $v_{\text{flame}} \sim 10^6$ cm s^{-1}, spreading over the neutron-star surface in a matter of seconds. If the nuclear burning ignites at a density $\rho \gtrsim 10^7$ g cm^{-3}, the burning front turns into a shock wave, forming a so-called detonation wave, and propagates at a velocity $v_{\text{deton}} \sim 10^9$ cm s^{-1}, spreading over the star in only a few hundredths of a second. However, if a very strong magnetic field is present ($B \gtrsim 10^{11}$ G), the burning front does not turn into a detonation wave, and convection and turbulence are suppressed. In this instance, the disturbance is mediated via a conduction wave propagating at a much lower velocity, $v_{\text{cond}} \sim 10^3$–$10^4$ cm s^{-1}, and takes several hundred seconds to spread across the surface.

As an aside, we mention that the physics of nuclear burning fronts is crucial to an understanding of how classical novae and Type Ia supernovae (both associated

---

[5] See, e.g., Strohmayer et al. (1998) and Galloway et al. (2006).
[6] Additional technical details on this topic may be found in Fryxell and Woosley (1982) and Bildsten (1995).

with white dwarfs) function as well. There, however, the burning takes place deep inside the star, obviating any possibility of actually seeing the process from the outside. In Type I X-ray bursts, on the other hand, the flame propagates within 10 m of the surface, and is easily observable. In addition, a typical neutron star rotates at 300–400 revolutions per second, providing hundreds of "snapshots" per burst to view all points on the stellar surface.

Plate 16 shows the results of one particular hydrodynamics calculation, designed to simulate the propagation of a nuclear burning front via a detonation wave. For simplicity, physical quantities have been averaged out in the azimuthal direction, so our view here is of a cross section in just two dimensions. To create the shock wave, the ignition was started inside a helium layer at a depth of about 90 m, with a density $\rho \approx 10^8$ g cm^{-3}. The detonation wave breaks through the surface of the neutron star at about $10 \, \mu$s, producing a rapidly expanding fireball. In less than a millisecond, the shock has extended the burning front to a height of over 1 km, and the outer layers of the photosphere are ejected.

Calculations such this, attempting to account for the thousands of X-ray bursts that have been observed by now, have considerably improved our view of physical processes in rather extreme conditions. Yet this is not the whole story. One can always do better, improving the temporal resolution even beyond RXTE's capabilities, and carrying out more sophisticated simulations. For example, one would like to see the results of a fully three-dimensional hydrodynamics simulation, analogous to the two-dimensional effort summarized in plate 16, in order to trace the detonation across the neutron-star surface while taking into account the strong Coriolis forces experienced by the burning material. Only then can one construct a realistic lightcurve for comparison with the observed brightness oscillations.

In the next section, we will study the "other" class of important bursting phenomena. Ironically, until the exquisite work of BATSE on the *Compton*-GRO, it was generally believed that γ-ray bursts were themselves also associated in some way with galactic neutron stars. The paradigm has shifted completely over the past decade and, in a way, γ-ray bursts now provide an excellent complement to X-ray bursts, in that they are not only associated with extragalactic sources but, unlike X-ray bursts, constitute the total destruction of the underlying object.

11.2 GAMMA-RAY BURST SOURCES

Several times per day, Earth is irradiated with the γ-ray glow of a distant explosion whose high-energy output—though lasting only a matter of seconds—has a brilliance during that fleeting moment as great as the entire visible universe. For several seconds, these bursts radiate a power equal to the total energy emitted by our entire Galaxy over a period of many years. The farthest such event seen to date,[7] known as GRB 050904,[8] had a redshift of $z = 6.39 \pm 0.12$, corresponding to an epoch when

[7] See Haislip et al. (2006).

[8] Gamma-ray bursts are named after the date of their discovery, so this particular event occurred on 9 May 2004.

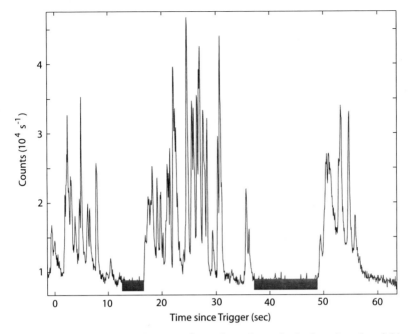

Figure 11.5 The duration of GRBs spans five orders of magnitude, from less than 0.01 s to more than 100 s. This is the lightcurve of GRB 920627, showing typical variability during the event. An often used measure of the duration T_{90} is the time during which 90% of the counts are made by the detector. GRBs are highly variable, showing 100% changes in flux on a timescale δt (sometimes just milliseconds) much shorter than T_{90}. (Image from Piran 2005)

the universe was only 900 million years old. Like the majority of other γ-ray bursts, it released many times the energy ($\sim 10^{51}$ ergs) of a typical supernova, creating conditions like those in the early universe within a region several hundred km across.

11.2.1 Gamma-Ray Burst Observations

Gamma-ray bursts (GRBs) are short, intense pulses of γ-rays lasting from a fraction of a second to several hundred seconds (figure 11.5). They arrive from random directions in the sky and (as we now know) from cosmological distances. The time-integrated flux, usually called the *fluence*, of these events ranges from $\sim 10^{-4}$ to 10^{-7} ergs cm^{-2}, corresponding to an *isotropic* luminosity of $\sim 10^{51}$–10^{53} ergs s^{-1}. However, the high-energy emission from these sources is believed to be beamed, lowering their actual power by one to two orders of magnitude below these estimates. This still makes them more powerful than a typical supernova, though sufficiently close in energy to warrant a study of any possible connection between the two.

Like many other important discoveries in astronomy (radio pulsars constituting a notable example), γ-ray bursts were first detected accidentally. Back in the 1960s, both the United States and the former Soviet Union had deployed military satellites designed to monitor compliance with the nuclear test-ban treaty. A tell-tale signature of a nuclear detonation is a brief, but intense, pulse of γ-rays. These devices did indeed measure such small bursts of high-energy emission, but the origin of the γ-rays was outside Earth's atmosphere, not on its surface. This became public information several years later, with the publication of results from the U.S. *Vela*[9] and Soviet *Konus* satellites.[10]

However, γ-ray bursts remained a complete mystery for several decades due to the fact that, until the 1990s, they were detectable only at γ-ray energies with instruments whose spatial resolution was (at best) a fraction of a degree. Source identification is impossible on this scale, and the growing mystery concerning the origin and nature of these elusive events spawned hundreds of theoretical models and interpretations, ranging from minor cometary phenomena in the solar system, to enormous supermassive black-hole convulsions in the distant cosmos.

But the 1990s saw the unfolding of two dramatic developments, heralding a new era in GRB research that would finally put an end to much of the speculation concerning this enigmatic phenomenon. The first was the launch of the *Compton* Gamma Ray Observatory (see section 1.5), carrying several instruments, including the Burst and Transient Experiment (BATSE). This detector was designed to be an all-sky monitor, specifically to record thousands of bursts for the express purpose of identifying their spatial distribution. Its earliest and most important result was a clear demonstration that γ-ray bursts are distributed isotropically in the sky (see plate 8), with no significant dipole or quadrupole moments, ruling out all possible origins other than a true cosmological population.[11]

The second major advance occurred with the launch in 1997 of the Italian–Dutch satellite *Beppo*-SAX.[12] The first of the X-ray missions with the capability of detecting photons over more than three decades in energy (from 0.1 to 200 keV), its principal strength was the fact that even with this broadband coverage, it could nonetheless still identify sources to within $\sim$1 arcmin accuracy in the 0.1- to 10-keV portion of the spectrum.

Soon after launch, *Beppo*-SAX obtained the first high-resolution X-ray images of the fading afterglow associated with the burst GRB 970228.[13] Now, after decades of uncertainty regarding the nature of GRBs, *Beppo*-SAX could finally identify γ-ray bursts with arcmin accuracy, at a rate of about 10 per year. This made it possible for other telescopes to follow the GRB afterglows at optical[14] and radio[15] wavelengths.

Of the many tens of afterglows observed with *Beppo*-SAX, the majority of cases have led to the identification of the host galaxy.[16] A spectacular example is illustrated

---

[9]See Klebesadel et al. (1973)

[10]See Mazets et al. (1974).

[11]See Meegan et al. (1992).

[12]Its name derives from Giuseppe "Beppo" Occhialini (1907–1993), a dominant figure in Italian physics and astrophysics in the latter half of the 20th century. SAX is an acronym for the Italian translation of "X-ray Astronomy Satellite."

[13]See Costa et al. (1997).

[14]See, e.g., van Paradijs et al. (1997).

[15]See, e.g., Frail et al. (1997).

[16]See, e.g., Hurley, Sari, and Djorgovski (2006).

in plate 15, which shows a *Hubble* Space Telescope image of the fireball accompanying the very first GRB discovered in this fashion, on 28 February 1997. The long-GRB[17] hosts are typically low-mass galaxies, with the blue colors and atomic-line spectra indicative of active star formation. Many of them are obscured, far-infrared luminous galaxies, some of which appear to be undergoing a tidal disruption.

Some afterglows have been followed for many months to over a year. Observations of longer-wavelength counterparts to GRBs have led to the measurement of redshift distances, and the confirmation that their distribution is undoubtedly cosmological. As we have seen, the current redshift record ($z = 6.39 \pm 0.12$) belongs to GRB 050904. The burst GRB 980425, with the nearest measured redshift ($z = 0.0085$) is important for another reason—its association with a supernova explosion—that we will examine later in this chapter (see also figure 11.10).

Actually, the bursts with measured intrinsic brightness would be detectable with current instrumentation out to much larger redshifts—perhaps as far as $z \sim 15$–20.[18] Indeed, within the first few minutes to hours after the γ-ray burst, the optical afterglow is far brighter than quasars. Thus, GRBs at very high redshifts could provide information, e.g., on the physical state of pregalactic gas, at much earlier epochs in the history of the universe than any other objects we know.

In examining the physics of γ-ray bursts, the first characteristic we deduce from their electromagnetic signal is that the typical GRB spectrum is nonthermal: the number of photons emitted per unit time is expressible as a function of photon energy ϵ in terms of two power laws ($N\epsilon \propto \epsilon^{-\alpha}$) connected at a *break* energy E_b. At low energies ($\epsilon \lesssim 0$–$1\,\mathrm{MeV}$), the spectral index α is ~ 1. The spectrum steepens considerably ($\alpha \sim 2$–3) toward higher energies.[19] Figure 11.6 shows the distribution of observed values of E_b for a sample of bright bursts. Most of them fall in the range 100–400 keV, with a clear maximum in the distribution around 250 keV.

In some cases, this phenomenological fit extends to at least GeV energies. For example, EGRET on the *Compton*-GRO detected seven GRBs with photon energies ranging from 100 MeV to 18 GeV. It is quite telling that this very high-energy emission is sometimes delayed by more than an hour after the main burst.[20] It is also quite revealing that no high-energy cutoff has ever been seen above a few MeV in any GRB spectrum. All these facts together suggest that the mechanism producing the GRB spectrum is not a common variety process we would have encountered with any of the other high-energy sources we have considered thus far. The radiation field produced by a GRB may be due to synchrotron emission (see section 5.5), or it may be a combination of synchrotron and inverse Compton scattering processes (see section 5.6), though with a unique source geometry and particle distribution.

The GRB lightcurve may be described as erratic, with a smooth, fast rise and quasi-exponential decay, through many peaks and substructure on a millisecond timescale (figure 11.5). Common measures for the burst duration include T_{90} and T_{50},

---

[17]We shall soon make an important distinction between long ($\gtrsim 2\,\mathrm{s}$) and short γ-ray bursts.
[18]See, e.g., Lamb and Reichart (2000).
[19]This phenomenological fit to the observed GRB spectrum was introduced by Band et al. (1993).
[20]See Dingus (2003).

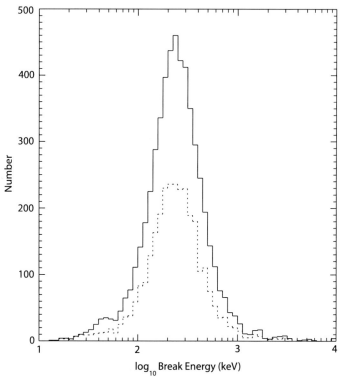

Figure 11.6 Gamma-ray bursts (GRBs) are short, intense pulses of soft γ-rays, lasting from a fraction of a second to several hundred seconds. They originate at cosmological distances and display a nonthermal spectrum, with a flux peaking at a few hundred keV, in some cases extending as a long high-energy tail up to GeV energies. A useful phenomenological fit for this spectrum comprises two power laws joined smoothly at a break energy, E_b. The distribution of GRBs as a function of E_b is shown here for two samples of bright events (solid and dashed lines). (Image from Preece et al. 2000)

corresponding to the time over which 90 and 50%, respectively, of the photon counts are received. The variability timescale δt is inferred from the width of the individual peaks, which are separated by a typical time interval Δt. The distributions for δt and Δt shown in figure 11.7 indicate that the average pulse interval ($\sim$1.3 s) is larger by a factor $\sim$1.3 compared to the average pulse width ($\sim$1 s).

The duration of bursts spans six orders of magnitude, from 10^{-3} s to about 10^3 s, with a well-defined bimodal distribution for events: those lasting longer than $\sim$2 s, and others ending earlier.[21] The short-burst spectrum is skewed toward higher photon energies than that of longer bursts—i.e., it is *harder*—so these events are usually termed "short hard bursts" (or SHBs) in the primary literature. The GRB hardness ratio is defined as the $\sim$100- to 300-keV fluence divided by that in the $\sim$50- to 100-keV energy band. A plot of this ratio versus the burst duration T_{90} is shown in

[21] See, e.g., Kouveliotou et al. (1993).

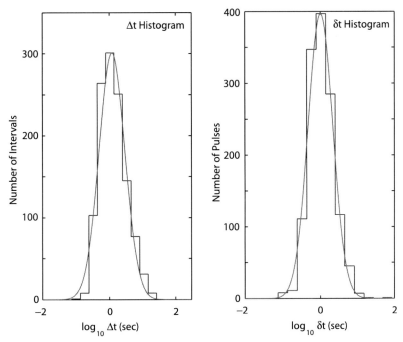

Figure 11.7 GRBs are composed of individual pulses, each of which displays a hard to soft evolution with peak energy decreasing exponentially with photon fluence. The pulse widths δt and the pulse separation Δt have similar log-normal distributions, though the pulse distribution seems to have an excess of long intervals. The relatively long segments of several dozen seconds with no activity may be classified as quiescent periods. The distributions for δt and Δt shown here indicate that the average pulse interval (~ 1.3 s) is larger by a factor ~ 1.3 compared to the average pulse width (~ 1 s). (Image from Nakar and Piran 2002)

figure 11.8, revealing an unmistakable division between the SHBs and their longer counterparts. We now know that these two different types of burst are associated with two very different kinds of catastrophic event.

One can demonstrate quite convincingly that the long and short bursts reflect different demographics by invoking another statistical test originally designed to study quasars.[22] Since the flux drops off as distance squared, and the volume within which sources may be found increases as the distance cubed, a plot of the number of sources with flux greater than S should follow a $-\frac{3}{2}$ power law versus S for a homogeneous distribution. The γ-ray burst data, however, indicate a much flatter profile, best characterized as a deficit in the number of weak bursts compared to what we would expect by extrapolating out to large distances.

An equivalent representation is the so-called V/V_{max} distribution, in which $V/V_{max} = (C_{max}/C_{min})^{-3/2}$, where C_{max} is the peak count rate for a particular event, and C_{min} is the detector's trigger threshold count rate. In this ratio, V is the volume contained within a sphere extending out to the location of the burst, and V_{max} is

[22]See Schmidt, Higdon, and Hueter (1988).

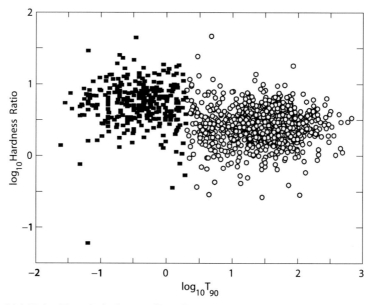

Figure 11.8 Using T_{90} as the indicator of burst length, we see a clear subdivision between hardness and softness in the observed γ-ray burst spectra, correlated with the event duration (solid symbols denote bursts with T_{90} less than 2 s; long bursts are represented by open symbols). Short bursts tend to be hard, and are therefore called SHB events. (Image from Qin et al. 2000)

the volume of a sphere extending out to the maximum distance at which this same burst would have still been detected given its observed count rate. The average value $\langle V/V_{max}\rangle$ thus provides a statistical test of the hypothesis that the parent population is spatially homogeneous, since in that case V/V_{max} would be uniformly distributed between 0 and 1, with an average value $\langle V/V_{max}\rangle = 0.5$.

The fact that the observed value of $\langle V/V_{max}\rangle$ is ~ 0.3, statistically different from 0.5, means that we are sampling out to the "edge" of the distribution,[23] yet according to BATSE, its angular dependence is uniform across the sky. Since we do not see any spatial correlation with objects within the Galaxy, or even with galaxies in the local group or the Virgo cluster, the parent distribution is evidently extended over much larger volumes than that, and must therefore have a cosmological origin. In addition, assuming that long and short bursts have the same spatial distribution, one can show from the burst catalogs that while $\langle V/V_{max}\rangle \approx 0.28$ for the former, this ratio is ≈ 0.39 for the latter, a statistically significant indicator that (the *observed*) short bursts are closer than the rest.[24] But note that this does not necessarily bear on the location of the various subcategories of burst. Rather, the different values

[23]Note, however, that other effects also contribute to the measured value of $\langle V/V_{max}\rangle$. For example, this ratio is 0.5 for a homogeneous distribution in a Euclidean space. Curvature effects can produce an average value different from this even though the distribution may be homogeneous in terms of local coordinates.

[24]See, e.g., Guetta and Piran (2005).

of $\langle V / V_{max}\rangle$ imply something about the difficulty of detecting short (intrinsically weaker) bursts.

The distinction between long and short bursts has been reinforced in recent years with a careful monitoring of their afterglow. With only rare exceptions, longer wavelength emission associated with GRBs is seen only from long bursts. There is strong evidence that this is more than a simple selection effect; short bursts do appear to be intrinsically much fainter across the spectrum. For example, *Chandra* observed the well-localized short hard burst GRB 020531 and detected no afterglow.[25] Its implied X-ray intensity must have been weaker by at least a factor $\sim$100–300 compared with that of typical long bursts.

When it is seen, the X-ray afterglow in long GRBs has the characteristic features indicated schematically in figure 11.9. It is the first and strongest follow-up emission to the γ-ray signal, and also the shortest. It seems to begin while the GRB is still active, and the lightcurve persists for several hours afterward. The overall energy released in the X-ray afterglow is generally a few percent of the GRB budget. *Beppo-SAX*, *ASCA*, *Chandra*, and *XMM-Newton* have sometimes also detected X-ray lines about 10 h after the main event, with typical line luminosities of around 10^{44}–10^{45} ergs s^{-1}, producing a total fluence of $\sim$$10^{49}$ ergs. Most of these lines are interpreted as emission lines of Fe K$_\alpha$.

Roughly half of the well-localized GRBs show an optical, infrared, and radio afterglow. The observed optical source a day after the burst is typically around the 19th–20th magnitude, though from here the signal decays over a timescale of months to a year. In all cases, the observed optical spectrum is itself a power law, with absorption lines created between the source and Earth superimposed on top of it. The highest redshift lines are typically associated with the host galaxy, providing a measurement of the GRB redshift.

This growing body of observational data suggests two different approaches to the task of building a theory of γ-ray bursts: a pseudo-phenomenological assessment of the ejecta producing the electromagnetic signal, and a more in-depth physical analysis of the engine itself. Such a dichotomy is driven in part by the constraints inherent in the types of observation we can make of the bursts and their afterglow. By the time the γ-ray emission has begun, the energy has long since been released, and certainly by the time the afterglow is measurable, we are witnessing the interaction of the explosion's ejecta with the surrounding medium. Phenomenological considerations can therefore provide us with an abundance of information concerning the evolution of the burst's "remnant." It is much more difficult to trace the early history of the central engine, since for that we must rely on an extrapolation to earlier times to understand the nature of the initial explosion.

But it is in the explosion itself that a compelling distinction may be made between the long and short bursts. So let us proceed with the general phenomenology of GRBs first, and then examine the pivotal observations and concordant theoretical developments that have provided us with believable models for the nature of these powerful events.

---

[25] See Klotz, Boér, and Atteia (2003) for a discussion of this burst's lack of an afterglow at several wavelengths, including X-rays.

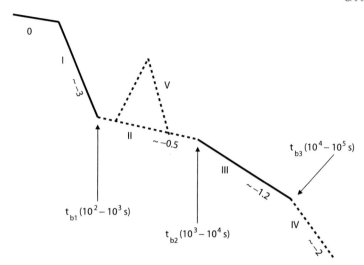

Figure 11.9 A synthetic diagram of a typical gamma-ray burst X-ray lightcurve, based on observations made with SWIFT, showing the five most commonly seen components: I, an initial steep decay, with a power-law index ~ -3 or steeper: II, a shallower-than-normal decay with slope ~ -0.5; III, a normal decay with slope ~ -1.2; and IV, a late steeper decay with slope ~ -2. Sometimes, flares are also seen (shown here as segment V). The phase "0" denotes the prompt (γ-ray) emission. Segments I and III are most common, and are shown here as solid lines. The other three components (dashed lines) are seen in only a fraction of the bursts. The typical temporal indices, labeled t_{b1}, t_{b2}, and t_{b3}, are shown at the corresponding breakpoints. (Image from Zhang et al. 2006)

11.2.2 Gamma-Ray Burst Theory

GRBs are inherently relativistic phenomena. At cosmological distances, the observed GRB fluxes would require energies up to $\sim 10^{54}$ ergs if the emission were isotropic. In addition, light-travel-time arguments based on the variability indicated in figure 11.7 suggest that this energy is liberated inside regions no bigger than ~ 100 km across. Thus, independently of the nature of the progenitor and the trigger for the explosion, we would expect such an intense and highly localized explosive release of energy to necessarily involve the rapid and extensive formation of e^+e^- pairs via $\gamma\gamma$ interactions.[26]

In fact, the optical depth to $\gamma\gamma \rightarrow e^+e^-$ interactions in such an environment would be much larger than one, and since the luminosity implied by the typical GRB flux is highly super-Eddington, even for a black hole source,[27] the exploding material must undergo rapid expansion. It then challenges us to understand why we even see radiation with energy $\epsilon \gg 1$ MeV, since such photons would presumably all have materialized into electron–positron pairs.

---

[26]This concept, originally framed within the context of a *fireball* scenario, was considered by Cavallo and Rees (1978).

[27]A black hole with a Schwarzschild radius of 100 km has a mass $\approx 33\ M_\odot$.

Although current thinking on the emission of γ-rays during the burst now focuses on the behavior of particles on the periphery of the expanding fireball, it is interesting to note that the solution to this problem (of seeing $\sim$10-GeV photons from a $\gamma\gamma$-thick medium) provided one of the earliest indications that we must be dealing with a relativistic outflow. To see why, let us consider the relativistic kinematics of the two-photon collision.

Let us suppose that the two incoming photons, with energies ϵ_a and ϵ_b, are approaching each other at an angle θ in the laboratory frame. To create an electron–positron pair in this interaction, the minimum energy of the incoming photons must be equivalent to just enough energy E^* in the center-of-momentum system (CMS) to produce the rest mass energy $(2m_e c^2)$ of the outgoing leptons. The CMS is characterized by

$$\mathbf{p}_a^* + \mathbf{p}_b^* = 0 , \tag{11.1}$$

where the asterisk indicates CMS quantities, and $\mathbf{p}$ is the three-momentum. From the Lorentz invariance of the four-momentum (see section 3.2), we therefore have

$$\frac{1}{c^2}(E^*)^2 = \frac{1}{c^2}(\epsilon_a^* + \epsilon_b^*)^2$$

$$= \frac{1}{c^2}(\epsilon_a + \epsilon_b)^2 - (\mathbf{p}_a + \mathbf{p}_b)^2$$

$$= \frac{2}{c^2}\epsilon_a \epsilon_b - 2\mathbf{p}_a \cdot \mathbf{p}_b . \tag{11.2}$$

Thus, since $\epsilon = pc$ for each photon, we find that the energies at threshold must satisfy

$$\epsilon_a \epsilon_b = \frac{2(m_e c^2)^2}{(1 - \cos\theta)} . \tag{11.3}$$

Clearly, two photons traveling in parallel ($\theta = 0$) could never produce an electron–positron pair, no matter what their energies were. For a head-on collision, on the other hand, one gets the minimum energies satisfying $\epsilon_a \epsilon_b = (m_e c^2)^2$.

According to equation (11.3), photons with 1-MeV and 10-GeV energies could survive and escape to an observer at infinity, but only as long as the angle θ were close to π. This would indeed happen if the hot plasma were moving relativistically, for then the radiation it emits would be beamed into the forward direction (relative to its bulk motion), within a narrow cone of half-opening angle $\sim 1/\gamma$ (see discussion in section 5.3). As long as $\theta < 1/\gamma$ for the given energies ϵ_a and ϵ_b, the photons would not materialize. Thus, the appearance of MeV and 10 GeV photons in a typical GRB spectrum suggests that the fireball expands with a bulk Lorentz factor

$$\gamma_{\text{bulk}}^2 \gtrsim \frac{\epsilon_a \epsilon_b}{4(m_e c^2)^2} \tag{11.4}$$

(using the expansion $\cos\theta \approx 1 - \theta^2/2$), or

$$\gamma_{\text{bulk}} \gtrsim 100 \left(\frac{\epsilon_a}{10\,\text{GeV}}\right)^{1/2} \left(\frac{\epsilon_b}{1\,\text{MeV}}\right)^{1/2} . \tag{11.5}$$

There is now ample evidence that the emitting plasma in a GRB is moving relativistically, though the simple fireball picture we have thus far described is untenable upon deeper scrutiny. The evidence includes radio scintillation measurements (see section 9.1), which indicate that the size of the afterglow is $\sim 10^{17}$ cm two weeks after the burst.[28] The implied speed of expansion is therefore $\sim c$.

The problem with a simple expanding fireball is that most of its internal energy would be converted into kinetic energy of the entrained baryons, rather than into a radiative luminosity, so the process would be very inefficient. In addition, since the medium is optically thick, the emergent spectrum would be quasi-thermal instead of the observed power-law (or combination of power-law) components.

A simple extension to this scenario improves the model considerably. The argument is based on the fact that a rapidly expanding outflow must eventually encounter an obstruction that causes it to shock. Often, this is simply the interstellar medium, perhaps dominated by the wind of the burst's progenitor. As we learned in section 4.3, shocks are efficient accelerators of particles. Thus, if shocks form in a GRB remnant once the fireball has become optically thin, they could reconvert the kinetic energy of the baryons back into nonthermal particles and, by extension, into radiation.

Fireball shocks come in two varieties. The easiest to anticipate are those produced when the GRB ejecta collide with the ambient medium and produce *external* shocks.[29] The synchrotron and combined synchrotron–inverse-Compton emission by particles accelerated in this environment can account for the general characteristics of the typical GRB spectrum. The accelerated electrons acquire the power-law distribution in equation (5.112) and, in the presence of turbulent magnetic fields built up behind the shock(s), produce synchrotron power-law radiation similar to that observed. Inverse-Compton scattering of these synchrotron photons then extends the GRB spectrum into the GeV range. The consensus view right now is that the much longer-lasting afterglow is indeed emitted by such external shocks. However, even with this evident success, we shall see below that the high-energy radiation in GRBs is probably produced by other means.

Internal shocks do even better than their external counterparts when it comes to the prompt emission itself—i.e., the γ- (and the earliest X-) radiation produced prior to the afterglow activity in the shocked external medium—because they also explain the rapid variability. Internal shocks arise when the plasma expands nonuniformly. Some of it may overtake a previous wave of ejection, and since the relative difference in velocities may still be supersonic, this collision can itself produce a shock.[30] In reality, several internal shocks may be present at any one time.

The observed GRB lightcurves are variable down to a timescale as short as a millisecond, even when the burst itself lasts tens of seconds (see figures 11.5 and 11.7). This is difficult to rationalize on the basis of a variable central engine, particularly since the evidence points to a catastrophic destruction of the progenitor. In addition, such variability would tend to get washed out within the optically thick fireball material, as we have seen. On the other hand, this rapid flickering could be

---

[28] See, e.g., Goodman (1997).

[29] See Rees and Mészáros (1992).

[30] See Rees and Mészáros (1994).

the radiative manifestation of multiple internal shocks jostling for preeminence in the expanding optically thin GRB remnant. We shall soon see why the variability timescale in this picture is expected to be so short.

At first glance, it would seem that even internal shocks would have difficulty producing millisecond variability, given that $c\delta t \sim 300\,\text{km}$, whereas the afore-mentioned radio scintillation measurements suggest an overall GRB remnant size $\sim 0.1\,\text{lyr}$, and images such as plate 15 reinforce the idea that GRBs are associated with supernova-type explosions, which cause a rapid, extensive expansion into the surrounding medium. But this naive view does not take into account the important effects of special relativity.

Suppose that the γ-ray emitting front moves at speed v from radius R_1 to R_2. The first photon emitted at R_1 then reaches the observer a time

$$\Delta T_{\text{obs}} = \frac{R_2 - R_1}{v} - \frac{R_2 - R_1}{c} \tag{11.6}$$

before the last photon emitted at R_2. Notice that when $v \approx c$, the first photon and the γ-ray emitting front move almost in parallel and reach R_2 at about the same time. Now, from the definition of the Lorentz factor in section 3.1, we have that

$$\frac{v}{c} = \frac{(\gamma_{\text{bulk}}^2 - 1)^{1/2}}{\gamma_{\text{bulk}}}, \tag{11.7}$$

from which we deduce that

$$\Delta T_{\text{obs}} \approx \frac{R_2 - R_1}{2c\gamma_{\text{bulk}}^2}. \tag{11.8}$$

The appearance of γ_{bulk}^2 in the denominator of this expression makes an enormous difference to the observed variability time when compared with the period of emission measured in the co-moving frame. Specifically, if we define $\Delta T_{\text{emiss}} \equiv (R_2 - R_1)/v$, then for $v \approx c$,

$$\Delta T_{\text{obs}} \approx \frac{1}{2\gamma_{\text{bulk}}^2} \Delta T_{\text{emiss}}. \tag{11.9}$$

That is, with $\gamma_{\text{bulk}} \sim 100$, even a fluctuation as long as $10\,\text{s}$ in the emitter's frame would appear as a mere millisecond variation to the stationary observer at infinity.[31]

The astute reader may accept this distinction between the observed and emitted variability timescales, but still wonder how it is that the overall burst duration can be so short when $10^{17}\,\text{cm}/c$ is about one month. Though in this case it is not the light-travel time factor that directly reduces the apparent burst duration, relativistic effects—in the guise of beamed emission—are responsible for this phenomenon as well.

Consider a relativistically expanding spherical shell or, more realistically, a locally spherical patch of shell (with radius R). Emission from parts of the shell moving at an angle θ from the line of sight to the observer arrives later than the radiation propagating directly along the reference direction, with a time delay $R(1 - \cos\theta)/c$.

---

[31] Incidentally, this so-called *light-travel time effect* is also responsible for the apparent superluminal motion of knots seen at the base of many quasar jets. We shall return to this issue in section 12.3.

For small angles θ, this time delay is just $R\theta^2/2c$. But since the radiation is beamed with an effective angle $\sim 1/\gamma_{\text{bulk}}$ (see section 5.3), the observer sees primarily the radiation emitted within the patch with opening angle $\theta \approx 1/\gamma_{\text{bulk}}$. The characteristic burst duration is therefore expected to be

$$\Delta T_{\text{burst}} \approx \frac{R}{2c\gamma_{\text{bulk}}^2} . \tag{11.10}$$

Again, the appearance of γ_{bulk}^2 in the denominator of this expression reduces the observed burst duration considerably from what it would otherwise be in the absence of relativistic effects. For example, with $\gamma_{\text{bulk}} \sim 100$, a γ-ray burst lasting ~ 1 month in the co-moving frame appears as a mere 300-s event to the observer at infinity.

The fact that both the typical GRB duration and variability timescale may be so easily reconciled with the observations, even for an explosive event extending out to a fraction of a light-year in size, constitutes one of the most compelling arguments in favor of the relativistically expanding shock model for GRBs. However, the fireball phenomenology of GRB emission does not address the most interesting question: what produces the explosion in the first place? Several clues delimit the range of possibilities. First, the total released energy needs to be $\gtrsim 10^{51}$ ergs, a significant fraction of the binding energy of a stellar compact object.

Second, most GRBs are collimated, with typical opening angles $1° < \theta < 20°$. We know this from a consideration of the burst afterglow, shown schematically in figure 11.9. The first critical downturn in the prompt emission is produced when the bulk Lorentz factor γ_{bulk} decreases to values for which $1/\gamma_{\text{bulk}} > \theta$. Thus, knowledge of γ_{bulk}, e.g., from the γ-ray and X-ray spectra, provides a reasonable estimate of the range in θ. Additional evidence that the fireball expansion is collimated is provided by the estimated GRB rate, inferred to be roughly one event every 300,000 years per galaxy. GRBs are therefore 3000 times rarer than typical supernova explosions. If these two phenomena are related, as now appears likely (see below), then significant anisotropy in the prompt high-energy emission could at least partially account for the observed rate differential.

The pivotal event that brought the supernova–GRB connection into focus, at least for the long bursts, occurred on 25 April 1998 (GRB 980425), nearly coincident with one of the most unusual supernova explosions ever seen, SN 1998bw (see figure 11.10). Core-collapse supernovae are the explosive deaths of massive stars that occur when their iron core collapses to a neutron star or black hole. They are generally not accompanied by highly relativistic mass ejection, but are visible from all angles and last from weeks to months. If their hydrogen envelopes have been lost prior to the collapse, they produce what are commonly known as Type Ib or Ic explosions; otherwise, the supernovae are labeled Type II.[32] Most of a supernova's emission, $\sim 10^{53}$ ergs s^{-1}, is known to occur in neutrinos, which are difficult to detect unless the event occurs nearby (within ~ 50 kpc). But even a fraction of this power, redirected into a collimated plasma outflow, is sufficient to drive a GRB.

A supernova origin for GRBs was confirmed in compelling fashion with the observation of another event (SN 2003dh) that occurred nearly simultaneously with GRB 030329 (see figure 11.11). In this case, the source spectrum evolved from a

---

[32]See, e.g., Filippenko (1997).

Figure 11.10 By circa 2004, host galaxies had been seen for all by 1 or 2 GRBs with arcsec precision of their optical, radio, or X-ray afterglow. The association of bursts with star-forming regions suggests that these events are related to supernova explosions. The first indication of such an association was made with the discovery of SN 98bw within the error box of GRB 980425. This was an unusually bright, type Ic SN, with an ejection velocity $\sim$20,000 km s^{-1}, corresponding to a kinetic energy $\sim$2 $\times$ 10^{52} ergs, more than ten times that of other supernova explosions. The optical transient seen in the New Technology Telescope R-band image of the spiral galaxy ESO 184-G82 on the left lies within the 8-arcmin error box of the *Beppo*-SAX Wide Field Camera image of this event, produced about one day after the GRB. The transient is absent in the older Digitized Sky Survey image on the right. (Image from Galama et al. 1998)

power-law continuum with narrow emission lines originating from HII regions in the host galaxy, to the development of broad peaks characteristic of a supernova. Many other GRB afterglows have also revealed "bumps" consistent in color, timing, and brightness with what would be expected from Type Ib or Ic supernovae of luminosity comparable to SN 1998bw. These broad features are produced by the high-velocity ejecta of the nascent remnant. Of course, observations such as these beg the question "Why should some stars produce ordinary supernova explosions, whereas others follow the GRB route?" Though this question is still open, it appears that *rotation* may be the distinguishing feature, GRBs may be produced only by the most rapidly rotating and most massive stars, whereas ordinary supernovae, which comprise about 99% of massive star deaths, occur in objects whose rotation plays a smaller role, or no role at all.

The model that best accounts for the inferred properties of the GRB explosion is the so-called *collapsar* scenario, in which a massive star endowed with fast rotation (let's say, a single Wolf-Rayet star) collapses and forms a small black hole that continues to accrete from a transient disk. The key ingredient in this picture is the relativistic jet that penetrates through the envelope of the collapsing star and breaks out into the surrounding medium.[33]

---

[33]The collapsar model was first proposed by Woosley and his collaborators (Woosley 1993). The collimation of a jet by the stellar mantle was shown to occur analytically by Mészáros and Rees (2001).

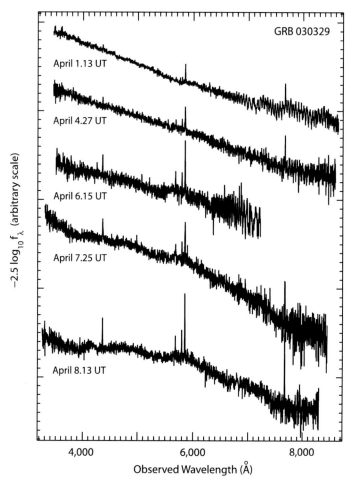

Figure 11.11 A SN 98bw-like bump in the GRB afterglow lightcurve, and its support for the GRB-supernova association, were confirmed dramatically with the very bright GRB 030329 and SN 2003dh. Following the evolution of the GRB 030329/SN 2003dh spectrum, starting from April 1.13 UT (2.64 days after the burst) to April 8.13 UT (6.94 days after the burst), one can see the transition from a power-law continuum ($F_\nu \propto \nu^{-0.9}$) with narrow emission lines originating from HII regions in the host galaxy at redshift $z = 0.168$, to the development of broad peaks characteristic of a supernova. (Image from Stanek et al. 2003)

According to this model, it is the massive iron core of the massive star ($M > 30\, M_\odot$) that collapses to a black hole, either directly or during the accretion phase that follows the core collapse. Due to the large angular momentum of the star's interior, a transient disk develops around the compact object, and a funnel emerges along the rotation axis, where the stellar material has relatively little rotational support. In numerical simulations of this process, the accretion disk has a mass of $\sim 0.1\, M_\odot$ and drains into the black hole over a period of several dozen seconds, powering the GRB.

However, there is a long-standing problem in core-collapse supernova models: it is not yet clear how the liberated energy is actually extracted and deposited into the surrounding matter, though neutrino transport and magnetic effects probably all contribute to this process. It is believed that once the proto-neutron star forms at the center of the collapse, a luminosity of $\sim 10^{53}$ ergs s^{-1} is radiated in neutrinos, 10% of which are captured on nucleons in the region between the nascent neutron star and the accretion shock that develops above it. This energy deposition shuts off any further accretion, and inflates a bubble of radiation and electron–positron pairs that push off the rest of the star, making a supernova. If this were not to happen, the collapse would proceed without any significant visual signal, leading to what is commonly termed a "failed" supernova.

Numerical simulations do not demonstrate an efficient energy deposition scenario very robustly. More often than not, the neutrino energy deposition, by itself, fails to launch and sustain an outbound shock of sufficient energy to power observed supernova explosions. The problem may be numerical, in the sense that full three-dimensional simulations must be carried out to capture the complex fluid flow in the convective region where the neutrinos deposit their energy. In addition, the neutrino transport must itself be handled very carefully and coupled to the hydrodynamics. The problem may also be due to inadequate physical modeling. Perhaps magnetic fields and rotation are not being handled properly, or the process may be affected by uncertainties in the high-density equation of state, changes in fundamental particle physics, such as neutrino flavor mixing, or other corrections.

The uncertainty regarding how a supernova explosion develops above the collapsing core is a lingering burden on our attempts to fully understand how the γ-ray burst engine releases its energy. But in the case of collapsars, once the energy is transferred (by whatever means), it presumably leaks out preferentially along the rotation axis, producing jets with opening angles $\lesssim 10°$. A snapshot of the star's envelope following its core collapse is shown in plate 17.

In these calculations, the process of core collapse, accretion along the polar column, and the jet propagation through the stellar envelope together take about ~ 10 s. The ensuing accretion onto the black hole takes tens of seconds longer. The timing of these events is therefore consistent with the measured properties of long bursts, whose observed afterglow is consistent with the breakout of the jet into the interstellar medium. But it would be quite a stretch to suggest that collapsars could also account for short bursts which, as we have seen, are observationally quite different from the former.

Instead, the short bursts appear to be associated with another class of progenitors that has received attention in recent years—neutron-star binaries or neutron star-black hole binaries that lose orbital angular momentum by gravitational wave radiation and undergo a merger (see section 9.3).[34] These catastrophic events also produce a black hole surrounded by a temporary debris torus whose accretion provides a sudden release of gravitational energy, though the duration of the burst in binary mergers is related solely to the fall-back time of matter flowing into the

---

[34]This class of models has been discussed extensively in the primary literature. Key papers include Paczyński (1986), Goodman (1986), Eichler et al. (1989), and Mészáros and Rees (1997).

black hole. The split between long and short bursts may therefore simply be the dichotomy between collapsar and merger explosions.

This suspicion was confirmed observationally in 2005, with the notable detection of two short bursts, by two different satellites. On 9 May of that year, the X-ray wide-field Burst Alert Telescope (BAT) on the newly deployed *Swift* satellite[35] sensed a pulse of γ-rays, which triggered a rapid slewing of the spacecraft to point its narrow-field X-ray telescope (XRT) and the ultraviolet-optical telescope (UVOT) in the direction of what turned out to be the short burst GRB 050509B. No optical afterglow was detected,[36] consistent with many earlier (failed) attempts to detect long-wavelength emission from such short events, but its X-ray afterglow allowed an accurate identification of the explosion with the periphery of a luminous, non-star-forming elliptical galaxy at a redshift of 0.225.

A few months later, another spacecraft built to study GRBs—the High Energy Transient Explorer 2 satellite (HETE)—detected its own short burst,[37] cataloged as GRB 050709. In this case, however, not only was there an X-ray afterglow, but (for the first time) an optical afterglow was seen following a short event. The burst occurred on the outskirts of a late-type galaxy at redshift $z = 0.16$ (see figure 11.12). At this redshift, the isotropic-equivalent energy released in γ-rays over the 25- to 2000-keV band was $\sim 6.9 \times 10^{49}$ ergs. The X-ray flare that followed 25–130 s after the burst had an energy fluence about twice this value. Thus, the total (isotropic) energy released during the first few hundred seconds of this event was only $\sim 10^{50}$ ergs, two orders of magnitude smaller than that seen during typical long-duration bursts.

Tellingly, however, though both X-ray and optical afterglows were detected in this event, no evidence of a supernova explosion emerged at any time before or after the prompt γ-ray emission,[38] supporting the growing suspicion that the short GRBs have a different origin than their longer counterparts. They are lower-energy explosions, with a correspondingly less-energetic relativistic blast wave, and they appear to occur at significantly shorter distances.

In addition, both of these well-identified short GRBs were localized far (between ~ 4 and 40 kpc) from the center of their respective host galaxy. Neutron star–neutron star binaries often receive significant kick velocities (100–1000 km s^{-1}) from the supernova that created each binary member. Thus, if one ignores the effects of the galactic potential, such a binary moving at 1000 km s^{-1} would be expected to travel as far as 100 kpc in 10^8 yr. The projected distance of GRBs produced by the catastrophic merger of compact binaries is therefore expected to be as large as ~ 40 kpc, consistent with what has been observed.

The nature of the central engine in short γ-ray bursts is now well constrained. The absence of any large-amplitude oscillations with a period in the range 1–10 s in their lightcurves and the large offset from the center of their host galaxy argue against an origin in soft γ-ray repeaters (see section 9.1). The time it takes collapsars

---

[35] See Gehrels et al. (2004).
[36] See Gehrels et al. (2005).
[37] See Villasenor et al. (2005).
[38] See Hjorth et al. (2005).

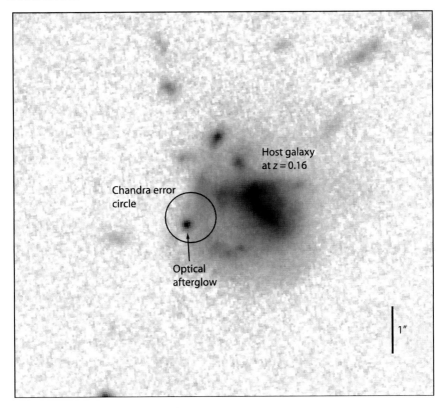

Figure 11.12 *Hubble* Space Telescope and *Chandra* X-ray observations of the afterglow and environs of GRB 050709. The circle, 0.5 arcsec in radius, represents the localization region of the X-ray afterglow source. Within this region, a prominent optical point source fades over one month, and is identified as the counterpart to the γ-ray transient. The irregular galaxy to its west is the $z = 0.16$ host galaxy. At this redshift, the afterglow's 1.38-arcsec offset from the brightest region of this galaxy corresponds to 3.8 kpc in projection. (Image from Fox et al. 2005)

to develop a relativistic jet and the absence of any observed supernovae associated with these events also argue against collapsars as the agents responsible for these bursts. Neutron-star binaries, on the other hand, merge in only a fraction of a second once the stars begin to touch and, given the relatively small mass of the temporary debris disk, it is not unreasonable to expect a lower average energy for short, hard bursts than for the collapsar-driven ones.

In the future, there is the exciting possibility of detecting gravitational waves directly from these events, using the second generation of laser interferometry detectors, the Laser Interferometer Gravitational-Wave detector (LIGO) and its sister facility VIRGO, located within the site of EGO (European Gravitational Observatory) at Cascina, Italy.[39] LIGO and VIRGO are facilities dedicated to the

---

[39] See Barish and Weiss (1999).

detection of cosmic gravitational waves. These instruments will detect ripples in spacetime by using interferometry to measure the time it takes light to travel between suspended mirrors with high precision. Two arms, perpendicular to each other, will produce an interference pattern by the two reflected beams if there's any perturbation caused by a passing gravitational wave. In years to come, we may learn a great deal more about GRBs, beyond the electromagnetic signals they produce. What awaits us is the likely prospect of detecting gravitational radiation directly from them, providing important new tests of Einstein's theory of general relativity.

SUGGESTED READING

The field of X-ray bursts is quite mature now, so relatively little new material appears in the primary literature. Most of the reviews are quite old compared to those in other areas, but their contents are, of course, still relevant. The recommended reading includes two of the most influential articles written on this topic and more recent articles highlighting the impact of modern computational resources on the numerical simulations. See Lewin and Joss (1981), Lewin, van Paradijs, and Taam (1993), Fryxell et al. (2000), and Heger et al. (2007).

For a review of gamma-ray bursts from a theorist's perspective, focusing on the fireball shock scenario, see Mészáros (2002).

A more recent review, highlighting progress on gamma-ray bursts in the Swift era, may be found at Zhang (2007).

The connection between supernova explosions and gamma-ray bursts is explored in Woosley and Bloom (2006).

Our understanding of gamma-ray bursts expanded rapidly once they could be associated reliably with emission at longer wavelengths. Read about the afterglow phenomenon in Costa et al. (1997), Mészáros and Rees (1997), Sari et al. (1998), and Rhoads (1999).

Chapter Twelve

Supermassive Black Holes

12.1 THEIR DISCOVERY AND IDENTIFICATION

Gamma-ray bursts may outshine the rest of the visible universe during their peak emission, but they are not even close to being the most powerful high-energy sources known. That honor belongs to the thousands (probably millions) of active galactic nuclei (AGNs) sprinkled throughout the cosmos, and to their more distant brethren—the quasi-stellar radio sources (or quasars, for short).[1]

Discovered in the early 1960s, the earliest members of this class were known principally as point-like radio emitters, identified with stellar-like sources possessing very unusual spectra, featuring several prominent emission lines. The recognition that these lines were highly redshifted led to the immediate realization that the objects producing them must be distant and extremely luminous,[2] typically producing radiation at a rate of $L_q \sim 10^{46}$ ergs s^{-1}. At this level, quasars are 10 times more luminous than the brightest galaxies.

Yet this radiative output is highly variable—in all observable wavelength bands—on timescales from days to years. Many of these sources vary even more quickly in X-rays, some showing a measurable change in a matter of hours. Light travel-time arguments therefore constrain the origin of their enormous luminosity to a highly compact volume, with a scale no bigger than that of the solar system.

This conclusion was actually reached very early on, because already by the end of the 1960s, the conversion of mass into radiation via accretion onto a black hole was recognized as the most efficient source of energy possible. A very elegant argument to this effect was provided by Lynden-Bell,[3] who demonstrated the implausibility of powering quasars via nuclear reactions alone.

To see this, suppose that a typical quasar shines for a time $\tau_q \sim 10^9$ years—a likely prospect, considering that these objects are sometimes observed at a redshift exceeding 6 (see below), when the universe was only a fraction of its current size. Then, the total energy output during its lifetime is $E_q \sim \tau_q L_q$, or roughly 3×10^{62} ergs. According to equation (1.3), the efficiency ($\epsilon_{nuc} \equiv \Delta E_{nuc}/c^2$) of producing energy through nuclear reactions is only ~0.7%, so the nuclear waste left behind by a quasar would be at least $M_q \approx 4 \times 10^{10} \, M_\odot$.

---

[1] We shall revist the AGN taxonomy in section 12.3, though we mention here that AGNs and quasars are commonly grouped together in the same family.

[2] See Schmidt (1963).

[3] See Lynden-Bell (1969). A related argument was also made earlier by Salpeter (1964), and Zel'dovich and Novikov (1965).

At the same time, quasar engines themselves cannot be very big. They are certainly smaller than $\sim 10^{15}$ cm, based on the aforementioned variability argument, given that their luminous output varies significantly over ~ 10 h. But the gravitational potential energy of $4 \times 10^{10} \, M_\odot$ compressed into a volume of spatial dimension $R = 10^{15}$ cm is $GM_q^2/R > 10^{65}$ ergs. Quoting directly from Lynden-Bell, "Evidently, although our aim was to produce a model based on nuclear fuel, we have ended up with a model which has produced more than enough energy by gravitational contraction. The nuclear fuel has ended as an irrelevance."

There are now several additional lines of evidence in support of the supermassive black-hole paradigm for these sources. For example, we shall see in section 12.3.4 that a relativistically deep gravitational potential well is unambiguously motivated by the detection of radio jets exhibiting knots moving across the plane of the sky at speeds exceeding c, a phenomenon known as "superluminal" motion.

In other observations, many AGN jets appear to remain well collimated and straight over Mpc scales (see, e.g., figure 12.4). Assuming the plasma within them is moving at close to the speed of light, this distance translates into an AGN temporal "memory" of up to $\sim 10^7$ yr. The most "natural" explanation for this occurrence is the gyroscopic stability associated with a single rotating body that launches the jet in a fixed direction.

Equally compelling evidence for the supermassive black-hole nature of AGNs and quasars is provided by direct probes of the spacetime near the central engine itself. For example, Very Large Baseline Array (VLBA) observations of water maser hotspots in material orbiting in a thin disk around the Seyfert 2 galaxy NGC 4258 reveal an almost perfect Keplerian motion about the central source of gravity.[4] Spanning a distance of 0.13–0.26 pc from the center, these masers indicate the presence of a mass $\sim 3 \times 10^7 \, M_\odot$ enclosed within a region no bigger than 0.13 pc.

The spectra of some AGNs permit us to sample the behavior of matter even closer to the central black hole than this. As we shall see in section 12.3, modern X-ray satellites, such as ASCA and XMM-*Newton* (plate 2), have uncovered many instances of relativistic motion in these systems, via the detection of a strong, broadened Fe K_α emission line in the energy range ~ 2–6.9 keV (see figure 12.16). These lines exhibit an enormous Doppler broadening, which, following equation (3.18), implies that the emitter is characterized by a wide range of Lorentz factors γ, and hence a large spread in velocities; in some cases, the spread reaches as high as 100,000 km s^{-1}, or roughly $0.3c$. In addition, these lines are asymmetric, skewed strongly toward the red end of the spectrum, suggesting emission by a disk extending down to just a few Schwarzschild radii (equation 3.88) above the black hole's event horizon.

Quasars are harder to study because of their great distance. Even so, the most widely accepted view today is that they too are found in the active nuclei of galaxies hosting a supermassive black hole (see figure 12.1). This coupling was not recognized immediately because the "host" galaxies appear very small and faint, and are hard to see against the much brighter quasar light emanating from the core.

Quasars actually reside in the nuclei of many different types of galaxy, from the normal to those highly disturbed by collisions or mergers. A supermassive black hole

---

[4]See Miyoshi et al. (1995).

Figure 12.1 This *Hubble* Space Telescope (HST) image reveals the faint host galaxy within which dwells the bright quasar known as QSO 1229+204. The quasar is seen to lie in the core of an ordinary-looking galaxy with two spiral arms of stars connected by a bar-like feature. (Photograph courtesy of John Hutchings, Dominion Astrophysical Observatory, and NASA)

at the nucleus of one of these distant galaxies "turns on" when it begins to accrete stars and gas from its nearby environment; the rate at which matter is converted into energy can be as high as $10\,M_\odot\,\mathrm{yr}^{-1}$ (see figure 12.6). So the character and power of a quasar depend in part on how much matter is available for consumption. Disturbances induced by gravitational interactions with neighboring galaxies can trigger the infall of material toward the center of the quasar host galaxy (see figure 12.2). However, many quasars reside in apparently undisturbed galaxies, and this may be an indication that mechanisms other than a disruptive collision may also be able to effectively fuel the supermassive black hole residing at the core.

12.1.1 The Supermassive Black Hole Census

By now, some 15,000 distant quasars have been found, though the actual number of supermassive black holes discovered thus far is much greater. Because of their intrinsic brightness, the most distant quasars are seen at a time when the universe

Figure 12.2 The collision between two galaxies begins with the unraveling of the spiral disks. This HST image shows the interacting pair of galaxies NGC 2207 (the larger, more massive object on the left) and IC 2163 (the smaller one on the right), located some 114 million light-years from Earth. By this time, IC 2163's stars have begun to surf outward to the right on a tidal tail created by NGC 2207's strong gravity. (Image courtesy of Debra M. Elmegreen at Vassar College, Bruce G. Elmegreen at the IBM Research Division, NASA, M. Kaufman at Ohio State U., E. Brinks at U. de Guanajuato, C. Struck at Iowa State U., M. Klaric at Bell South, M. Thomasson at Onsala Space Observatory, and the Hubble Heritage Team based at the Space Telescope Space Institute and AURA)

was a mere fraction of its present age, roughly one billion years after the Big Bang. The current distance record is held by an object found with the Sloan Digital Sky Survey (SDSS), with a redshift of ∼6.42, corresponding to a time roughly 700–800 million years after the universe was born.[5]

The SDSS has shown that the number of quasars rose dramatically starting from this epoc to a peak around 2.5 billion years later, falling off sharply toward the present time. Quasars turn on when fresh matter is brought into their vicinity, and then fade into a barely perceptible glimmer not long thereafter.

We may get a better idea of how many supermassive black holes may be lurking in the cosmos by reexamining plate 6—the *Chandra* Deep Field North image. This is

[5] See White et al. (2005).

one of the two deepest images ever made of the distant universe at X-ray energies—the other being its counterpart in the southern hemisphere. As we noted earlier in section 2.1, based on the number of suspected supermassive black holes in this image, one infers an overall population of ~300 million spread throughout the cosmos.

And yet, these X-ray detections speak only of those particular supermassive black holes whose orientation facilitates the transmission of their high-energy radiation toward Earth. Their actual number must be higher than this; indeed, there is now growing evidence that many—perhaps the majority—of these objects are obscured from view. This inference may be drawn from a consideration of the faint X-ray background pervading the intergalactic medium, which has been a puzzle for many years. (We shall return to this in chapter 13.) Unlike the cosmic microwave background radiation left over from the Big Bang, the photons in the X-ray haze are too energetic to have been produced at early times. Instead, this radiation field suggests a more recent provenance associated with a population of sources whose overall radiative output may actually dominate over everything else in the cosmos. Stars and ordinary galaxies simply do not radiate profusely at such high energy, and therefore cannot fit the suggested profile.

A simple census shows that to produce such an X-ray glow with quasars alone, for every known source there must be ten more obscured ones. This would also mean that the growth of most supermassive black holes by accretion is hidden from the view of optical, UV, and near infrared telescopes. We now have some direct evidence that this must be true, following the recent discovery of an object termed a Type-2 quasar.[6] Invisible at optical wavelengths, the nucleus of this otherwise normal looking galaxy betrays its supermassive guest with a glimmer of X-rays. The implication is that many more quasars, and their supermassive black-hole power sources, may be hidden in otherwise innocuous-looking galaxies.

And so, the all-pervasive X-ray haze, in combination with the discovery of gas-obscured quasars, now points to supermassive black holes as the agents behind perhaps *half* of all the universe's radiation produced after the Big Bang. Ordinary stars no longer monopolize the power as they had for decades prior to the advent of space-based astronomy and the development of high-energy astrophysics.

Between the quasar realm (extending out to distances ~11 Glyr) and the nearby galactic nuclei we shall consider in the next section (restricted to distances of only a few Mlyr or less) dwell the supermassive black holes accreting at a rate somewhere between the two extremes we see near us and at the edge of the universe. An archetype of this AGN group, Centaurus A, graces the southern constellation of Centaurus at a distance of 11 Mlyr (see figure 12.3). At the center of the dark bands of dust, HST recently uncovered a disk of glowing, high-speed gas, swirling about a concentration of matter with a mass of $\sim 2 \times 10^6\ M_\odot$. This large mass in the central cavity cannot be due to normal stars, which would shine brightly, producing an intense optical spike toward the middle, unlike the rather tempered look of the infrared image shown here.

Another well-known AGN, Cygnus A (see figure 12.4), demonstrates the strong collimation and consistency of direction exhibited by the jets produced at the core

[6]See Fabian et al. (2000).

Figure 12.3 This image of Centaurus A is a composite of three photographs taken by the European Southern Observatory. The dramatic dark band is thought to be the remnant of a smaller spiral galaxy that collided, and ultimately merged, with a large elliptical galaxy. At radio wavelengths, one sees two jets of plasma spewing forth from the central region in a direction perpendicular to the dark dust lanes. (Photograph courtesy of Richard M. West and the European Southern Observatory)

of such an object. The luminous extensions in this field project out from the nucleus of Cygnus A, an incredible distance three times the size of the Milky Way. Yet located 600 Mlyr from Earth, they cast an aspect only one-tenth the diameter of the full moon.

12.1.2 Their Formation

Clearly, we have a rather good idea now of how these objects are distributed around the cosmos, but an increasingly important question being asked in the context of supermassive black holes is which came first, the central black hole or the surrounding galaxy? Quasars seem to have peaked 10 Gyr ago, early in the universe's existence. The light from galaxies, on the other hand, originated much later—after

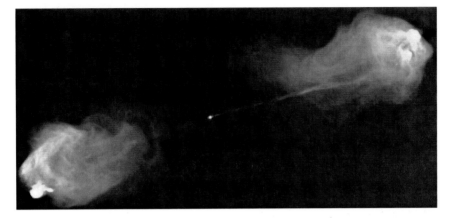

Figure 12.4 A VLA image of the powerful central engine and its relativistic ejection of plasma in the nucleus of Cygnus A. Taken at 6 cm, this view reveals the highly ordered structure spanning over 500,000 lyr, fed by ultrathin jets of energetic particles beamed from the compact radio core between them. The giant lobes are themselves formed when these jets plow into the tenuous gas of the intergalactic medium. Despite its great distance from us (over 600 Mlyr), it is still by far the closest powerful radio galaxy and one of the brightest radio sources in the sky. The fact that the jets must have been sustained in their tight configuration for over half a million (possibly as long as ten million) years means that a highly stable central object—a rapidly spinning supermassive black hole acting like an immovable gyroscope—must be the cause of all this activity. (Photograph courtesy of Chris Carilli and Rick Perley, NRAO, and AUI)

the universe had aged another 2–4 Gyr. Unfortunately, both measurements are subject to uncertainty, and no one can be sure we are measuring *all* of the light from quasars and galaxies, so this argument is not quite compelling. But we do see quasars as far out as we can look, and the most distant among them tend to be the most energetic objects in the universe, so at least *some* supermassive black holes must have existed near the very beginning. At the same time, images such as figure 12.2 provide evidence of mergers of smaller structures into bigger aggregates, but without a quasar. Perhaps not every collision feeds a black hole or, what is more likely, at least some galaxies must have formed first. Several scenarios for the formation of supermassive black holes are currently being examined. In every case, growth occurs when matter condenses following either the collapse of massive gas clouds or the catabolism of smaller black holes in collisions and mergers.

All of the structure in the universe traces its beginnings to a brief era shortly after the Big Bang. Very few "fossils" remain from this period; one of the most important is the cosmic microwave background radiation. The rapid expansion that ensued lowered the matter density and temperature, and about one month after the Big Bang, the rate at which photons were created and annihilated could no longer keep up with the thinning plasma. The radiation and matter began to fall out of equilibrium with each other, forever imprinting the conditions of that era onto the radiation that reaches us to this day from all directions in space.

We now know that the temperature anisotropies are smaller than one part in 10^3, a limit below which density perturbations associated with ordinary matter would not have had sufficient time to evolve freely into the nonlinear structures we see today. Only a gravitationally dominant dark-matter component could then account for the strong condensation of mass into galaxies and supermassive black holes. The thinking behind this is that whereas the cosmic microwave background radiation interacted with ordinary matter, it would retain no imprint at all of the dark matter constituents in the universe. The nonluminous material could therefore have condensed unevenly (i.e., it could have been clumpy) all the way back to the Big Bang.

The first billion years of evolution following the Big Bang must have been quite dramatic in terms of which constituents in the universe would eventually gain primacy and lasting influence on the structure we see today. The issue of how the fluctuations in density, mirrored by the uneven cosmic microwave background radiation, eventually condensed into supermassive black holes and galaxies is currently an active area of research. The evidence now seems to be pointing to a coeval history for these two dominant classes of object—supermassive black holes and galaxies— though as we have already noted, at least some of the former must have existed quite early.

The possibility we already suggested above[7] is that the first supermassive objects formed from a condensation of dark matter alone; only later would these seed black holes have imposed their influence on baryonic matter. But this dark matter has to be somewhat peculiar, in the sense that its constituents must be self-interacting, i.e., they must be able to exchange heat energy with each other. As long as this happens, a fraction of its elements evaporate away from the condensation, carrying with them the bulk of the energy, while the rest collapse to create an event horizon. The net result is that the inner core of such a clump forms a black hole, leaving the outer region and the extended halo in equilibrium about the central object. Over time, ordinary matter gathers around it, eventually forming stars, and planets.

Ordinary matter could not have achieved this early condensation because it simply wasn't sufficiently clumped to begin with. Perhaps this material formed the first stars, followed by more stars, eventually assembling a cluster of colliding objects. Over time, the inner core of such an assembly would have collapsed due to the evaporation of some of its members and the ensuing loss of energy into the extended halo, just as the dark matter would have done.[8] A seed black hole might have formed in the cluster's core. Estimates show that once formed, such an object could have doubled its mass every 40 million years, so over the age of the universe, even a modestly appointed black hole could have grown into a billion-solar-mass object. The problem is that this could not have happened in only 700–800 million years, when the first supermassive black holes appeared.

The case for a coeval growth of supermassive black holes and galaxies is therefore quite compelling—especially now that a remarkable correspondence has been found between the two classes of object.[9] The data displayed in figure 12.5 show the relationship between the mass of the black hole and the velocity dispersion of stars within the spheroidal (or bulge) component in the host galaxy. Based on the very tight

---

[7]This was first proposed by Balberg and Shapiro (2002).

[8]See, e.g., Haehnelt and Rees (1993).

[9]See Ferrarese and Merritt (2000) and Gebhardt et al. (2000).

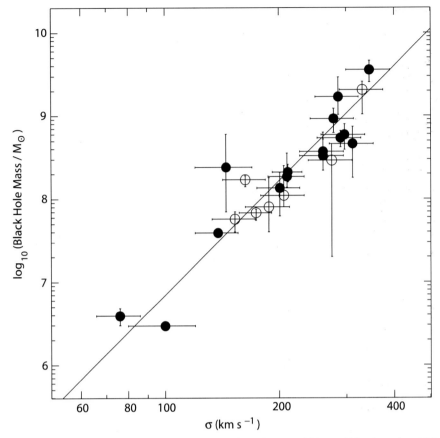

Figure 12.5 Plotted against the velocity dispersion of stars orbiting within the spheroidal component of the host galaxy, the mass of a supermassive black hole is tightly correlated with the former, hinting at a coeval history. The stars are simply too far away from the nucleus to be directly affected by the black hole's gravity at the present time. (Image from Ferrarese and Merritt 2000)

correlation evident in this graph, the mass of the central black hole can be predicted with remarkable accuracy simply by knowing the velocity of stars orbiting so far from it that its gravitational influence could not possibly be affecting their motion at the present time.

This result is one of the most surprising correlations yet discovered in the study of how the universe acquired its structure. Supermassive black holes, it seems, "know" about the motion of stars that are too distant to directly feel their gravity, suggesting an entangled history between a central black hole and the stellar activity in the halo. Although they may not be causally bonded today, they must have had an overlapping genesis in the past.

Additional evidence of this may be extracted from statistical analysis of source distributions like those evident in plate 6. If we adopt a (generous) canonical efficiency

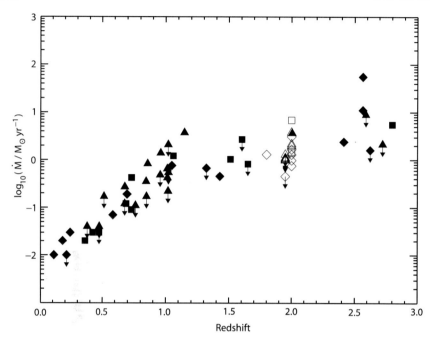

Figure 12.6 The black-hole accretion rate, in units of solar masses per year, versus redshift, assuming a canonical efficiency rate of 10% conversion of rest mass energy into radiation. To generate a 10^9 $M_\odot$ black hole over an accretion period of order 0.5 Gyr, a rate of order 2 $M_\odot$ yr^{-1} is required. (Image from Barger et al. 2001)

of 10% for the conversion of rest mass energy into radiation (see section 3.4.1), we obtain the mass inflow rate versus redshift relation shown in figure 12.6. This rate was as high as 10 $M_\odot$ yr^{-1} at $z \sim 3$, and has declined monotonically to its much lower value of around 0.01 $M_\odot$ yr^{-1} in the current epoch. An even more interesting evolutionary trait is obtained by integrating these data over the black-hole spatial distribution, which produces the redshift dependence of the so-called accretion rate density (figure 12.7). This quantity effectively gives us the rate at which black hole mass is increasing with time. The apparent $(1 + z)^3$ dependence of this measure is identical to the universe's star formation history, as deduced from the comoving UV intensity as a function of redshift.

Integrating this curve over time, we infer that about 0.006 of all the mass contained within the bulge of any given galaxy should be in the form of a central supermassive black hole. This is close to the value of $\sim$0.002 found in the local neighborhood of our Galaxy. With these facts in hand, and satisfied that most supermassive black holes and their host galaxies grew interdependently, one may hypothesize that once created, a primordial condensation of matter continues to grow with a direct feedback on its surroundings. This may happen either because the quasar heats up its environment and controls the rate at which additional matter can fall in from its cosmic neighborhood, or because mergers between galaxies affect the growth of

Plate 14 Artist's impression of the Cygnus X-1 binary system, containing a ~ 10-$M_\odot$ black hole in orbit about HDE 226868, a blue supergiant star. The temperature of the companion is 31,000 K. The gas flowing from its surface toward the accretion disk orbiting the black hole heats up further as it compresses, producing X-rays and γ-rays within tens of Schwarzschild radii (see chapter 3) of the compact object. (Image courtesy of Martin Kornmesser, ESA/ECF)

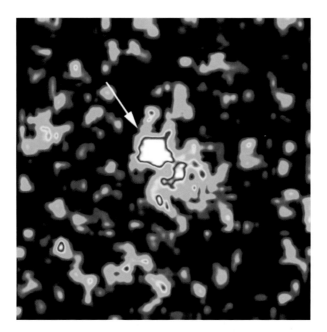

Plate 15 False-color *Hubble* Space Telescope image of the fireball accompanying gamma-ray burst GRB 970228. (The name is based on the date of observation, in this case 28 February 1997.) This is the first optical image ever taken that associates a gamma-ray burst source with a potential host galaxy. Observations such as this provide compelling evidence that gamma-ray bursts occur in distant galaxies. The arrow points to the fireball, a white blob immediately to the upper left of the image center. The extended object to the lower right is the host galaxy. Follow-up images taken just one month later showed that the fireball had faded. The fact that the fireball is offset from the center of the galaxy means that supermassive black holes, which reside at the nucleus of the host galaxies, cannot be the objects producing gamma-ray bursts. A more likely explanation, as we shall see in figure 11.10, is that most gamma-ray bursts are associated with very energetic supernova explosions. (Image courtesy of K. Sahu, M. Livio, L. Petro, D. Macchetto, STScI and NASA)

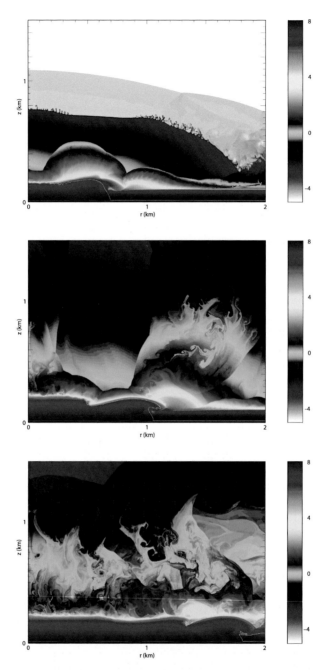

Plate 16 Simulated helium detonation on the neutron-star surface, showing the front breaking through the accreted envelope and propagating laterally: $60\,\mu s$ (top), $90\,\mu s$ (middle), and $150\,\mu s$ (bottom) after ignition. The vertical axis extends to a height of 1.5 km above the surface. A 2-km portion of the stellar surface is shown in the horizontal direction. The thin green line marks a helium mass fraction of 0.95 (roughly separating the ashes from the fuel), the light blue line marks a nickel mass fraction of 0.95 (roughly the interface between the accreted envelope and the surface), and the dark blue line marks a density of $10\,\mathrm{g\,cm^{-3}}$. The vertical color scale indicates mass density, from $10^{8}\,\mathrm{g\,cm^{-3}}$ at the base to the tenuous atmosphere with $10^{-5}\,\mathrm{g\,cm^{-3}}$ at the top. (Images from Zingale et al. 2001)

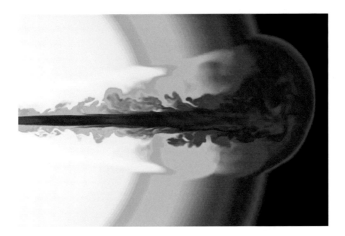

Plate 17 Numerical simulation of a relativistic GRB jet breaking out of a 15-$M_\odot$ Wolf-Rayet star, 8 seconds after its launch with energy 3×10^8 ergs s^{-1}. The stellar radius is 8.9×10^{10} cm. After breaking out, the core jet equilibrates to a Lorentz factor of $\sim$200. The mildly relativistic material surrounding the jet expands to larger angles with a Lorentz factor $\sim$15–30. (Image from Zhang, Woosley, and Heger 2004)

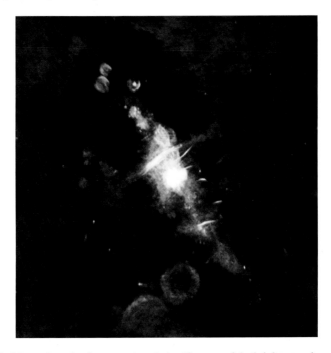

Plate 18 At 90 cm, the galactic center reveals itself as one of the brightest and most intricate regions of the sky. This VLA radio image, spanning an area of about 1000 light-years on each side, reveals a rich morphology encompassing supernova remnants (the circular features), wispy, snake-like synchrotron filaments, and highly ionized hydrogen gas. (Produced at the U.S. Naval Research Laboratory by N. E. Kassim and collaborators from data obtained with the National Radio Astronomy's Very Large Array Telescope, a facility of the National Science Foundation operated under cooperative agreement with Associated Universities, Inc. This image originally appeared in LaRosa et al. 2000)

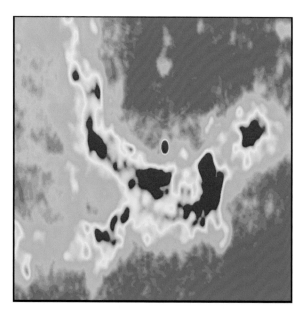

Plate 19 The central portion of the spiral pattern of Sagittarius A West dominates the inner-most 2×2-light-year region of the Galaxy at 2 cm. The bright point-like source near the middle of the image is Sagittarius A*, the radiative manifestation of the supermassive black hole, with a mass of $\approx 3.7 \times 10^6 \ M_\odot$. To the north of Sagittarius A*, the gas blown upward from the envelope of the luminous red giant star IRS 7 creates a cometary-like feature (in light blue against the dark blue background). The distance between Sagittarius A* and the red giant is approximately $\frac{3}{4}$ of a light-year. (Image courtesy of F. Yusef-Zadeh at Northwestern University, and the National Radio Astronomy Observatory)

Plate 20 Full-band (2- to 8-keV) image of the inner $8.'5 \times 8.'5$ field (roughly 20×20 square parsecs at the $\sim$8-kpc distance to the galactic nucleus) centered on Sagittarius A*. The image has been adaptively smoothed to permit the point sources to be easily distinguished from the diffuse emission. The colors indicate intensity on a logarithmic scale, with white the highest and red the lowest. (Image from Muno et al. 2003)

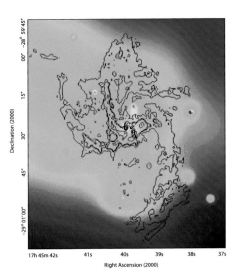

Plate 21 Smoothed *Chandra* 0.5- to 7-keV image showing the central $1.'3 \times 1.'5$ of the Galaxy (roughly $\frac{1}{6}$ the size of the field of view in plate 20). Overlaid on the X-ray image are VLA 6-cm contours. The X-ray emission from Sagittarius A* itself appears as a red dot at $17^h45^m40.0^s$, $-29°00'28''$. Bright diffuse emission from hot gas is visible throughout the region and appears to be produced primarily via the collisions of winds from the young, early stars surrounding the supermassive black hole. (Image from Baganoff et al. 2003)

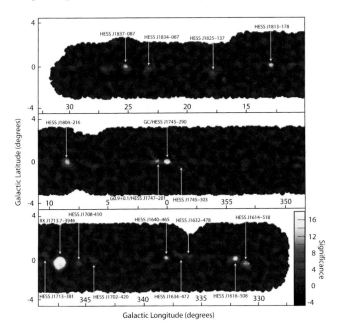

Plate 22 Significance map of the H.E.S.S. Galactic Plane survey with a typical energy threshold of 250 GeV. The on-source counts for each grid point are integrated in a circle of radius $0.22°$. The labels indicate specific γ-ray sources, including RX J1713.7–3946 (HESS J1713–397), G 0.9 + 0.1 (HESS J1747–281), and the galactic Center TeV source HESS J1745–290. The color scale indicates the post-trial significance of the γ-ray sources, on a scale truncated at 18σ. The signals from the galactic Center source and from RX J1713.7–3946 exceed this level. (Image from Aharonian et al. 2006)

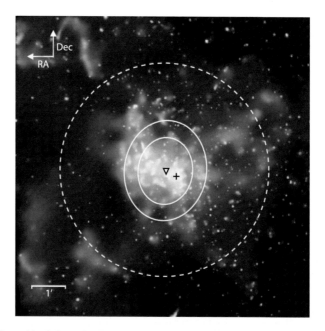

Plate 23 Centroid of the galactic center source (see plate 22) superimposed on the 2- to 8-keV 8.′5 × 8.′5 image seen earlier in plate 20. The location of the radio source Sagittarius A* is shown as a cross. The 68 and 95% confidence limits for the source position are shown as solid ellipses, taking into account the 20″ pointing error. The dashed curve shows the 95% confidence limit on the rms source size. The TeV source appears to be offset by ∼10″ toward positive galactic longitude relative to Sagittarius A*. (Image from Aharonian et al. 2004)

Plate 24 Artist's concept of the young, blue stars in the stellar peak P3 encircling the supermassive black hole at the nucleus of M31 (close to P2 in figure 12.13). The background stars, older and redder, are more typical of those inhabiting the cores of most galaxies. The new, younger component, revealed recently by the *Hubble* Space Telescope (HST), comprises more than 400 stars, formed in a burst of activity about 200 million years ago. These objects are tightly packed in a disk only a light-year across. Under the black hole's gravitational influence, they orbit at speeds exceeding 1000 km s^{-1}. (Image courtesy of NASA, ESA, and A. Schaller)

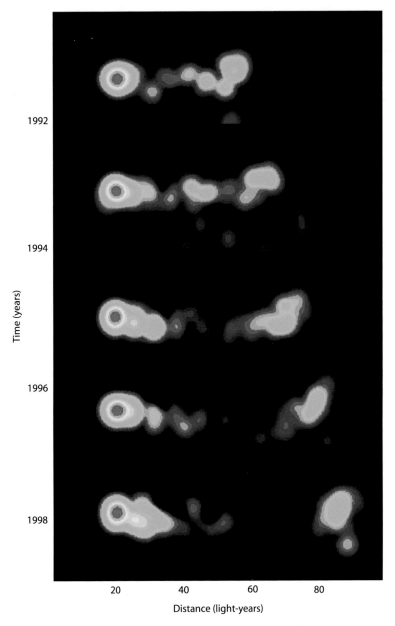

Plate 25 The blazar 3C 279 is also a member of the class of quasars exhibiting superluminal motion, as evidenced in this sequence of five radio images produced over a seven-year baseline. The stationary core, marking the location of the central supermassive black hole, is the bright red spot to the left of each image. The observed location of the rightmost blue-green clump of radiating plasma moved about 25 light-years (in projection) from 1991 to 1998. To us, the changes therefore appear to occur faster than the speed of light. This is a special relativistic effect (see also figure 12.17), and the motion of the clump is not really faster than light. The blue-green emitter is part of a jet, and the inferred speed is due to light-travel-time effects for a source moving near light-speed (actually at $0.997c$) close to (within $2°$ of) the line of sight. (Image courtesy of NRAO and Anne Wehrle)

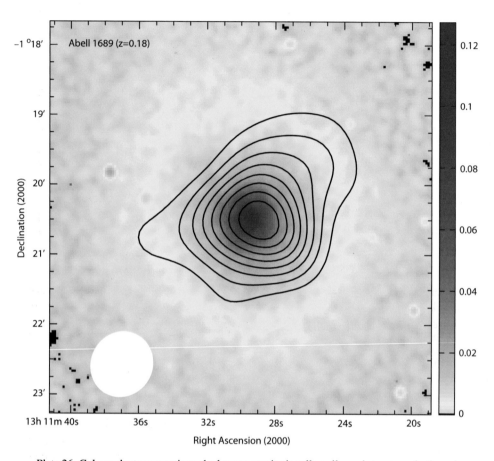

Plate 26 Galaxy clusters constitute the largest gravitationally collapsed structures in the universe. The intracluster medium (ICM) is filled with a hot diffuse plasma at a temperature $\sim 10^8$ K that radiates primarily via thermal bremsstrahlung. Additional evidence for the existence of this plasma is provided by cosmic microwave background photons passing through the hot ICM; some of them inverse-Compton scatter off the energetic electrons, causing a small ($\sim$1-mK) distortion of the microwave background spectrum (known as the Sunyaev-Zel'dovich effect). The image shown here was produced with *Chandra*, and it shows the X-ray surface brightness in the 0.7- to 7-keV band in units of counts pixel^{-1} (1.″97 pixels) for the cluster Abell 1689 (at a redshift of 0.18). Overlaid is the Sunyaev-Zel'dovich effect decrement contours, with the full width at half-maximum of the synthesized beam (the effective point-spread function) shown in the lower left corner. (Image from Bonamente et al. 2006)

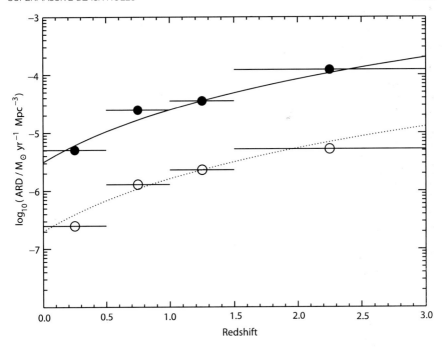

Figure 12.7 The time history of the accretion rate density (effectively, the rate at which black hole matter is increasing, in units of solar masses per year per megaparsec cubed), integrated from the data exhibited in figure 12.6. The upper (solid) and lower (open) bounds in this figure correspond to the accretion rate density calculated from the bolometric and X-ray luminosities, respectively. The curves illustrate a $(1 + z)^3$ dependence, normalized to the $1 < z < 1.5$ redshift bin. (Image from Barger et al. 2001)

colliding black holes in the same way that they determine the energy (and therefore the average speed) of the surrounding stellar distribution. In addition, protogalactic star formation would have been influenced significantly by the quasar's extensive energy outflows.[10] Thus, in addition to the reasons we suggested earlier for the tight correlation between the mass of the central object and the stellar velocities much farther out, this coupling could also be due to the ensuing feedback on the galaxy's spheroidal component.

In either case, we cannot ignore the importance of galaxy mergers to the hierarchical construction of elaborate elliptical or disk-plus-bulge profiles. Most large galaxies have experienced at least one major merger during their lifetime. Larger galaxies have grown with a succession of collisions and mergers, a process contributing significantly to the variety of shapes encompassed by the Hubble sequence. Meanwhile, the turbulence generated in the core of the collision drives most of the gas into the middle, where it forms new stars and feeds a central black hole or a pair of black holes (figure 12.8). This cosmic cascade must be at least partially

[10] See Silk and Rees (1998).

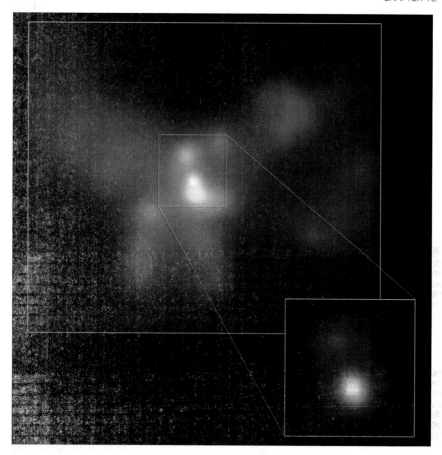

Figure 12.8 NGC 6240 is a butterfly-shaped galaxy, believed to be the product of a collision between two smaller galaxies some 30 million years ago. This *Chandra* X-ray image (34,000 lyr across) shows the heat generated by the merger activity, which created the extensive 10^7–10^8 K gas. We see here for the first time direct evidence that the nucleus of such a structure contains not one, but two active supermassive black holes, drifting toward each other across their 3000-lyr separation; they're expected to merge into a bigger object several hundred million years hence. (Image courtesy of Stephanie Komossa, Günther Hasinger, and Joan Centrella, and the Max-Planck-Institut für Extraterrestrische Physik and NASA)

responsible for the tight correlation between the black-hole mass and the halo velocity dispersion.

12.2 NEARBY OBJECTS

However, not all members of the AGN (or AGN/quasar) class are necessarily very powerful, or very distant (i.e., extending to redshifts of 2 or more). The arguments we have made thus far are less compelling for the more abundant low-luminosity objects, whose energy requirements are less demanding, and where long jets or superluminal motion are seen less frequently or not at all. The nearby galactic nuclei, including the center of our own Galaxy, fall into this group. Yet the evidence for the existence of a supermassive black hole is even stronger in these systems than for those we have been describing. Certainly their proximity is largely responsible for this, since we can study them at a level of detail we cannot even hope to achieve with the others. And so we will begin our study of the high-energy characteristics of AGNs with two of the closest nuclei—the galactic center and the center of the M31 galaxy in Andromeda.

12.2.1 The Galactic Center

Plate 18 shows a wide-field, high-resolution 90-cm image centered on the radio complex known as Sagittarius A, covering an area of $4° \times 5°$ with an angular resolution of $43''$. This map of the galactic center is based on archival data originally acquired with the Very Large Array (VLA) 333 MHz system in all four array configurations between 1986 and 1989.[11]

The bright, central source in plate 18 may be magnified further by tuning the receivers to a higher frequency, and therefore a better resolution. At a wavelength below 6 cm (see plate 19), the material within several parsecs of the nucleus shines forth as a three-armed spiral consisting of highly ionized gas radiating a thermal continuum (see section 5.3). Each arm in the spiral is about 3 lyr long, but one or more of these may be linked to the overall structure merely as a superposition of gas streamers seen in projection. At a distance of 3 lyr from the center, the plasma moves at a velocity of about 105 km s^{-1}, requiring a mass concentration of just over $3.5 \times 10^6 \, M_\odot$ inside this radius. The hub of the gas spiral corresponds to the very bright point-like radio source known as Sagittarius A*, which defines the dynamical center of our Galaxy. We shall see shortly that most of its emission is due to a combination of thermal and nonthermal synchrotron processes (see sections 5.5 and 5.6), and to self-Comptonization of this radiation by the same particles (section 5.7).

A rather different—though no less interesting—view of the galactic center emerges with progressively sharper images of this region in the X-ray band. X-ray emission has been observed on all scales, from structure extending over kiloparsecs down to a fraction of a light-year, with contributions from thermal and nonthermal, pointlike and diffuse sources.

The most detailed X-ray view of the galactic center has been made with *Chandra's* Advanced CCD Imaging Spectrometer (ACIS) detector (see plate 1), which combines the wide-band sensitivity and moderate spectral resolution of ASCA and *Beppo*-SAX with the much higher spatial resolution ($\sim 0.''5$–$1''$) of the High-Resolution Mirror Assembly (HRMA). Figure 12.9 shows the field (oriented along

---

[11]The observations were carried out by Pedlar et al. (1989) and Anantharamaiah et al. (1991). But it was only the use of a wide-field algorithm that properly compensates for the nonplanar baseline effects seen at long wavelengths that permitted LaRosa, Kassim, and Lazio (2000) to properly image such a large field of view and obtain increased image fidelity and sensitivity.

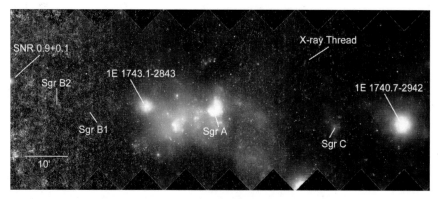

Figure 12.9 X-ray intensity map of the galactic center, formed from a mosaic covering a $\sim 2 \times 0.8$ square-degree band in galactic coordinates centered at $(l^{II}, b^{II}) = (-0.1°, 0°)$. The galactic plane runs horizontally along the center of this map. (Image courtesy of Q. Daniel Wang at the University of Massachusetts, Amherst, and NASA)

the galactic plane) mapped out in the 1- to 8-keV range by the most complete *Chandra* survey to date.[12]

The high spatial resolution of the *Chandra* X-ray Observatory allows for a separation of the discrete sources from the diffuse X-ray components pervading the galactic center region. One may get a better sense of this capability by inspecting the magnified view of the central $8.'5 \times 8.'5$ portion of this field, shown in plate 20. These observations have led to a detection of roughly 1000 discrete objects within the inner 2×0.8 square degrees, very few of which were known prior to this survey. Their number and spectra indicate the presence of numerous accreting white dwarfs, neutron stars, and solar-size black holes, with a luminosity $\sim 10^{32}$–10^{35} ergs s^{-1} in the 2- to 10-keV band.

But it is the region closest to the supermassive black hole that interests us here, and the innermost $1.'3 \times 1.'5$ of plate 20 is magnified and rendered in plate 21, showing both the X-ray point sources and the diffuse X-ray emission within a few light-years of Sagittarius A*. For comparison, the color-coded X-ray emission in this image is superposed with the VLA 6-cm contours, which reveal the counterclockwise orbiting mini-spiral of ionized gas we encountered earlier in plate 19.

A fit to the X-ray emission[13] from the hot gas within $10''$ of Sagittarius A* yields an inferred 2- to 10-keV diffuse flux of $(1.9 \pm 0.1) \times 10^{-15}$ ergs cm^{-2} s^{-1} arcsec^{-2} and a corresponding luminosity of $(7.6^{+2.6}_{-1.9}) \times 10^{31}$ ergs s^{-1} arcsec^{-2}. Based on the parameters of the best-fit emissivity model, which assumes optically thin bremsstrahlung radiation (see section 5.3), the local, hot diffuse plasma is estimated to have a RMS electron density $\langle n_e^2 \rangle^{1/2} \approx 26$ cm^{-3}. Thus, assuming twice solar abundances (typical for this part of the Galaxy), the total inferred mass of

---

[12]The saw-shaped boundaries of this map, plotted in galactic coordinates, result from a specific roll angle of the observations. See Wang, Gotthelf, and Lang (2002).

[13]See Baganoff et al. (2003)

this gas (i.e., the product of this density with the spatial volume of this region) is $M_{local} > 0.1\ M_\odot$.

The hot plasma within a few parsecs of Sagittarius A* appears to be injected into the interstellar medium via stellar winds, and the diffuse X-ray emission therefore constitutes an excellent probe of the gas dynamics near the black hole. There is ample observational evidence in this region for the existence of rather strong outflows in and around the nucleus, obtained via the measurement of emission-line Doppler shifts and equivalent widths, which reveal the presence of 500- to 1000-km s^{-1} winds and number densities $\sim 10^{3-4}$ cm^{-3} near the mass-ejecting stars. The implied total mass injection rate into the Galaxy's central region is $\sim$3–4 $\times 10^{-3}\ M_\odot$ yr^{-1}.

At least as far as our Galaxy is concerned, we can now begin to understand the data plotted toward the left-hand side of figure 12.6. On this graph, the Milky Way occupies the region at $z = 0$, where the average black-hole accretion rate is $\sim 10^{-2}\ M_\odot$ yr^{-1}. If the medium surrounding the central black hole contains little gas, then the accretor clearly cannot grow at rates like those seen in the universe's early history, when as much as 10 $M_\odot$ of material could fall into the black hole each and every year.

With modern computers, it is now feasible to carry out comprehensive, high-resolution hydrodynamic (see section 6.2) simulations of the wind–wind interactions, using a detailed suite of stellar wind sources and their inferred wind velocities and outflow rates. A simulation of the wind structure within the inner 2-lyr region of the Galaxy is shown in figure 12.10, $\sim$ 10,000 yr after the beginning of the calculation. These simulations were carried out using a Smooth Particle Hydrodynamics (SPH) code developed from a basic algorithm written and tested over a ten-year span at Los Alamos's Theory Division.[14]

The wind–wind collisions create a complex configuration of shocks (see section 4.3) that efficiently convert the kinetic energy of the outflows into internal energy of the gas. In figure 12.10 the dark surfaces indicate regions of gas with low specific internal energy; these tend to lie near the wind sources themselves. The gray surfaces mark regions of high specific internal energy, where gas has passed through multiple shocks. From simulations such as this, it is evident that about $\frac{1}{4}$ of the total energy in the central parsec is converted to internal energy via multiple shocks. The total kinetic energy of material there is $\sim 8 \times 10^{48}$ ergs, while the total internal energy is $\sim 3 \times 10^{48}$ ergs.

The bremsstrahlung emissivity from these wind–wind collisions within 10″ of Sagittarius A* produce the entire diffuse X-ray flux detected from this region by *Chandra* (see plate 21). It is also important to emphasize the fact that these outflows bring the environment near the black hole into steady state within only $\sim$ 4000 years. It is unlikely, therefore, that supernova explosions occurring near a supermassive black hole could create an environment with anomalously low densities by sweeping out most of the material from the central parsecs. What we are seeing at the center of our Galaxy may be quite typical of nearby galactic nuclei in which the gas content of the central medium has been largely depleted due to black-hole accretion and star formation.

---

[14]This work was carried out jointly at Los Alamos and the University of Arizona by Rockefeller et al. (2004), using the tree-based algorithm described in Warren and Salmon (1993, 1995).

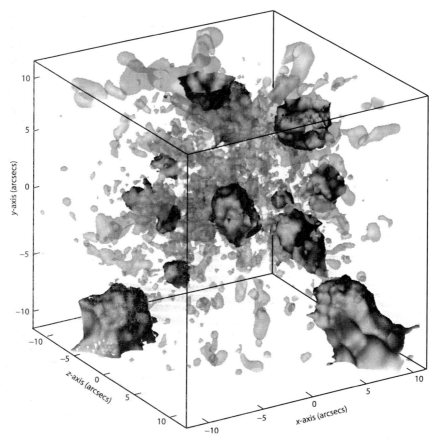

Figure 12.10 Within a few light-years of Sagittarius A*, the strong winds from massive stars collide and shock, converting kinetic energy into heat, some of which is radiated as a diffuse X-ray glow (see plate 21). This figure shows the turbulent gas structure roughly 10,000 years after the winds are "turned on" in a three-dimensional hydrodynamic simulation. The 2- to 10-keV luminosity from the central 3 parsecs of the Galaxy appears to be due mostly to this process. Shown here are the isosurfaces of specific internal energy of the shocked gas within the inner 20″-cube of the Galaxy ($1'' \approx 0.04$ pc). The line of sight is along the z-axis. The darkest surfaces correspond to a specific internal energy of 2.5×10^{12} ergs g^{-1}; the gray surfaces correspond to 3.8×10^{15} ergs g^{-1}. (Image from Rockefeller et al. 2004)

At the galactic center, the supermassive black hole is accreting from the turbulent medium simulated in figure 12.10. This presents quite a different picture than the one we developed for binary systems in chapters 9 and 10. There we learned that the gas flowing toward the compact object from its companion carries a steady supply of angular momentum, leading to circularization inside a steady disk. For those systems, we even know how to estimate the disk's size, and concluded in section 6.4 that R_{circ} may be a thousand times larger than a neutron star. The disk itself is even bigger.

On the other hand, the material accreting onto a supermassive black hole from its environment carries a variable angular momentum, which at times may even cancel some of the angular momentum already in the disk. One would not be surprised, therefore, to observe significant variability in a source like Sagittarius A*. In addition, when the medium surrounding the supermassive black hole has a low density, one would not expect a large viscous, accretion disk to even form. Inspection of figure 12.10 suggests that accretion in such a case is instead initiated by the stochastic infall of gaseous clumps whose specific angular momentum is sufficiently small to allow them to circularize within the black hole's gravitational sphere of influence.

Indeed, several observational lines of evidence suggest that the dynamically most important region of the accretion flow is restricted to the inner $\sim$20 Schwarzschild radii—essentially the distance between Earth and the Sun, for a black-hole mass $3.7 \times 10^6 \, M_\odot$. This appears to be where Sagittarius A*'s variable X-ray emission is produced.

We know this in part because of remarkable observations carried out by *Chandra* and XMM-*Newton*, an example of which is highlighted in figure 12.11. Compared with its quiescent emission, X-ray (and also IR) flares such as this reflect a significant (though transient) change in the system's physical state. The biggest events produce fluxes tens of times greater than those from the nonflaring configuration. Some have speculated that perhaps there is no real "steady" state in this source (and others like it), and that all of its emission is associated with transient events spanning a range of durations and amplitude; we just happen to identify the biggest of these as "flares."

The observed short timescale (about 40 min) for the variation, together with light travel time arguments, delimits the emitting region to a size no bigger than about 5 AU, or equivalently about 70 Schwarzschild radii. As we shall see, the X-ray modulation—evident even by eye in figure 12.11—appears to restrict the possibilities even further, since a Keplerian origin for the associated periods seems to be unavoidable.

But before we discuss the very exciting possibility that we are witnessing phenomena hovering just outside Sagittarius A*'s event horizon, let us first consider how these flares are produced. A telling feature in the spectrum of events such as this is that the observed IR photon distribution is so much steeper than that of its X-ray counterpart. This excludes a direct extrapolation of the same power law from one to the other. For example, during the peak of the IR flare[15] observed 15 July 2004, a power-law fit $\nu F_\nu \propto \nu^{-\alpha}$ (with ν the emission frequency) to its power spectrum yields $\alpha = 2.2 \pm 0.3$. During its rising and decaying phases, a similar fit requires an index $\alpha = 3.7 \pm 0.9$. By comparison, the best-fit photon index during a typical X-ray event is $1.3^{+0.5}_{-0.6}$, representing a significant flattening of $\sim$1 of the spectrum compared to that in Sagittarius A*'s quiescent state.[16]

For this reason, it is thought that the mm/sub-mm to IR portion of the spectrum is probably due to synchrotron, whereas the X-rays are produced by synchrotron self-Compton (in which the same particles that initially emit the synchrotron photons also subsequently upscatter some of them to much higher energy; see sections 5.6

---

[15] See Eisenhauer, Genzel, Alexander et al. (2005).
[16] See Baganoff et al. (2001) and the accompanying News and Views article by Melia (2001b).

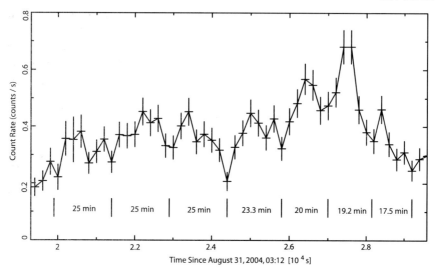

Figure 12.11 Light curve of the 31 August 2004 X-ray flare detected with XMM-*Newton*, binned in 200-s intervals. Markers and labels are used to show the gradual decrease in the period from 25 to 17.5 min. (Image from Bélanger et al. 2006b)

and 5.7). This constraint is empirically motivated, and is independent of the specific features one might introduce into a fully fledged theory describing how the emission occurs. From it, we infer that the flare region must be infused with relativistic particles embedded within a tenuous, highly magnetized plasma.

The transient azimuthal asymmetry giving rise to the modulated flux may be due to either a dynamical event or to a viscous process near the marginally stable orbit.[17] It may be a wave pattern co-rotating with the gas, or perhaps it is a hot spot where matter has fallen in from larger radii, impacting the small disk.[18] In either case, the pattern of modulation over one complete cycle is the result of several relativistic effects, including a Doppler shift, light bending, and area amplification near the black hole's event horizon.

Our discussion in section 3.4.1 provides a framework for analyzing the behavior of this material close to the black hole's event horizon, in terms of $\tilde{l}$, the conserved angular momentum per unit mass of the orbiting gas, and its energy-at-infinity per unit mass, $\tilde{E}$, expressed in equations (3.95) and (3.96) for the case of a Schwarzschild object. However, this formulation won't be sufficient if the black hole is spinning. Indeed, the monotonic decrease of the period evident in figure 12.11 (and in similar modulated lightcurves of Sagittarius A* observed in the IR) suggests that we may be witnessing the evolution of an event moving inward through the last portion of the accretion disk, very near the marginally stable orbit (see discussion following equation 3.96). As such, the effects of frame dragging (section 3.4.2)

---

[17]Some earlier considerations on how the timing data may be interrogated for information on the orbital properties at this radius may be found in Melia, Bromley, Liu, and Walker (2001).

[18]See Tagger and Melia (2006).

may be critical to a correct interpretation of the periods inferred from these observations.

Fortunately, it is not difficult to generalize the formulation of section 3.4.1 for material orbiting a spinning black hole. Qualitatively, we expect that the same range of orbits should be accessible in the Kerr metric as in the Schwarzschild case, though the frame dragging effect can introduce quantitative differences. As we saw in Section 10.4 for motion in the equatorial plane of the black hole, the expressions in (3.95) and (3.96) generalize to the form[19]

$$\tilde{l}_\phi^2 = \frac{GM\,(r^2 \mp 2ar_G^{1/2}r^{1/2} + a^2)^2}{r^{3/2}(r^{3/2} - 3r_G r^{1/2} \pm 2ar_G^{1/2})} \tag{12.1}$$

and

$$\tilde{E} = \frac{c^2(r^{3/2} - 2r_G r^{1/2} \pm ar_G^{1/2})^2}{r^{3/2}(r^{3/2} - 3r_G\,r^{1/2} \pm 2ar_G^{1/2})}, \tag{12.2}$$

when the spin parameter $a \neq 0$ (see equation (3.101)). The upper sign in these equations refers to direct orbits (i.e., co-rotating with $\tilde{l} > 0$), while the lower sign refers to retrograde orbits (counterrotating with $\tilde{l} < 0$). The coordinate angular velocity of a circular orbit is evidently (see equation 10.18)

$$\Omega \equiv \frac{d\phi}{dT} = \pm \frac{cr_G^{1/2}}{r^{3/2} \pm ar_G^{1/2}}. \tag{12.3}$$

From this expression, we infer a radius $\langle r \rangle \approx 2.7\,r_S$ from the average period in figure 12.11, assuming a black-hole mass $3.7 \times 10^6\,M_\odot$, with $a = 0$. This is interesting for several reasons. First, since this location is very near the marginally stable orbit ($3r_S$) of a Schwarzschild black hole, we could be observing an event caused by the sudden reconfiguration of matter flowing across this critical orbit toward the event horizon. Thus, if the shortest observed period (i.e., 17.5 min seen toward the end of this flare) were to actually correspond to this orbit, the black-hole spin could not be zero since $\langle r \rangle < 3r_S$. Indeed, from equation (12.3) we see that in this case[20] $a/r_G \approx 0.4$ (where as usual $r_G \equiv GM/c^2$) for a prograde orbit, or $a/r_G \approx 0.3$ in the case of retrograde disk rotation.

Second, this X-ray flare (and others detected earlier by *Chandra*) features strong variability near the middle of the event; the X-ray flux drops by a factor of 40% or more in 10–15 min. Simple light travel time arguments therefore constrain the X-ray emitting region to be no bigger than ~ 17–$34r_S$—much smaller than the size we estimated earlier on the basis of the total event duration. This size is consistent with the tight orbit we inferred from the quasi-periodic modulation of the lightcurve.

However, one must always be careful with the interpretation of a modulation using equation (12.3), for the inner edge of the accretion disk is unlikely to be

[19]The full details of this derivation may be found in Bardeen, Press, and Teukolsky (1972).
[20]See also Genzel, Schödel et al. (2003).

truncated cleanly, as we have assumed thus far. The question of where exactly the disk terminates has been asked in a broader context with the help of extensive MHD simulations of the plunging region in a pseudo-Newtonian potential (see equation 8.75) to identify several characteristic inner radii in black-hole accretion.[21] The above analysis has assumed that either the average X-ray period ($\sim$21.4 min) or the last (and smallest) period ($\sim$17.5 min) ought to correspond to the marginally stable orbit.

For magnetized accretion, however, matter flowing past the marginally stable orbit may still remain "magnetically" coupled with the outer disk even below the marginally stable orbit, so the constraint $\langle r \rangle < 3r_S$ does not necessarily mean that a/r_G is quite as large as 0.3–0.4. It makes more sense to consider a dynamically more meaningful radius—the so-called stress edge—where plunging matter loses dynamical contact with the material farther out. From a practical standpoint, this location may simply be defined as the surface at which the speed first becomes supermagnetosonic, that is, where it first exceeds the sound speed associated with fluctuations in the magnetic field. The MHD simulations indicate that this transition occurs somewhere between $2.3r_S$ and $3r_S$.

So, although the orbital interpretation of the modulation seen in the IR and X-ray flares from Sagittarius A* is now fairly robust, it is far from clear how one must disentangle the various effects related to the marginally stable orbit in order to correctly "measure" the spin parameter a. Still, the fact that we are even discussing this issue now is quite remarkable, considering that a decade or two ago we weren't even sure whether the dark matter condensation at the galactic center ought to be concentrated in a single object. Events sampled by observations such as that shown in figure 12.11 are as close to a supermassive black hole as we have ever seen in X-rays. As far as high-energy astrophysics is concerned, Sagittarius A* is therefore a very useful representative of the AGN class. But it is also the faintest member of this group and, perhaps not coincidentally, also lacks several key characteristics that help to define these sources, such as relativistic jets and a prodigious output of gamma-ray energy.

It was quite surprising, therefore, when air Cerenkov telescopes (see figure 1.11) detected a significant source of TeV radiation coming from the direction of Sagittarius A*. To date, the galactic center has been identified in the $\sim$TeV range by three different instruments[22]: Whipple, CANGAROO, and, most recently and significantly, HESS. The Hegra ACT instrument[23] has also put a (weak) upper limit on

---

[21] See Krolik and Hawley (2002).

[22] As we learned in chapter 1, air Cerenkov telescopes incorporate Earth's atmosphere into the overall detector design. Energetic photons create electron–positron pairs at high altitude; these continue to interact on their way down, emitting secondary energetic photons via bremsstrahlung and Compton scattering. The ensuing cascade grows until the charges eventually run out of energy. Particles in this shower travel faster than the speed of light in air and consequently emit a faint, bluish light known as Cerenkov radiation, after the Russian physicist who made comprehensive studies of this phenomenon. The photons in this glow typically form a pancake-like front, some 200 m in diameter and only about a meter thick. The mirrors in the telescope on the ground capture this light and feed it to a central detector. See Kosack et al. (2004), Tsuchiya et al. (2004), and Aharonian et al. (2004).

[23] See Aharonian et al. (2002).

galactic center emission at 4.5 TeV and the Milagro water Cerenkov extensive air-shower array[24] has released a preliminary finding of a detection at similar energies from the region defined as $l \in \{20°, 100°\}$ and $|b| < 5°$.

The galactic center survey conducted by HESS is shown in plate 22, revealing several point emitters of TeV radiation, in addition to the central source HESS J1745–290, which is shown with greater magnification in plate 23. The HESS data acquired over two epocs (June/July 2003 and July/August 2003) can be fitted by a power law (see figure 12.12) with a spectral index of 2.21 ± 0.09 and normalization of $(2.50 \pm 0.21) \times 10^{-8}$ m^{-2} s^{-1} TeV^{-1} with a total flux above the instrument's 165-GeV threshold of $(1.82 \pm 0.22) \times 10^{-7}$ m^{-2} s^{-1}. There is also a 15–20% error from energy resolution uncertainty.

It is too early to tell whether this TeV source should really be identified with Sagittarius A* itself. There is no obvious mechanism by which the black hole could radiate in this energy range. However, we do know that relativistic particles must be present within ~20 Schwarzschild radii of the central object to produce Sagittarius A*'s variable IR and X-ray flux. It has been suggested that the most energetic among them escape into the surrounding medium and initiate a series of interactions with the ambient plasma (figure 12.10).

Once they are out of the acceleration zone, these particles diffuse through the interstellar gas and dust, random-walking several parsecs away from the black hole. But on their way out, the protons undergo a series of interactions, including $pN \rightarrow pN$, m_{meson}, and $m_{N\bar{N}}$, where N is either a p or a neutron n, m_{meson} denotes the energy-dependent multiplicity of mesons (mostly pions), and $m_{N\bar{N}}$ is the multiplicity of nucleon/antinucleon pairs (both increasing functions of energy). Since $m_{N\bar{N}}/m_{\text{meson}} < 10^{-3}$ at low energy and even smaller at higher energies,[25] one may safely ignore the antinucleon production events. The charge exchange interaction $(p \rightarrow n)$ occurs around 40% of the time at accelerator energies, and this fraction is predicted to be only very weakly energy dependent. Other possible interactions of accelerated p's—all potentially important for cooling—are $p\gamma \rightarrow p\pi^0\gamma$, $p\gamma \rightarrow n\pi^+\gamma$, $p\gamma \rightarrow e^+e^-p$, and $pe \rightarrow eNm_{\text{meson}}$.

The neutral pions produced in this cascade decay into a pair of photons ($\pi^0 \rightarrow 2\gamma$), which inherit most of the energy transferred to them from the parent particles. The fact that the observed TeV spectrum is a well-defined power law with index ~2.2 suggests that Fermi acceleration, or something like it (see section 4.3), may ultimately be responsible for producing the implied nonthermal distribution. And since the TeV luminosity of HESS J1745–290 ($\sim 10^{35}$ ergs s^{-1}) is well within the energy budget available to the black hole, it may very well be that some of these ~10- to 100-TeV protons accelerated by Sagittarius A* could ultimately be the source of the ~1-TeV radiation from this object. Aside from identifying the galactic supermassive black hole as an important particle accelerator in the central region of the Galaxy, the viability of this picture therefore has some bearing on the more general question of relativistic particle ejection in the broader class of AGNs and quasars.

---

[24]See Fleysher and the Milagro Collaboration (2002).
[25]See Crocker et al. (2005).

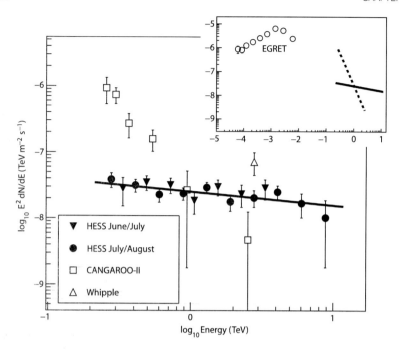

Figure 12.12 Spectrum ($E^2 dN/dE$) of the galactic center TeV source shown in figure plate 23. Full circles represent the HESS July/August 2003 data set, and full triangles represent the HESS June/July 2003 data set. The straight line is the power-law fit to the July/August spectrum. Data from other instruments are as follows: open squares represent the CANGAROO-II spectrum from Summer 2001 and 2002: the open triangle shows the Whipple measurements from 1995 through 2003, converted to a differential flux at the peak detection energy assuming a Crab-like spectrum. A comparison between the HESS TeV spectrum with that at ~GeV energies is shown in the inset, where the EGRET data (circles) are contrasted with fits to the CANGAROO-II (dashed line) and HESS (solid line) spectra. The disparity between these two energy ranges is probably accounted for by the relatively poor spatial resolution of EGRET (~1°), which suggests that the EGRET and HESS spectra correspond to two different sources. (Image from Aharonian et al. 2004)

12.2.2 Other Weak Nuclei

Several other nearby galactic nuclei also have much to offer in terms of providing us with a picture of the physical processes taking place near a supermassive black hole. Sagittarius A* is the closest, but it is also one of the least massive of its class. Moreover, our line of sight toward the galactic center lies in the midplane of the Galaxy, where attenuation by the interstellar medium is severe,[26] restricting our ability to observe the central region at all wavelengths.

At a distance of 780 kpc, the black hole known as M31* at the nucleus of the Andromeda galaxy is the nearest analog to Sagittarius A*. But its surroundings

---

[26]For example, in the visible band, the reddening corresponds to $A_V \sim 30$ for Sagittarius A*.

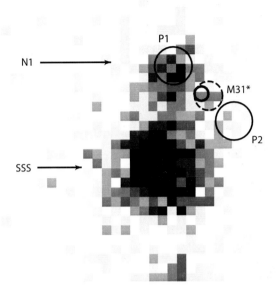

Figure 12.13 At a distance of 780 kpc, M31* is the nearest analog to Sagittarius A* (see Melia 1992b). The nucleus of M31 (the Andromeda Galaxy) houses three distinct stellar peaks, known as P1, P2, and P3 (the recently discovered concentration of blue, young stars— see plate 24), and a $\sim 1.4 \times 10^8 \ M_\odot$ supermassive black hole spatially coincident with the radio point source M31*. None of the bright X-ray sources in the central region, including N1 (also known as CXOM31 J004244.3+411608) and the central super-soft source SSS (also called CSOM31 J004244.3+411607) are identifiable with M31*. However, recent *Chandra* observations have uncovered a faint X-ray source, with a luminosity $L_X \approx 3 \times 10^{35}$ ergs s^{-1}, whose error circle overlaps the 0."1 heavy circle representing the radio point source, making it a likely candidate for the high-energy counterpart of M31*. (Image from Garcia et al. 2005)

(see figure 12.13) are quite different from those we encountered at the core of the Milky Way in the previous section. The nucleus of Andromeda comprises a central dark-matter distribution and three concentrations of starlight. Two of these, commonly labeled P1 and P2 (peaks 1 and 2), have been known for many years. The conventional view is that these are merely the opposite ends of an elliptical distribution of old stars orbiting the central distribution, which is near P2 at one focus of the ellipse.[27]

Recent observations[28] with the *Hubble* Space Telescope (HST), however, have confirmed the existence of a third stellar component, P3, a tiny nucleus of hot, blue stars embedded near (or within, depending on how one defines the region) P2 (see plate 24). Despite its small size (barely one lyr across), P3 contains stars with the highest average circular rotation velocity—almost 1700 km s^{-1}—measured so far in

[27] See Tremaine (1995).
[28] See Bender, Kormendy, Bower, et al. (2005).

any galaxy. According to Newton's law of gravity, the central mass required to corral such fast stars this close to the nucleus exceeds $10^8 \, M_\odot$, rendering Andromeda's black hole some 30 times more massive than Sagittarius A*.

The age of the roughly 400 stars that form P3 presents somewhat of a challenge. They are bright and blue, and therefore very young compared with the age of the Andromeda galaxy; most of them must have formed less than 200 Myr ago. A similar situation is present at the galactic center, where observations reveal a small cluster of young, massive stars surrounding Sagittarius A* as well. It appears that a new theory of star formation in the chaotic environment surrounding a supermassive object must be developed to account for these surprising discoveries. Nonetheless, the discovery of P3 has provided us with an accurate indication of Andromeda's center and the mass of the black hole being hosted there.

These measurements have turned M31* into an ideal accretor with which to test the ideas on Bondi-Hoyle accretion from section 6.2, using the exquisite imaging capability of the *Chandra* X-ray Observatory. M31* is a point source of radio waves,[29] apparently coincident with the concentration of dark matter near P2. Though unresolved at the $\sim 0.''35$ (~ 1 pc) level, it is believed to be the radiative manifestation of Andromeda's supermassive black hole.

None of the bright X-ray objects in M31's nuclear region are coincident with M31*, but a 50-ks *Chandra* image has shown 2.5σ evidence for a faint ($\sim 3 \times 10^{35}$ ergs s^{-1}) X-ray counterpart to the radio source.[30] The most natural interpretation is that we are witnessing the faint X-ray glow produced by matter captured via Bondi-Hoyle accretion from the medium surrounding the black hole.

The existence of hot, diffuse emission in the core of M31 was first identified in *Einstein* (see chapter 1) observations,[31] and later confirmed with several more recent satellite detectors, culminating with *Chandra*. But before we can use these data to examine whether M31* is a Bondi-Hoyle accretor, we must first estimate the temperature and density of the X-ray-emitting gas. Fitting a combination of XMM-*Newton* and *Chandra* observations of this region with an optically thin bremsstrahlung model (see section 5.3), one identifies three distinct components, each at a temperature ~ 0.26–0.5 keV, spread throughout the inner several hundred-parsec core of M31. The density of the diffuse gas near the supermassive black hole may be estimated from the integrated emissivity (equation 5.56) along several lines of sight, which yields an electron number density $n_e \sim 0.1$ cm^{-3} within 200 pc of M31*.

Thus, using equation (6.49) to estimate the sound speed c_s, we can calculate the accretion radius $r_{\rm acc}$ with the help of equation (6.73), which yields a value $r_{\rm acc} \approx 3.4$ pc, or approximately $0.''9$ at the 780-kpc distance to M31. This becomes a very tantalizing number once we recall that *Chandra*'s angular resolution is $\sim 0.''5$. Quite remarkably, we conclude that *Chandra* is actually resolving the Bondi-Hoyle accretion region in M31*.

Resolving the Bondi-Hoyle radius allows us to securely measure the size of the accretion region, but it provides only half of the story; the rest of it develops

[29] See Crane, Dickel, and Cowan (1992).
[30] See Garcia, Williams, Yuan et al. (2005).
[31] See Trinchieri and Fabbiano (1991).

as we attempt to estimate the accretion rate (from equation 6.68) and compare that with M31*'s observed luminosity. Under optimal conditions for converting accreting mass into radiant energy, we would expect roughly 10% efficiency (see section 3.4). With the above conditions and accretion radius r_{acc}, we would thus expect a maximum luminosity of $\sim 3 \times 10^{40}$ ergs s^{-1} from M31*. By comparison, its measured X-ray power is five orders of magnitude smaller.

This surprising result is actually not unique to M31*—it is a common trait among all the nearby weak nuclei, including Sagittarius A* and (as we shall see shortly) M87. Theoretically, it is still an open question why the quantity of gas captured at r_{acc} does not provide an accurate indication of its emissivity as it descends into the maw of the black hole. We are faced with two possibilities, it seems: either $\dot{M}$ changes with radius, and much of the captured matter escapes or is ejected before reaching the region where X-rays are produced, or the radiative efficiency of the plasma is very low.[32] This would certainly require that the medium be optically thin and, more importantly, that the captured gas have very little angular momentum, allowing it to descend deep into the potential well without forming a large accretion disk.[33] Only then would we expect viscous dissipation to be relatively ineffective at converting gravitational potential energy into heat, and thence radiation.

The nucleus of M87 provides our final illustration of a weak AGN, but its mass is significantly larger than that of either Sagittarius A* or M31*, so it stretches our survey in important new directions for comparative study. The giant elliptical galaxy M87 (also known as NGC 4486) contains a black hole of mass $M \sim 3 \times 10^{9} M_\odot$, measured from the kinematics of stars in its vicinity using HST observations.[34] It contains a one-sided jet (see figure 12.14) and a large-scale radio structure (not shown in this figure). Like Sagittarius A* and M31*, however, the activity displayed by its nucleus if far less than what is predicted by simple Bondi-Hoyle accretion from its hot interstellar medium.

Chandra observed the nucleus of M87 on several occasions in 2000, and spectra were extracted from circular annuli centered on the black hole,[35] yielding the temperature kT of the gas in these regions. In this source, kT decreases from 1.83 keV at 10 kpc from the nucleus, to 0.8 keV within the central kpc. As before, one can also estimate the density by interpreting the integrated emissivity along various lines of sight in terms of a simple bremsstrahlung model, which in this case yields an electron number density $n_e \approx 0.17$ cm^{-3}.

Thus, from equation (6.73) we estimate a Bondi-Hoyle accretion radius $r_{acc} \approx 5 \times 10^5 r_S$ (where $r_S \approx 3 \times 10^{14}$ cm, or roughly 20 AU) for this object. At the distance to M87 (18 Mpc), the angular resolution of *Chandra* corresponds to a radius of less than 100 pc, which is a few times 10^5 Schwarzschild radii—a good match for resolving the Bondi-Hoyle radius of this supermassive black hole, which we shall henceforth call M87* for consistency with the notation for Sagittarius A* and

---

[32] See, e.g., Blandford and Begelman (1999).

[33] This possibility was first mooted by Melia (1992a) in the context of Sagittarius A* and its faintness relative to other active nuclei.

[34] See Ford et al. (1994).

[35] See Di Matteo et al. (2003).

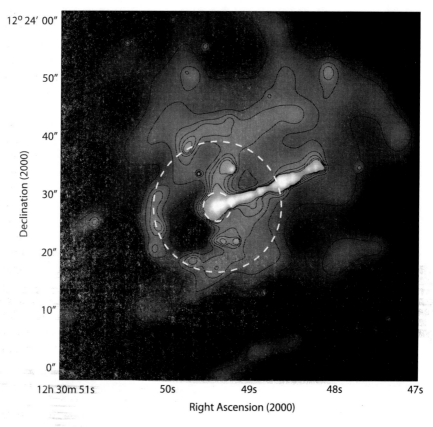

Figure 12.14 *Chandra* image of M87 and the core of the Virgo cluster in the 0.5- to 5-keV energy band. The overlaid contours (solid lines) are logarithmically spaced (from 5–30 counts per 0."5 pixel). The emission from the region bounded by the innermost dashed circle (which has a radius of 4 detector pixels, approximately 2", or 0.2 kpc at the distance to the source) is dominated by the central AGN, with a significant contribution from the jet. This emission is characterized by a power law with photon index 2.23 ± 0.04. The overall power produced by this region is more than four orders of magnitude smaller than that expected from Bondi-Hoyle accretion through the annulus bounded by the inner and outer dashed circles, i.e., between 0.2 and 1 kpc. (Image from Di Matteo et al. 2003)

M31*.[36] The corresponding accretion rate (from equation 6.68) is $\dot{M} \approx 0.1\, M_\odot$ yr^{-1}, which translates into a maximum luminosity of about 5×10^{44} ergs s^{-1}, if we adopt the standard 10% efficiency.

---

[36] Strictly speaking, this designation is reserved for the radio point source associated with the black hole. Brown (1982) named the unusual radio source at the center of our Galaxy Sagittarius A* (often simply shortened to Sgr A*) to distinguish it from the extended emission of the Sagittarius A complex and to emphasize its uniqueness and importance. Ten years later, this nomenclature was also used to denote the central radio point source, now known as M31*, in the nucleus of M31 (Melia 1992b) and has since been generalized to identify all such sources in the nuclei of nearby galaxies.

The central X-ray point source (presumably the counterpart to M87*) can be described by a power-law model with a photon index ≈ 2.23 and a flux density at 1 keV of $\approx 8 \times 10^{-13}$ ergs cm^{-2} s^{-1} keV^{-1}. The implied X-ray power is therefore $\approx 7 \times 10^{40}$ ergs s^{-1}, about four orders of magnitude less than expected. One may draw the same conclusions for this source as for M31*—that either $\dot{M}$ is not constant all the way down to the X-ray emitting region, or that the infall is radiatively inefficient.

However, in this case, there is an additional piece of the puzzle that may ultimately provide a satisfactory solution for most, or all, of these nearby weak nuclei. Unlike Sagittarius A* and M31*, the nucleus of M87 clearly has a jet, and though we will not be considering the physics of these structures until the next section, we will preempt that process just slightly by introducing the concept of a jet power here. As we shall see shortly, most of the radiation in these relativistic outflows appears to be produced by incoherent synchrotron and synchrotron–self-Compton emission, the former in the radio and the latter in X-rays and (when seen) in gamma rays. It is straightforward[37] to estimate from the measured luminosity what the kinetic power in the jet must be for it to sustain the observed radiative output over its full extent; for M87*, that value is roughly 10^{44} ergs s^{-1}.

We find, therefore, that the Bondi-Hoyle accretion rate, corresponding to a maximum luminosity of $\sim 5 \times 10^{44}$ ergs s^{-1}, matches the overall energetics of M87* quite well. In other words, though only a tiny fraction of this power comes out as X-rays, the rest of it accounts for the mass loss rate in the jet and the power required to sustain it.

Some have wondered then, whether this outflux of power may itself provide feedback to the interstellar medium, possibly heating it and altering the mass capture rate at $r_{\rm acc}$. From the expression in equation (6.73), we know that the accretion radius decreases as the temperature (and hence the sound speed) of the ambient medium rises. Physically, this is due to the fact that a more energetic interstellar medium is less bound to the accretor, so the gas has to approach closer to the central object before the gravitational potential can overwhelm the specific kinetic (or heat) energy of the gas.

It is interesting to note that ideas such as this echo the proposals we considered earlier in this chapter in attempts to explain the M-σ correlation shown in figure 12.5. Unraveling the physics of accretion onto supermassive black holes, particularly the faint ones, is work in progress. This is one of the most exciting areas of research in high-energy astrophysics and we will no doubt learn more about these objects in the years to come.

12.3 SUPERMASSIVE BLACK HOLES IN AGNS

For historical reasons, and largely because these systems are inherently quite complex, astronomers have tended to subdivide active galactic nuclei into groups defined primarily by their specific observational characteristics. As our knowledge of these systems has improved, so too has our realization that many of the differences we see

---

[37] See Bicknell and Begelman (1996).

may ultimately be attributed to observational selection effects; chief among these is the inclination angle of the object relative to our line of sight. Not surprisingly, attempts have been made to "unify" the various groups into a dominant class, with the hope of finding a single explanation for the whole AGN phenomenon. We now know, however, that aside from these selection effects, there are indeed several intrinsic differences that remain from subclass to subclass, and so there is still some benefit to be derived from a detailed classification.

12.3.1 AGN Taxonomy

Seyfert galaxies[38] constitute one of the largest subgroups of AGNs, and had been under consideration for their unusual nuclear characteristics long before quasars were discovered in 1963. These objects tend to be early-type spiral galaxies with radio-quiet nuclei at the faint end of the AGN magnitude scale (with $M_V > -23$). Spectroscopically, they are characterized by strong, high-ionization emission lines.

It is useful to subdivide these further into two groups, known as Type 1 and Type 2 Seyfert galaxies. The first of these display two distinct sets of emission lines—broad and narrow—superimposed on top of one another. The less luminous Seyfert 2's have predominantly only the narrow lines. Interpreting the line width as a Doppler effect (see equation 3.18) due to the motion of the emitting gas about the central object, we infer from the broad features a speed (at FWHM) as high as $\sim 10^4$ km s^{-1}; the narrow lines imply much lower speeds, typically $< 10^3$ km s^{-1}.

The link between these two subgroups of Seyfert galaxies has been identified as an orientation effect, coupled to the geometry of the core, which includes a supermassive black hole surrounded by a thick molecular torus. Seyfert 1's, the argument goes, are oriented such that we can see directly down to the black hole itself, thus exposing the rapidly moving gas that produces the broad emission lines. In contrast, Seyfert 2's are galaxies whose active nucleus contains a torus aligned edge on to our line of sight. This blocks our view of the ionized environment close to the accretor, so we see predominantly the narrow lines produced much farther out, above and below the torus.[39]

Strong observational support for this interpretation has been provided by certain "intermediate" Seyfert galaxies that seem to fit neither the Type 1 nor the Type 2 classification. Some evidence for highly polarized broad emission lines in certain Type 2's indicates that the light produced deep in the funnel of the torus is being scattered toward us by ionized plasma above the opening, which has the effect of attenuating one component of polarization relative to the second, thereby producing a net polarized signal in the post-scattered wave.

With the exquisite imaging capability of *Chandra*, we now also have compelling visual evidence that this is taking place in Seyferts such as NGC 1068. X-ray images of this galaxy reveal the presence of gas blowing away at high speeds from the vicinity of the central black hole.[40] X-rays originating from a hidden disk partially

---

[38] See Seyfert (1943).
[39] See, e.g., the review article by Antonucci (1993).
[40] See Ogle et al. (2003).

scatter off the torus and partially evaporate its outer layers to produce the outflow, but *Chandra* does not directly see the black hole and its nearby environment, thus confirming that this galaxy has a high inclination angle typical of Seyfert 2's.

The second large category of AGNs is *quasars*, the most luminous ($M_V < -23$) members of this class, showing a bluish ($U - B < 0$) continuum, with strong broad optical emission lines. As we noted earlier in this chapter, these objects were distinguished from Seyfert galaxies by being spatially unresolved in optical photographs, implying an angular size smaller than $\sim7''$. This situation changed with the advent of high-resolution instrumentation, such as HST, which produced images like figure 12.1, revealing the faint galaxy hosting the quasar in its core. We now know that most quasars are associated with elliptical galaxies, consistent with our view that major collisions between these stellar aggregates (which dissolves their spiral structure to produce a spheroidal remnant) are often necessary to channel high doses of material into the core, where the ensuing prodigious rate of accretion onto the nucleus turns the supermassive black hole into a powerful beacon.

Quasars are themselves subdivided yet again—into the radio-quiet and radio-loud categories,[41] depending on whether their ratio of radio (5 GHz) to optical (680 THz) flux densities is smaller or larger than 10. Since only about 15–20% of all quasars are radio loud, the majority of them are detectable primarily through optical surveys (such as the Sloan Digital Sky Survey; see chapter 2), or in X-ray maps, such as those produced by ROSAT (plate 5) and *Chandra* (plate 6).

Within the radio-loud membership, one sees another dichotomy based on spectral index. There are the steep-spectrum radio quasars (often simply called SSRQs), and the flat-spectrum objects (FSRQs), according to the value of their radio spectral index α (appearing in the expression for flux density, $S_\nu \propto \nu^{-\alpha}$). The former have $\alpha > 0.5$, while the latter have $\alpha < 0.5$. Since (as we shall see) jets are the dominant source of emission at radio wavelengths, this division ultimately says something about the jet's orientation relative to our line of sight.

Additionally, radio-loud AGNs that reside in elliptical galaxies and are underluminous ($M_V > -23$) are called *radio galaxies*. This category includes two of the sources we have already discussed in this chapter: M87 at the core of the Virgo cluster (figure 12.14), and Centaurus A (figure 12.3).

At the risk of letting this subdivision get out of hand, we mention several additional observationally motivated refinements that include the *broad-emission-line* radio galaxies (BLRGs), which show both broad (asymmetric) and narrow emission lines in their spectra, and the *narrow-line* radio galaxies (NLRGs), which display only narrow lines. The BLRGs and NLRGs may be regarded as the radio-loud (elliptical galaxy) counterparts of the Seyfert 1 and Seyfert 2 designations.

Galaxies with *low-ionization nuclear-emission-line regions* (LINERs) constitute the subclass of AGNs that are similar to Seyfert 2's, i.e., displaying only the narrow emission lines, though with relatively strong low-ionization lines, such as OI and NII. Almost half of all spiral galaxies show LINER activity in their cores.

Blazars comprise a very interesting subclass of radio-loud AGNs (see figure 12.15), characterized by their unusually rapid variability (with changes in magnitude

---

[41]This was first done by Kellermann et al. (1989).

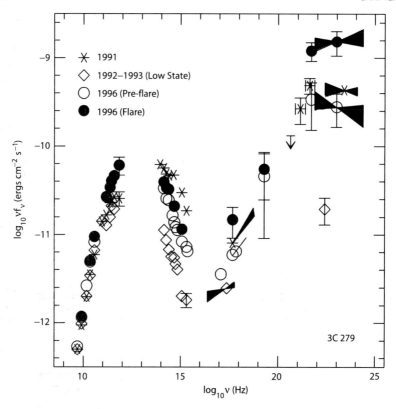

Figure 12.15 3C 279 (at a redshift of 0.538) is a member of the class of blazars, radio-loud active galactic nuclei, whose polarized, luminous and rapidly variable nonthermal continuum emission extends from radio to GeV and TeV energies. Its emission is dominated by a relativistic jet oriented close to the line of sight. Open circles represent the spectrum in the low state; filled circles are for the flaring state. The spectral energy distribution of blazars is typically double-humped; the first component peaks at IR/optical wavelengths in some cases, and at UV/X-rays in others. There is little doubt that this emission originates from relativistic electrons in the jet. The second component extends from X-rays to γ-rays. It may be due to inverse Compton scattering by electrons within the jet (giving rise to synchrotron-self-Compton), or perhaps outside the jet. (Image from Wehrle et al. 1998)

greater than 0.1 mag day^{-1}), their strong and variable optical linear polarization (with polarization fractions greater than $\sim$3%; see section 5.4), and their flat radio spectrum and featureless broad nonthermal continuum (see section 5.6). Very Large Baseline Interferometry (VLBI) observations[42] have revealed that many blazars are

---

[42] In radio astronomy, the technique known as Very Long Baseline Interferometry blends together the radio signals gathered simultaneously by many telescopes, often separated by many miles, or even thousands of miles. A major component in the arsenal of VLBI work is known as the Very Large Baseline Array (VLBA), which was commissioned in 1993. It constitutes a system of ten 82-foot-diameter dish antennas located across the United States, from Hawaii to the Virgin Islands. All ten antennas work as a single instrument, controlled from an operations center in Socorro, New Mexico.

superluminal sources (see section 12.3.4, especially plate 25), showing apparent transverse velocities with magnitudes greater than the speed of light.

These supermassive black holes are believed to have their jets oriented almost exactly along our line of sight. Their emission is therefore greatly enhanced by Doppler boosting effects, and their observed variability timescale greatly shortened by special relativistic transformations, as we shall see in section 12.3.4. Blazars encompass several subgroups, including the *optically violently variable* (OVVs) quasars, characterized by rapid and large amplitude optical continuum variations; the *highly polarized* quasars (HPQs), which exhibit a high percentage of optical linear polarization; and the lower luminosity *BL Lac* objects, which show no, or only weakly, detectable emission lines. Again, all blazars seem to be hosted within elliptical galaxies.

Attempts at unifying the various classes of AGNs, e.g., via the invocation of an orientation effect, also introduce into the mix many extragalactic radio sources that apparently represent quasars being viewed (by us) at inopportune angles. These too are subdivided into two groups according to their radio morphology and luminosity. Members of the first class, called Fanaroff-Riley type I (or FR I, for short), are compact, unresolved radio sources with flat spectra. The second category, called Fanaroff-Riley type II (or FR II), includes resolved radio sources with steep spectra, extended over regions (e.g., in lobes) more than 1 kpc from the center of the associated galaxy.[43]

The FR I's are lower luminosity, edge-darkened radio sources that show no prominent hot spots in their outer lobes and have only weak optical emission lines. On the other hand, FR II's are very luminous, edge-brightened sources, whose radio lobes are dominated by bright hot spots, probably where shocks are produced as the jet material impacts the surrounding medium. In contrast to the FR I's, these sources also exhibit strong optical emission lines. In unification schemes, BL Lac objects are related to FR I's, whereas quasars, OVVs, and blazars are somehow connected to the FR II's.

12.3.2 Radio Synchrotron Emission

It is generally believed that AGNs produce two relatively distinct electromagnetic signals, one due to an isotropic process, the other due to a relativistic jet; depending on the object and its orientation relative to our line of sight, one or the other of these may dominate the observed flux.

At the center of the AGN lies a supermassive black hole whose gravitational influence provides the ultimate source of power for the high-energy activity, just as we have been discussing for the nearby, weaker nuclei. Matter captured from the surrounding medium loses angular momentum through viscous or turbulent processes within an accretion disk, which sometimes glows brightly at ultraviolet and soft X-ray wavelengths (see sections 7.2.2 and 8.1). In some cases, hard X-ray emission is also produced by a pervasive sea of energetic electrons above the disk, in the form of a tenuous corona. And if the black hole is spinning, power may

---

[43]As the naming convention suggests, this classification was introduced by Fanaroff and Riley (1974).

be extracted electromagnetically from the black hole's rotational energy, which contributes to the expulsion of relativistic particles in the form of a prominent jet extending hundreds of kpc out from the nucleus (see Cygnus A in figure 12.4).

The radio interferometric techniques we discussed in the previous section provide images of AGN cores with angular scales approaching 10^{-3} arcsec. Even for an object as distant as Cygnus A, this spatial resolution corresponds to a size as small as ~3 lyr. VLBA images produced in this way reveal jet-like structures that sometimes vary from year to year, implying superluminal expansion of the inner regions. We shall return to this fascinating phenomenon in section 12.3.4.

From the beginning, the development of a theoretical framework for radio emission in AGNs has advanced with a single fundamental postulate,[44] that the dominant radiative mechanism in most radio sources is incoherent synchrotron emission, produced by nonthermal, relativistic leptons gyrating in a mostly randomized magnetic field (see section 5.6). There is now ample evidence for this, including the ubiquity of nonthermal spectra and the measurable degree of linear polarization (see section 5.4). For example, synchrotron processes are clearly evident in the radio to soft X-ray spectrum of (core-dominated) blazar sources,[45] manifested as a bump on the left-hand side of figure 12.15.

With the exquisite spatial resolution available today, we know that there exist at least four distinct regions where the radio emission is produced: the nucleus, a jet, a radio lobe (often with hots spots, or an edge brightening), and cocoons that seem to enshroud part of the jet. Which of these are seen depends on the class of object (see section 12.3.1).

AGN jets were recognized as dynamically important structures linking the various emission regions soon after the nonthermal-synchrotron model was proposed. It had been shown as early as 1958 that a source such as Cygnus A (figure 12.4) must contain at least 10^{59} ergs in the energy of relativistic particles and magnetic fields within the giant radio lobes.[46] Coupled with evidence that the nucleus is variable, this led to the idea that the internal energy throughout the radio-emitting region must be supplied continuously from the center through channels or jets. Adiabatic energy losses are minimized because most of the energy is transported outwards in the form of bulk kinetic energy.[47]

The giant radio lobes generally have a size > 3 kpc, loosely correlated with their absolute power. The most powerful objects with a luminosity $> 10^{42}$ ergs s^{-1} are typically 100–300 kpc in linear extent and, as already noted above, tend to be edge brightened. This morphology finds a natural explanation in terms of an energy transport via the jet, which results in a collision between the advancing head of the outflowing plasma and the surrounding medium.

The impact slows down the plasma considerably, so that, whereas the jet emission is apparently beamed—the physics of which we will return to shortly—the

[44]This postulate was proposed by Alfvén and Herlofsen (1950) and Shklovsky (1953 and translated into English in a volume edited by Lang and Gingerich in (1979).

[45]See Blandford and Königl (1979).

[46]See Burbidge (1959).

[47]See Rees (1971), Longair, Ryle, and Scheuer (1973), Scheuer (1974), and Blandford and Rees (1974).

radio lobes are more or less stationary with respect to the observer's frame. With knowledge that the radio luminosity from these sources is produced by synchrotron processes, it is straightforward to use the framework developed in sections 5.4, 5.5, and 5.6 to determine the physical conditions in the emission region.

To begin with, let us consider a scenario in which the relativistic plasma,[48] confined to a uniform spherical region of radius R, is observed to emit a total luminosity L, with a power-law spectrum $F_\nu \propto \nu^{-\alpha}$ between the frequencies ν_0 and ν_1. (We will consider an illustrative set of parameter values shortly.) The particle distribution producing this radiation must have the form given in equation (5.112), with $x = 1 + 2\alpha$.

There are two unknowns here—the density of relativistic particles and the magnetic field. But we can relate them by using the observed synchrotron power, under the (reasonable) assumption that the medium is optically thin between ν_0 and ν_1.

The total energy density, including relativistic particles and magnetic field, is

$$u_{\text{tot}} = u_e + \frac{B^2}{8\pi} \tag{12.4}$$

where, according to equation (5.112),

$$u_e = \int_{m_e c^2}^{\infty} K E^{1-x}\, dE = \frac{K m_e c^2}{x - 2} \tag{12.5}$$

(remembering that $x > 2$). This combination of particles and magnetic field produces a luminosity L derived by integrating the emissivity in equation (5.120) over the whole volume ($V = 4\pi R^3/3$), and frequencies from ν_0 to ν_1. The result is

$$L \approx \frac{3.4 \times 10^{-21} a(x)\, K V B^{(x+1)/2}}{x - 1} \left(\frac{6.26 \times 10^{18}\ \text{Hz}}{\nu_0} \right)^{(x-1)/2}, \tag{12.6}$$

where we have also assumed that $\nu_1 \gg \nu_0$. This equation defines the particle normalization constant K in terms of B, which we write as

$$K \equiv K_0 B^{-(x+1)/2}, \tag{12.7}$$

with

$$K_0 \equiv \frac{(x-1)L}{3.4 \times 10^{-21} a(x) V} \left(\frac{\nu_0}{6.26 \times 10^{18}\ \text{Hz}} \right)^{(x-1)/2}. \tag{12.8}$$

We can often argue that the system settles into its minimum energy configuration. Thus, B may be estimated by minimizing the energy density u_{tot}, which leads to

$$B^{(x+5)/2} = \frac{x^2 - 1}{x - 2} \frac{\pi m_e c^2 L}{1.7 \times 10^{-21} a(x) V} \left(\frac{\nu_0}{6.26 \times 10^{18}\ \text{Hz}} \right)^{(x-1)/2}. \tag{12.9}$$

For typical parameter values $\alpha = 0.8$, $R = 200\,\text{kpc}$, $L = 10^{44}$ ergs s^{-1}, $\nu_0 = 400$ MHz, and $\nu_1 = 5$ GHz, equation (12.9) then gives $B \approx 10^{-5}$ G, and the corresponding particle density may be estimated from equation (5.112) with the normalization constant given in equation (12.7).

---

[48] By this we mean that the particles individually have relativistic energies, but their motion is random, producing a very hot gas whose bulk motion, if any, is nonetheless still nonrelativistic.

Under these conditions, particles lose their energy on a relatively short timescale compared to the presumed age of the giant structures. From equation (5.113), setting $\nu_1 \sim \nu_c$, we infer a maximum particle Lorentz factor $\gamma_{max} \sim 2 \times 10^4$. Thus, defining the synchrotron cooling time

$$\tau_{synch} \equiv \frac{\gamma_{max} \, m_e c^2}{P}, \qquad (12.10)$$

in terms of the power P given by equation (5.117), we infer a relativistic-particle lifetime $\tau_{synch} \approx 15$ million years. The radio lobes, however, may exist for hundreds of millions (if not several billions) of years, confirming the suspicion that jets must be funneling energy from the central black hole to the outer giant structures to sustain their prodigious radiative output.

However, one must remember that the issue of energy transport into the radio lobes is only weakly connected to how the energy is actually transferred to the relativistic particles. Each of the four distinct emission regions we introduced above seem to have their own particular spectral indices, so the particle populations in the various acceleration sites are quite distinct. Nuclei of compact sources have flat spectra, $-0.3 < \alpha < 0.5$. These result from a superposition of a continuum of self-absorbed regions.[49] The spectral indices of jets typically fall in the range $0.5 < \alpha < 0.7$, and those of radio lobes are typically $0.5 < \alpha < 1$. Low-surface-brightness cocoons have very steep spectra ($\alpha > 1.2$), interpreted as the result of radiative losses.[50]

Almost certainly different acceleration mechanisms are at work in each of these four regions. In section 4.3, we learned that an acceleration based on the Fermi process is ubiquitous in nature, but the acceleration rate and the particle spectral index depend on whether first-order or second-order scatterings dominate. Extended jets are unlikely to be permeated by very strong shocks, otherwise they would be easily disrupted. However, supersonic jets can be "noisy" and subject to small-scale Kelvin-Helmholtz instabilities, which generate long-wavelength ($\lambda > 1$ kpc) waves that cascade down to shorter wavelengths to form a turbulent magnetic field.[51] In such an environment, particles bounce continuously off randomly moving fluctuations, gaining energy via a second-order Fermi process.

Some shocks may also be present within jets (contributing particles to the relativistic pool via first-order Fermi acceleration), but it is far more likely that these would be generated in the hot spots at the end of a jet, where the material collides with the surrounding medium. Thus, since the resulting spectral index is a sensitive function of the ratio of acceleration and escape timescales, it is quite likely that the four different radio-emitting regions in AGNs produce particles with varied distributions based on which of the Fermi acceleration schemes dominate the energy transfer in that particular location.

In sources (such as blazars) whose jet is aligned close to our line of sight, the spectrum is dominated by emission from the jet itself. Their flux measured at Earth

---

[49] See, e.g., Wittels, Shapiro, and Cotton (1982).
[50] See Jenkins and Scheuer (1976).
[51] See, e.g., Bicknell and Melrose (1982) and Henriksen, Bridle, and Chan (1982).

is so high that relativistic beaming must play a significant role in focusing what would otherwise be an isotropic radiation field into a directed cone of energy.

Observations indicate that, at least on the smallest scales, energetic particles flow out from the central engine along the poles of the disk or torus at velocities approaching the speed of light. They form collimated radio-emitting structures that sometimes end in the giant radio lobes we have been discussing (e.g., for Cygnus A in figure 12.4).

The Doppler boosting of radiation emitted by plasma moving with an inferred bulk Lorentz factor ~ 10 can result in an observed flux thousands of times brighter than it would appear in the emitter's rest frame. It is therefore not a coincidence that quasars detected by the EGRET instrument on the *Compton* Gamma Ray Observatory, sensitive in the ~ 100-MeV to 10-GeV range (see section 1.5), also happen to be predominantly jet-dominated radio-emitting AGNs. Several of these sources are routinely also detected up to the TeV γ-ray range via ground-based Cerenkov telescopes (see section 1.4). It is generally believed that blazars produce the entire electromagnetic signal we detect from them in their Doppler-boosted jet aligned close to our line of sight.

To understand how this comes about, we need only consider the Doppler shift formula (equation 3.18), which relates the frequency of the radiation emitted in the plasma's rest frame to that observed with our detector, and the Lorentz invariant quantity I_ν/ν^3 (equation 5.42), which allows us to transform the intensity from one frame to the next.

When an AGN jet is aligned close to our line of sight, the source of radiation is approaching us, and time intervals measured by the observer are shortened relative to those in the emitter's frame, even allowing for time dilation. This happens because the emitting plasma is trailing its photons, so the effective region over which the radiation is produced is shortened in our frame. Thus, a time interval $\Delta t'$ in the emitting plasma[52] becomes

$$\Delta t = \delta^{-1} \Delta t' \tag{12.11}$$

in our coordinate system, where

$$\delta \equiv \gamma (1 - \beta \cos \theta)^{-1} \tag{12.12}$$

is known as the kinematic Doppler factor. In these expressions, β is the bulk velocity in units of the speed of light, and θ is the angle between the velocity vector and the line-of-sight. Correspondingly, the radiation is also blue-shifted, so the observed frequency is

$$\nu = \delta \nu'. \tag{12.13}$$

But the most dramatic effect in jet emission is the intensity enhancement arising from Doppler boosting. From equation (5.42), we infer that

$$I_\nu = \delta^3 I'_{\nu'}. \tag{12.14}$$

---

[52] In this section, a prime is used consistently to denote quantities measured in the emitter's frame.

One power of δ in this expression comes from the compression of the time interval (equation 12.11); the other two are due to the transformation of the solid angle (see the discussion in section 5.2, particularly equation 5.40).

Finally, the broadband flux is obtained by integrating I_ν over frequency (and then solid angle), and since $d\nu = \delta \, d\nu'$, the overall boost corresponds to a net factor δ^4. That is,

$$I = \int d\nu \, I_\nu = \delta^4 \int d\nu' \, I'_{\nu'}. \tag{12.15}$$

Note, however, that a proper interpretation of the observations may be further complicated by the spectrum itself, which may shift toward higher frequencies due to blue-shifting effects. For example, if the intrinsic spectrum is a power law with index α (i.e., $I_\nu \propto \nu^{-\alpha}$), then

$$I_\nu = \delta^{4+\alpha} \, I'_\nu, \tag{12.16}$$

where both intensities are now measured at the *same* frequency ν. In simple terms, this happens because the shifting of the spectrum toward higher frequencies also brings a higher (preboosted) intensity into the observation window.

As we shall see in section 12.3.4, there is now ample evidence for the existence of bulk Lorentz factors $\gamma \sim 10$ (or more) in relativistic jets. Thus, even if the emission is more or less isotropic in the emitter's frame, by the time the electromagnetic signal reaches the observer, its measured flux may be enhanced by as much as a factor of 10,000.

12.3.3 High Energy Emission

High-energy observations of quasars and AGNs seem to confirm the theoretical expectation of a Doppler-boosted signal. The third (and last) catalog of EGRET point sources[53] includes 271 resolved objects of which 93 have been identified (either confidently or with reservation) as blazars. Not only are these objects bright at energies above $\sim$100 MeV, in many cases their observed γ-ray luminosity actually dominates their spectrum in other wavebands by factors of 1 to 1000. For example, the multiwavelength spectrum of the superluminal quasar 3C 279 (figure 12.15) shows that its γ-ray luminosity increases relative to its bolometric output as the overall radiative power rises, and typically dominates the emission at other wavelengths by a factor 10.

Additional evidence that the γ-rays from AGNs are beamed comes from their observed rapid variability at high energy, where the flux changes measurably over the course of a day.[54] The intensity in 3C 279, for example, has on (at least one) occasion declined by a factor 4–5 in under 3 days.[55]

---

[53] See Hartmann et al. (1999).

[54] A suitable definition of the variability timescale is how long it takes for the γ-ray flux to double, or to be cut in half.

[55] See Kniffen et al. (1993).

Consider that for the γ-rays to escape the source, the optical depth $\tau_{\gamma\gamma}$ to pair production[56] must be of order unity or less. Let us first suppose that the source radiates isotropically, with a γ-ray luminosity L and a characteristic size R. We estimate L from the observed flux and the size R from the measured variability using a light-travel time argument. At any radius r, the density of photons with energy of order the electron rest mass is then the flux $L/4\pi r^2$ divided by $m_e c^2$, and divided again by c. Thus, since the cross section for $\gamma\gamma$ scattering near threshold is approximately σ_T, we may write

$$\tau_{\gamma\gamma} \approx \int_R^\infty dr\, \sigma_T \frac{L}{4\pi r^2 m_e c^3}, \tag{12.17}$$

which easily integrates to

$$\tau_{\gamma\gamma} \approx \left(\frac{L}{R}\right) \frac{\sigma_T}{4\pi m_e c^3}. \tag{12.18}$$

But for 3C 279, the luminosity and size estimated under the assumption of isotropic emission imply a value of $\tau_{\gamma\gamma}$ greater than 1000, far above the level where the γ-rays could easily escape. For us to even see γ-rays from this blazar, the true γ-ray luminosity must therefore be much smaller than this, and its size must be much larger than that implied by its variability without the inclusion of the shrinkage factor appearing in equation (12.11).

Relativistic beaming alleviates this problem in several ways. To begin with, the actual luminosity would be a factor δ^4 smaller than L (see equation (12.15)). Furthermore, according to equation (12.11), the size R estimated from the observed variability timescale (let's call this Δt) is a factor δ smaller than the actual distance traversed by the emitter during a time $\Delta t'$. Together, these two modifications reduce $\tau_{\gamma\gamma}$ by δ^5.

We can use this important result in at least two ways. We may either choose to adopt the value of δ ($\sim$10) implied by the superluminal motion (see section 12.3.4) in sources such as 3C 279 to show that $\tau_{\gamma\gamma} \ll 1$, or take the reverse approach of arguing that since we see γ-rays from these objects, $\tau_{\gamma\gamma}$ must be less than one, which places a lower limit on δ. In that case, δ must be greater than one, showing again that the blazar high-energy emission must therefore be relativistically beamed.

It is generally believed that the second bump (e.g., from hard X-rays to γ-rays in 3C 279's spectrum shown in figure 12.15) is produced via inverse Compton scattering of lower-energy photons by the same electrons responsible for the synchrotron emission.[57] Some of the seed photons originate within the jet itself (giving rise to a synchrotron-self-Compton component; see section 7.2.2); the rest are produced in the environment surrounding the supermassive black hole, particularly by the accreting plasma.

In other AGNs whose jet is not aligned close to our line of sight, or in which the jet is weak or nonexistent, it is in fact the central disk that accounts for most of the

---

[56]The physics of pair creation via $\gamma\gamma$ interactions in black-hole systems is relatively independent of the central object's mass. The reader may find a more complete discussion of this effect in section 10.2.

[57]See Melia and Königl (1989).

high-energy emission seen from these objects. The largest amplitude and most rapid variability is observed in the X-ray band. The European EXOSAT satellite uncovered variability in sources with an X-ray luminosity $\sim 10^{43}$ ergs s^{-1} on timescales as short as a few hours. Follow-up observations with subsequent X-ray instruments confirmed that large amplitude X-ray variability within only a few hours is actually commonplace amongst AGNs.[58] It is therefore reasonable to suppose that most of this high-energy emission is produced within a solar system-sized region, only tens of Schwarzschild radii from the black hole.[59]

The X-ray appearance of the disk depends on several relativistic effects, including Doppler shifts (equation (3.18)) associated with very rapid orbital motion, aberration (section 5.2), and the effects of gravitational redshift and light bending (section 3.4).[60] Indeed, one of the most exciting discoveries in recent years has been the identification, within the 2- to 10-keV spectrum, of a dominant iron line with a broad skewed profile that bears the signature of the spacetime in the immediate vicinity of the central black hole.[61] One of the most famous examples is shown in figure 12.16.

More generally, the X-ray spectra of these accreting black holes exhibit a power-law component extending into the hard X-rays, up to 200 keV or more, on top of a softer thermal component believed to originate from within the Keplerian region. Almost certainly, this nonthermal emission is due to inverse Comptonization of the soft X-rays by energetic particles within a hot corona above the disk, possibly energized by magnetic fields from the flaring plasma in the equatorial plane.[62] In turn, this hard X-ray component then irradiates the dense disk material, which gives rise to a characteristic reflection spectrum, including a so-called Compton bump at $\sim$20–30 keV, and the fluorescent FeI K$_\alpha$ line.

This iron line is actually a doublet comprising the K$_{\alpha 1}$ and K$_{\alpha 2}$ lines at 6.404 and 6.391 keV, respectively. Their energy separation is too small to have been resolved thus far, and a weighted mean at 6.4 keV is generally assumed in all the fits. However, the resulting line in the emitter's frame is symmetric and intrinsically much narrower than the spectral resolution of X-ray detectors available today, so the observed line shape and broadening can be used to probe the spacetime and disk dynamics near the black hole.

Examining the line profile in figure 12.16, we should notice a hint of a double peak, though the one at $\sim$4 keV is much weaker than the other. These two features correspond to emission from the approaching (blue) and receding (red) sides of the disk, the former being much stronger due to the effects of Doppler boosting. The

---

[58] See, e.g., Markowitz and Edelson (2001).

[59] According to equation (3.88), the Schwarzschild radius for an object with mass 10^8 $M_\odot$ is approximately 2 AU, or about 1000 light-seconds. Thus, variability on a timescale of 2 h corresponds to a (light-travel) distance of only 6–7 Schwarzschild radii for such a system.

[60] Early papers on this topic include those by Page and Thorne (1974) and Cunningham (1975). A more recent treatment may be found in Melia (2007).

[61] See Fabian et al. (1989) and Laor (1991).

[62] This situation therefore bears some resemblance to the energetic activity seen in the Sun, which, in fact, provided some of the early motivation for developing this kind of model. See, e.g., Haardt and Maraschi (1991), Zdziarski et al. (1994), and Nayakshin and Melia (1997).

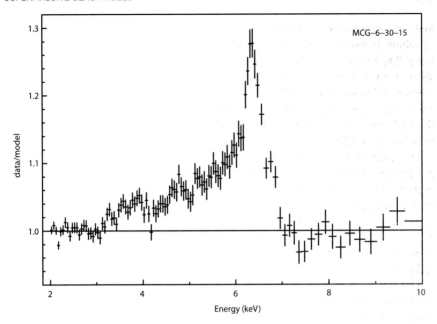

Figure 12.16 The broad (asymmetric) iron line in MCG-6-30-15 observed with XMM-*Newton* in 2001. The red wing provides compelling evidence of emission from a disk extending down to just a few Schwarzschild radii above the black hole's event horizon. (Image from Fabian et al. 2002)

transverse Doppler effect (due to time dilation), shifts both peaks to energies lower than they would have in a Newtonian framework, in which the two peaks would appear on either side of the line center at 6.4 keV.

However, the dominant effect is clearly the gravitational redshift, which not only skews the line shape toward the red, but also decreases the line intensity on that side. This is easy to understand because I_ν/ν^3 is a Lorentz invariant (see equation 5.42), so a gravitational redshift must alter *both* the frequency and the intensity of the radiation produced in the inner region of the disk.

As the energy resolution of X-ray instruments continues to improve, the impact of further refinements in this kind of investigation will push the limits to which we can test the physics of strong fields. One should not underestimate the extraordinary potential for using the fluorescent iron emission line in AGNs to study in detail the structure of spacetime—particularly the Kerr metric—near a supermassive black hole.

12.3.4 Superluminal Motion

One of the quasars we featured in the previous two sections—3C 279—also happens to be a superluminal source, meaning that certain features at the base of its jet move across the plane of the sky at apparent velocities exceeding c (see plate 25).

Though quasars had been known since the early 1960s, the experimental verification of the superluminal effect could not be obtained until radio astronomy extended the resolution of the measuring instruments from arcsecs to mas. A mas corresponds to a length of 1.2 light-years—the scale over which the superluminal motion occurs—at a distance of 100 Mpc, roughly the near edge of the quasar population. This need for higher angular resolution motivated groups in Canada, the United States, England, and the former Soviet Union to develop Very Long Baseline Interferometry (VLBI), in which independent local oscillators are used to obtain and record signals on magnetic tape for later correlation.

3C 279 was monitored by an American–Australian team during the period between 1968 and 1970, when its visibility and total flux density changed due to the fact that a radio emitting component had reached a diameter of about 1 mas, implying expansion (given its distance) at an apparent velocity of at least twice the speed of light. These measurements[63] provided evidence for the relativistic expansion of the emitter, confirming the basic prediction offered several years earlier.[64]

The reaction to the discovery of apparent superluminal motion was generally one of skepticism, inducing some to refute special relativity and/or the concept of cosmological redshifts.[65] However, in subsequent years, the development of more sophisticated analysis methods for VLBI and the establishment of extensive observing programs have led to the generation of true images of the superluminal motion of the sources (as we see in plate 25), allowing us to see not only the birth of these components, but also their evolution.

Like many of the other effects we have been studying in special relativity, the phenomenon of superluminal motion is based on a well-understood physical principle. Suppose we observe a blob of relativistic electrons emitting synchrotron radiation as they move from points 1 to 2 over a time Δt in the laboratory frame (figure 12.17). Because 2 is closer to the observer than 1, the apparent duration of the light pulse received at Earth is Δt minus the light travel time from point 1 to the line running horizontally out from point 2:

$$\Delta t_{app} = \Delta t - \frac{\Delta t v \cos \theta}{c}$$

$$= \Delta t \left(1 - \frac{v}{c} \cos \theta \right). \tag{12.19}$$

The apparent transverse velocity is therefore

$$v_{app} = \frac{v \, \Delta t \, \sin \theta}{\Delta t_{app}}$$

$$= \frac{v \, \sin \theta}{1 - \beta \cos \theta}. \tag{12.20}$$

---

[63] See Moffet et al. (1972).

[64] The relativistic expansion model was proposed by Rees (1966, 1967), who also anticipated the fact that superluminal velocities might be observed under special circumstances.

[65] See, e.g., Stubbs (1971).

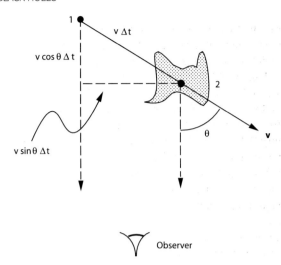

Figure 12.17 A blob of plasma emits radiation while moving with very high velocity **v** from point 1 to point 2 over a time interval Δt. A distant observer infers a different duration for the emission because of the reduced light travel time from point 2 compared to point 1. Therefore, the blob appears to move superluminally because it covers the projected distance from 1 to 2 over an inferred time shorter than Δt.

To find the maximum permissible value, we now differentiate v_{app} with respect to θ and set the result equal to zero. This yields the critical angle θ_c, where

$$\cos \theta_c = \frac{v}{c} \equiv \beta. \tag{12.21}$$

Correspondingly, $\sin \theta_c = \sqrt{1 - \beta^2} = 1/\gamma$. Therefore, the maximum apparent velocity is

$$v_{max} = \frac{v\sqrt{1 - \beta^2}}{1 - \beta^2} = \gamma v. \tag{12.22}$$

Though this optimal condition is not met in every instance, the apparent velocity v_{app} can nonetheless be greater than c for a large range of angles. A systematic study of blazars over a seven-year period[66] has shown that most of these objects produce an outflow away from the active core, though the direction and speed of the superluminal components sometimes change as they propagate along the jet. In other sources, the plasma blobs appear to follow a ballistic trajectory.

The results of this very informative survey are summarized in figure 12.18, whose sample includes active galaxies, BL Lac objects, and quasars (see section 12.3.1 for a definition of these categories). Most of the sources are quasars, and the speeds of their jet components are between zero and ten times c, but there is a clear tail extending out to $v/c \sim 34$. Features associated with active galaxies have motions

---

[66] See Kellermann et al. (2003).

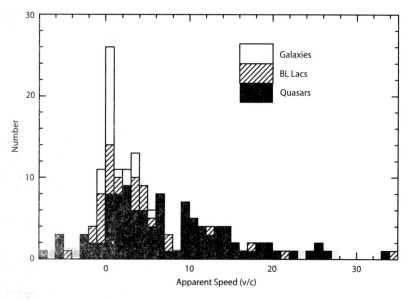

Figure 12.18 Distribution of linear speeds for 156 individual radio features seen in a sample of 96 sources that have well-determined motions. Negative velocities mean that the jet components appear to be moving toward the nucleus. However, these are consistent with zero motion within the errors. (From Kellermann et al. 2003)

in the range $0 < v/c < 8$, whereas BL Lac objects appear to be more uniformly distributed over the entire range from zero to 35.

The process of jet formation is still not completely understood, though several possibilities are quite credible. We are clearly dealing with bulk speeds very close to c, so a hydrodynamic acceleration cannot, by itself, achieve this kind of expansion. In such a scenario, one would expect a motion at close to sound speed (see equation 6.49), but this would require gas temperatures of order 10^{13} K or more. Even in the case of Sagittarius A* at the galactic center (section 12.2.1), one of the most radiatively inefficient supermassive black holes we know, the temperature of the infalling plasma is several orders of magnitude lower than this. So most workers in this field believe that magnetic fields are somehow involved in the particle acceleration process. But it is not known whether this field is anchored in the plasma approaching the event horizon of the black hole, or whether it is threading the disk. Perhaps both a spinning black hole and an accretion disk are necessary to produce a relativistic jet. We do not yet have a compelling answer.

The subject we have covered in this chapter is so extensive that our treatment can only be viewed as an introduction. Still, we have described (at least) two of the most important characteristics of supermassive black holes that are likely to command our attention over the coming years, as the sensitivity and spatial resolution of high-energy detectors continue to improve. Certainly the study of the skewed, broad 6.4-keV line seen in many sources will be the focus of our attempts to understand the spacetime surrounding these enigmatic objects, and the phenomenon of superluminal motion in the jets of other quasars and AGNs will no doubt feature prominently in our eventual understanding of how supermassive black holes eject their relativistic plasma into the surrounding medium.

SUGGESTED READING

Our discussion of the black-hole census has focused on recent observations and analysis, but population studies began in the early 1980s, albeit with the limited number of data available back then. The key publications that established the initial mass and luminosity functions for these objects include the following: Avni et al. (1980), Zamorani et al. (1981), and Soltan (1982).

In more recent times, the Sloan Digital Sky Survey has revolutionized our view of the quasar distribution. One of the most surprising results has been the discovery of quasars at redshifts greater than 6. In the standard model of cosmology, that corresponds to only $\sim$800 million years after the Big Bang. It is not yet clear how such large condensations of mass could have occurred so quickly in the rapidly expanding universe. See Fan et al. (2001).

Another remarkable property of supermassive black holes is that the velocity dispersion within the galaxies that host them somehow traces the central object's mass, even though the stars are currently too far from the black hole to be influenced by its gravity. Supermassive black holes and their host galaxies must therefore have evolved together, probably influencing each other as they grew. The reader may find more details on this paradigm-changing discovery in Ferrarese and Merritt (2000) and Gebhardt et al. (2000).

At a distance of only 25,000 light-years, the supermassive black hole at the center of our Galaxy is by far the closest member of its class. Observations of this object, known as Sagittarius A*, are clarifying its nature at such a rapid pace that theorists have a difficult time keeping up. An extensive review of the primary literature is now available in book form. See Melia (2007).

The discovery of a periodic modulation in Sagittarius A*'s infrared and X-ray lightcurves, though still needing confirmation, has heralded a new phase in the study of accretion onto supermassive objects. The implied emission region hovers barely outside the event horizon, offering the real possibility of studying the Kerr spacetime within 2–3 Schwarzschild radii of the black hole. Read about these tentative discoveries in Genzel et al. (2003) and Bélanger et al. (2006b).

The painstaking analysis that led to the discovery of 400 young, blue stars at the nucleus of the Andromeda galaxy is a lesson in perseverance and a careful blending of multiwavelength data. See Bender et al. (2005).

The unification of various classes of AGNs is based in part on the idea that orientation restricts what we see from any given object. Read more about some of the observational evidence for this important concept in Antonucci (1993).

Broad Fe lines observed in several black-hole systems suggest that a range of gravitational redshifts creates their asymmetric profile. A study of this effect is growing in prominence and importance, both observationally and theoretically. Some of the principal papers on this topic include Nandra and Pounds (1994), Reynolds et al. (1997), and Reynolds and Nowak (2003).

Superluminal motion in some AGNs was anticipated in the 1960s, and observed soon thereafter. Read about these developments in Rees (1966, 1967), Moffet et al. (1972), Pearson and Readhead (1988), and Kellermann et al. (2003).

For a comprehensive review of AGN variability, see Ulrich, Maraschi, and Urry (1997).

Chapter Thirteen

The High-Energy Background

Having surveyed the various classes of high-energy objects in the previous chapters, one might understandably come away with the impression that this branch of astronomy is restricted to compact sources, i.e., condensations of matter possessing extreme equations of state. Certainly their deep potential well facilitates the dissipation of gravitational energy into random particle motion, and (when present) the strong magnetic fields anchored to the surface (or near an event horizon) can facilitate the transfer of this liberated energy to a subpopulation of relativistic particles, whose radiative signature includes the X-rays, γ-rays, and other messenger particles we detect with our instruments.

However, not all of the high-energy activity we perceive in the cosmos is necessarily associated with unresolved point sources. In fact, the genesis of high-energy astrophysics could rightly be traced back to the early 1900s (see chapter 1), when Victor Hess and his contemporaries worried about the nature of a pervasive field of ionizing radiation near Earth's surface. Point sources would not be discovered until much later (in the 1960s), so the diffuse high-energy field was actually recognized first.

But just as compact objects form distinct groups characterized by their observed morphology and theoretical interpretation, so too must the various diffuse sources of emission be considered separately. Not only do these components represent a variety of spatial scales, but their origins are traceable to rather diverse physical processes. We will begin this final chapter with a discussion of cosmic rays, whose existence has been known (or suspected) for over a century, and whose early influence gave birth to the discipline of high-energy astrophysics.

13.1 COSMIC RAYS

Cosmic rays are shielded from direct view by Earth's atmosphere, though their existence has been inferred through indirect means, such as the pervasive low-level ionization implied by the gradual discharging of an electroscope in the laboratory (chapter 1). Direct observation of these particles is possible only with instruments flown in space or with high-altitude balloons. These detectors, however, are very restricted in size and weight, so observations with them are always limited by poor statistics. The differential cosmic-ray spectrum (see figure 13.1) is a steeply falling function of energy, roughly in accord with a power-law index -2.7 up to an energy $\sim 4 \times 10^{15}$ eV and then steepening toward higher energy (more on this below). Current instrumentation can provide a significant detection up to only about 100 Tev ($\approx 10^{14}$ eV). The neutral component (i.e., γ-rays) of cosmic rays has a flux several

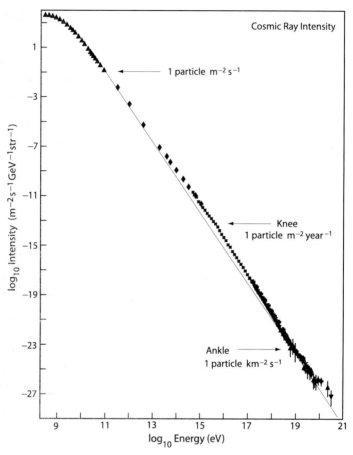

Figure 13.1 A compilation of the cosmic-ray all-particle spectrum observed over the whole range of energies accessible by the various experimental techniques now in use (see text). The distribution is roughly a power law over a wide range of energies, though comparison of the data to a single power law (the thin solid curve) reveals significant breaks at the "knee" (around 4×10^{15} eV) and, to a lesser extent, at the "ankle" (at $\sim 5 \times 10^{18}$ eV). Cosmic-ray particles with energies $> 10^{20}$ eV have also been detected, but the sources and the physical mechanism(s) responsible for producing such enormous energies are unknown. Current theories of cosmic-ray production generally fall into two main camps: "bottom-up" acceleration, in which charged particles are accelerated to high energies in supernova shocks, active galactic nuclei, powerful radio galaxies, or the strong electric fields generated by magnetized, rotating neutron stars; and "top-down" scenarios, in which energetic particles are created from the decay of more massive particles originating in the early universe. (From Bhattacharjee and Sigl 2000)

orders of magnitude lower than that of the charged particles, so this statistical limit occurs at even lower energies for them, typically around 100 GeV for instruments like those flown aboard the *Compton* Gamma Ray Observatory (see figures 1.8 and 1.9).

Above about 100 TeV, the showers of secondary particles created by the interaction of an incoming cosmic ray with the upper layers of Earth's atmosphere are extensive enough to be detectable from the ground (see section 1.4 and figure 1.11). There are at least two methods employed for this purpose. The first is based on the detection of charged particles in the shower using standard instrumentation, such as water Cerenkov detectors, in which the forward cone of light is produced as the hadrons, electrons, and muons plow through the medium faster than the local speed of light. The ground array of detectors allows one to measure the lateral cross section of the particle cascade, from which one may then infer the energy of the incipient particle. The second, complementary technique involves the measurement of Cerenkov light produced in the atmosphere itself. Other sources of light, such as the fluorescence of air nitrogen atoms excited by the charged particles in transit, add to the overall signal and its detectability.

The cosmic-ray spectrum compiled from all these different experimental appro-aches (figure 13.1) exhibits a power-law dependence on energy, though a careful comparison with a single function (the dashed line in this figure) clearly shows significant deviations at the so-called "knee" at $\sim 4 \times 10^{15}$ eV (~ 4 PeV) and at the "ankle", near $\sim 5 \times 10^{18}$ eV (~ 5 EeV). The cosmic rays of Extremely High Energy (EHE)—those above $\sim 10^{20}$ eV—pose a serious challenge to conventional theories of particle acceleration, so their study is heavily featured in observational campaigns by the current suite of detectors, including the Pierre Auger Observatory.[1]

Actually, though we now understand the general aspects of how cosmic rays are energized, major gaps remain *throughout* the spectrum, with the level of uncertainty tending to increase with energy. The total cosmic-ray energy density measured above Earth's atmosphere is dominated by particles in the range 1–10 GeV. Below ~ 1 GeV, the intensities are correlated with solar activity, so these particles clearly originate with the Sun. At higher energies, however, the cosmic-ray flux is anticorrelated with the strength of the solar wind, whose screening effects therefore point to a cosmic-ray origin outside the solar system.

Conventional wisdom has it that the bulk of the cosmic rays between 1 GeV and the knee are confined to the Galaxy, and are probably produced in supernova remnants. Between the knee and the ankle, the situation is murkier, though the ankle is sometimes interpreted as a crossover from a Galactic to an extragalactic population. Beyond ~ 10 EeV, the cosmic rays are generally expected to have an extragalactic origin due to their isotropy and the fact that galactic magnetic fields would not be able to confine them to the Galaxy.

Setting aside the most energetic particles for the moment, let us suppose that the bulk of the cosmic-ray population is due to galactic accelerators. These photons and relativistic hadrons and leptons diffuse throughout the interstellar medium, but they don't survive forever because they interact with gas along the way and

---

[1] The Pierre Auger Observatory was designed to measure the flux, arrival direction distribution, and mass composition of cosmic rays from EeV to EHE with high statistical significance over the whole sky (see Abraham et al. 2004). The design for this observatory includes several thousand water-Cerenkov detectors arranged on a triangular grid, with each side 1.5 km in length, overlooked by optical sta-tions, each containing six telescopes designed to detect air-fluorescence light. The water tanks respond to the muons, electrons, and positrons, whereas the fluorescence cameras measure the emission from atmospheric nitrogen. This hybrid arrangement provides several significant advantages over the more traditional approaches, including angular coverage and sensitivity.

produce secondary particles, very much as they do when they strike Earth's upper atmosphere.

We can use the observed secondary to primary abundance ratios of galactic cosmic rays—a direct consequence of this interaction—to estimate the (energy-dependent) mean column density $N(E)$ they traverse. Evidently, these observations[2] indicate that

$$N(E) \approx 6.9 \left(\frac{E}{20Z \text{ GeV}} \right)^{-0.6} \text{ g cm}^{-2}, \tag{13.1}$$

where Z is the mean charge of the cosmic ray and E is its energy. If we define $\tau_{\text{res}}(E)$ to be the mean residence time of a cosmic ray with energy E in the Galaxy, and remembering that they travel at speeds very close to c, we infer that

$$\tau_{\text{res}}(E) \approx \frac{N(E)}{c\rho_{\text{ism}}}, \tag{13.2}$$

where ρ_{ism} is the density in the interstellar medium, and therefore

$$\tau_{\text{res}}(E) \approx 7 \left(\frac{E}{20Z \text{ GeV}} \right)^{-0.6} \left(\frac{\rho_{\text{ism}}}{10^{-24} \text{ g cm}^{-3}} \right)^{-1} \text{ Myr}. \tag{13.3}$$

Thus, if we assume that the cosmic-ray population is in equilibrium, the accelerators must replenish the energy lost to the secondaries at a rate

$$L_{\text{cr}} = \int d^3x \, dE \frac{E \, 4\pi \, j_{\text{cr}}(E)}{c\tau_{\text{res}}}, \tag{13.4}$$

where $j_{\text{cr}}(E)$ is the cosmic-ray intensity displayed in figure 13.1, and the integral over solid angles has been carried out for an isotropic distribution, which produces the factor of 4π. It is straightforward to evaluate this remaining integral over energy and volume, which produces a cosmic-ray replenishment rate

$$L_{\text{cr}} \sim 1.5 \times 10^{41} \text{ ergs s}^{-1}. \tag{13.5}$$

This is quite a remarkable number when we start to consider what process (or processes) within the Galaxy could sustain this power over several hundred million years or more, because it represents about 10% of the estimated total power output in the form of kinetic energy ejected into the interstellar medium by galactic supernovae. Therefore, from an energetics point of view, supernova remnants could well account for most of the cosmic rays. This is the main reason many suspect that cosmic rays, at least up to the knee, originate predominantly from first-order Fermi acceleration (see section 4.3) in the shocks between expanding supernova ejecta and the surrounding medium.

Another reason for believing that most of the cosmic rays originate within the Galaxy is that if we assume the cosmic-ray density to be the same throughout the universe, then the production of γ-rays from the decay of neutral pions produced in interactions between the cosmic rays and ambient protons (see discussion near the end of section 12.2.1) exceeds the observed limits. This is not the case within the

---

[2]See, e.g., Swordy et al. (1990).

Galaxy itself, where pion decays actually contribute to the overall γ-ray background (see figure 13.4), but would clearly be a problem outside the Galaxy. Consider, for example, what would happen in the Small Magellanic Cloud (SMC). The observed upper limit to the γ-ray flux from this part of the sky turns out to be a factor of a few below that predicted from pionic decays assuming a universal cosmic-ray density. At least in the SMC, then, the cosmic rays should have a density well below that of our local neighborhood.[3]

Past these general energy arguments, however, the theory behind cosmic-ray acceleration becomes less well defined and, within a galactic context, even unworkable at the highest observed energies. There are basically two kinds of energizing mechanisms considered by workers in this field: (1) direct acceleration of charged particles by an electric field (e.g., like that considered in section 4.2), and (2) statistical acceleration (either first- or second-order Fermi) in a magnetized plasma (see section 4.3).

In the direct method, the details of how the particles are actually accelerated and the maximum energy they attain depend on the particular physical situation under consideration. But for a variety of reasons, these mechanisms are now disfavored. The principal complaint levied against them is that they do not produce a characteristic power-law spectrum in any natural way. As we saw in section 4.3, the existence of a power-law particle distribution almost certainly means that any fractional gain in energy by a subgroup of the population must be accompanied by a significantly large fractional loss in the number of accelerating particles. Otherwise, it is difficult to see why the system would not produce a simple peaked distribution with a characteristic energy set by the acceleration time.

The statistical acceleration we developed in section 4.3 works much better in accounting for a power-law distribution of cosmic rays, though only for a segment of the population. For any given acceleration site, there is a maximum energy achievable, E_{max}, limited either by the acceleration time required to reach this level, given that the shock has a finite lifetime, or by the size of the shock, which must be larger than the gyroradius of the particle being accelerated to keep it in the acceleration zone.

Under the best of conditions, it is the second factor that delimits E_{max}, so let us see how well a supernova remnant can do in producing cosmic rays. According to our general formulation of the Lorentz force (equation 4.18), the equation of motion for a proton moving under the influence of a magnetic field $\mathbf{B}$ is

$$\mathbf{F} = q\frac{\mathbf{v}}{c} \times \mathbf{B}. \tag{13.6}$$

Thus, given that $\mathbf{p} = \gamma m_p \mathbf{v}$ (see equation 3.53) and that $|\mathbf{v}| \approx c$ for all these particles, we see that

$$\gamma m_p \frac{d\mathbf{v}}{dt} \approx q\frac{\mathbf{v}}{c} \times \mathbf{B} \tag{13.7}$$

for a proton with mass m_p.

---

[3] See Sreekumar et al. (1993).

For the sake of estimating the gyroradius of this proton, let us simply take $\mathbf{v}$ to be more or less perpendicular to $\mathbf{B}$, so that

$$\gamma m_p \frac{v^2}{r_{\text{gyr}}} \approx q \frac{v}{c} B, \qquad (13.8)$$

where, in obvious notation, $v = |\mathbf{v}|$ and $B = |\mathbf{B}|$, and r_{gyr} is the gyroradius. Rearranging, we find that

$$\gamma m_p c^2 \approx q B r_{\text{gyr}}. \qquad (13.9)$$

The maximum attainable energy from Fermi acceleration therefore emerges in a region with the largest possible field $\mathbf{B}$ and the widest shock. In the interstellar medium, $B \sim 10^{-6}$ G. With compression inside the shock, this field may grow to 10^{-5} G, perhaps even 10^{-4} G. Certainly at the galactic center and, more generally, within molecular clouds, fields of this magnitude (and greater) are not uncommon.[4] Typically, the expanding supernova ejecta terminate in a shock whose width is ~ 1–10 pc. All told, we would therefore expect from equation (13.9) that supernova remnants should accelerate hadrons up to an energy

$$E_{\text{max}} \equiv \gamma m_p c^2|_{\text{max}} \sim 6 \times 10^{15} \text{ eV}, \qquad (13.10)$$

i.e., essentially up to the knee in figure 13.1.

This argument, coupled to a favorable comparison of the energetics, is rather compelling, at least up to intermediate energies. However, equation (13.9) can also be used to demonstrate that the Galaxy cannot confine cosmic rays much beyond the EeV range, since for an average magnetic field intensity $B \sim 10^{-6}$ G these particles have a gyroradius greater than the galactic scale height. The EHE cosmic rays must therefore constitute a separate component, and almost certainly originate outside the Galaxy. Alternative models involving other energetic sources, such as pulsars and neutron stars in close binary systems, appear to be ruled out by the complete absence of detectable TeV radiation from Cygnus X-3 and Hercules X-1.[5]

The Pierre Auger Observatory, which we introduced earlier in this chapter, was conceived primarily to solve the EHE cosmic-ray puzzle, sustained not only by the lack of an obvious energizing mechanism, but also by the absence of any clear underlying source population. Current theories of the EHE cosmic-ray origin can broadly be categorized into two distinct groups: the "bottom-up" acceleration types, in which charged particles are accelerated from lower energies to the requisite high energies in certain astrophysical (presumably extragalactic) environments, and the "top-down" scenarios, in which the energetic cosmic rays are the decay products of more massive particles produced in the early Universe. Fortunately, both camps identify observable signatures associated with their respective positions, and we shall see over the next few pages that the Pierre Auger Observatory is indeed now making definitive measurements to resolve this long-standing problem.

In bottom-up models, even if particles can accelerate to energies beyond EeV, they generally find it difficult to escape from the acceleration zone without losing

---

[4] See, e.g., Sarma et al. (2002).
[5] See Borione et al. (1997).

most of their energy. The one exception appears to be the powerful radio galaxies (see figure 12.4), whose giant radio lobes are tenuous, though sufficiently large to trap accelerating particles long enough to reach the EHE domain. Still, even for them the requisite physical parameters appear to be rather extreme.[6]

Regardless, all extragalactic sources of cosmic rays are constrained by an important effect arising from the propagation of these energetic particles through the intergalactic medium. If the EHE cosmic rays are conventional, such as nucleons or heavy nuclei, then the most relativistic among them lose their energy rapidly due to the so-called Greisen-Zatsepin-Kuzmin (GZK) process, namely, photo-production of pions from the collision between cosmic rays and photons in the microwave background.[7] So if the sources of EHE cosmic rays seen at Earth are extragalactic, they cannot reside at arbitrarily large distances from us.

We can understand this limitation in the following way. Let us suppose that our nucleon (with mass m_n) is highly relativistic (with energy well above the PeV scale) in the cosmic rest frame—i.e., the frame in which the cosmic microwave background radiation is isotropic. Then, in the nucleon's own rest (primed) frame, the four-momentum vector (equation 3.52) describing a nucleon–photon collision may be written

$$P'^\mu = \left(m_n c + \frac{\epsilon'_\gamma}{c}, \frac{\epsilon'_\gamma \hat{\mathbf{x}}}{c} \right), \tag{13.11}$$

where ϵ'_γ is the photon's energy in this frame, and $\hat{\mathbf{x}}$ is its direction of motion.

Since P'^μ is a four-vector, its contraction with itself is a Lorentz scalar, permitting us to relate energies and momenta in one frame to those in another, in this case specifically those in the center-of-momentum (starred) frame, where the four-momentum vector right after the collision (in which only the nucleon and a newly formed pion remain) is

$$P^{*\mu} = \left(\frac{E_n^*}{c} + \frac{E_\pi^*}{c}, 0, 0, 0 \right). \tag{13.12}$$

Thus, since $P'^\mu P'_\mu = P^{*\mu} P^*_\mu$, we have

$$m_n^2 c^2 + \frac{\epsilon'^2_\gamma}{c^2} + 2 m_n \epsilon'_\gamma - \frac{\epsilon'^2_\gamma}{c^2} = \frac{E_n^{*2}}{c^2} + \frac{E_\pi^{*2}}{c^2} + \frac{2 E_n^* E_\pi^*}{c^2}. \tag{13.13}$$

At threshold, all of the energy in the center-of-momentum frame is in the mass of the particles, so $E_n^* = m_n c^2$ and $E_\pi^* = m_\pi c^2$, and, therefore,

$$\epsilon'_\gamma = m_\pi c^2 + \frac{m_\pi^2 c^2}{2 m_n}. \tag{13.14}$$

The value of this threshold energy, at which the cross section exhibits a pronounced resonance associated with single pion production, is roughly 160 MeV.

According to the Doppler shift formula (equation 3.18), the nucleon Lorentz factor required to convert a microwave photon (with energy ϵ_γ) in the cosmic rest frame

---

[6]See Norman, Melrose, and Achterberg (1995).
[7]This effect was first described by Greisen (1966) and Zatsepin and Kuzmin (1966).

into one of these γ-rays in the nucleon's rest frame is therefore

$$\gamma = \frac{\epsilon'_\gamma}{\epsilon_\gamma (1 - \beta \cos \theta)}, \tag{13.15}$$

where $\theta = \pi$ for a head-on collision. With $\epsilon_\gamma \sim 10^{-3}$ eV, this gives $\gamma \approx 8 \times 10^{10}$, which corresponds to a cosmic ray with threshold energy $8 \times 10^{10} \, m_n c^2 \sim 8 \times 10^{19}$ eV in the laboratory frame.

Knowing the cross section for this interaction, and the density of cosmic microwave photons, one can then easily estimate the distance over which pion photo-production will occur.[8] Near threshold, the nucleon interaction length is roughly 6 Mpc, and one therefore expects to quickly run out of extragalactic sources that can contribute to the cosmic-ray flux seen at Earth for distances greater than this value, when the particle energy exceeds $\sim 10^{20}$ eV. Notice in figure 13.1 that the high-energy limit to the cosmic-ray spectrum is very close to this energy.

Energetic photons in the cosmic-ray distribution are also subject to attenuation as they propagate through the cosmic microwave background. For them, the dominant interaction is pair creation, $\gamma \gamma_{\text{cmb}} \rightarrow e^+ e^-$. In the cosmic rest frame, the four-momentum vector for this interaction is

$$P^\mu = \left(\frac{\epsilon_\gamma + \epsilon_{\gamma,\text{cmb}}}{c}, \frac{\epsilon_\gamma - \epsilon_{\gamma,\text{cmb}}}{c} \hat{x} \right), \tag{13.16}$$

where ϵ_γ is the energetic photon, and $\epsilon_{\gamma,\text{cmb}}$ is a typical photon in the microwave background.[9] Correspondingly,

$$P^{*\mu} = \left(\frac{E^*_+ + E^*_-}{c}, 0, 0, 0 \right), \tag{13.17}$$

where E^*_+ and E^*_- are, respectively, the energy of the positron and electron in the center-of-momentum frame. Thus, with $P^\mu P_\mu = P^{*\mu} P^*_\mu$, we easily find that at threshold (where $E^*_+ = E^*_- = m_e c^2$)

$$\epsilon_\gamma = \frac{m_e^2 c^4}{\epsilon_{\gamma,\text{cmb}}}. \tag{13.18}$$

With $\epsilon_{\gamma,\text{cmb}} \approx 10^{-3}$ eV, this energy is roughly 2.6×10^{14} eV.

In the EHE domain, the attenuation is mostly due to collisions between the cosmic-ray photons and background photons with energy $< 10^{-6}$ eV (≈ 100 MHz), i.e., with radio photons. Unfortunately, the universal radio background is poorly known due to the difficulty of disentangling the galactic and extragalactic components, so we cannot yet be certain how serious the attenuation of EHE photons is during their propagation to Earth.

Nonetheless, the fact that nucleons (and nuclei), at least, cannot originate from distances greater than ~ 50 Mpc (conservatively speaking) means that unless the

---

[8]See, e.g., Berezinsky and Gazizov (1993).

[9]This expression for P^μ assumes the photons are colliding head-on. However, our analysis is not strongly affected by this assumption, since the momentum $\epsilon_{\gamma,\text{cmb}}/c$ of the microwave photon is insignificant compared to that of the energetic photon, so we could just as easily have approximated the spatial components of P^μ as the vector $(\epsilon_\gamma/c)\hat{x}$.

extragalactic magnetic field B_{eg} is large, they should experience very little deflection on their journey to Earth. Currently, the best estimates we have for B_{eg} come from Faraday rotation measurements,[10] which yield a value $\sim 10^{-9}$–10^{-8} G.

From equation (13.9), we infer that the radius of gyration for a nucleon with energy $E = \gamma m_n c^2$ in the extragalactic magnetic field is

$$r_{gyr} \approx \frac{E}{q\,B_{eg}}. \tag{13.19}$$

Thus, the angle θ_D through which the nucleon deflects as it propagates a distance D is

$$\theta_D \sim \frac{Dq\,B_{eg}}{E}. \tag{13.20}$$

For an energy $E = 10^{20}$ eV and a distance of 50 Mpc, this is roughly 1°—not very promising for distinguishing one galactic object from another, but definitely workable for extragalactic sources, such as AGNs[11] and radio galaxies, whose density in the plane of the sky is far smaller than that of the foreground stars.

Indeed, as we alluded to earlier in this chapter, the best candidates for producing EHE cosmic rays via the bottom-up approach appear to be hot spots within the giant termination lobes of radio galaxies, such as Cygnus A (see figure 12.4). Not only are these features large enough to contain the energetic particles with their growing gyroradius as they approach 10^{20} eV, but the ambient density of soft photons in the hot spots is sufficiently small that the cosmic rays to do not suffer catastrophic energy losses (e.g., via inverse Compton scattering) as they escape into the intergalactic medium. Correlations between the arrival directions of the most energetic cosmic rays and distant radio galaxies have already been reported in the literature,[12] though these sources appear to lie at distances > 100 Mpc from Earth, which would be inconsistent with the GZK limit we discussed earlier. So radio galaxies could very well turn out to be the source of EHE cosmic rays, but a full resolution of this issue must await the acquisition of better EHE cosmic-ray statistics with the Pierre Auger Observatory.

Alternatives to the bottom-up approach for the origin of EHE cosmic rays may be grouped into two main classes of top-down models. The first one involves the decay or annihilation of topological defects produced through a phase transition

---

[10]These measurements have been carried out by many groups. Some of the better known results have appeared in Kronberg (1994) and Blasi, Burles, and Olinto (1999). Faraday rotation causes the plane of linear polarization to rotate as the wave propagates through a magnetized region. In simple terms, one can understand this by considering the linear polarization as a superposition of two oppositely circularly polarized waves. These waves in turn cause charges within the magnetic region to rotate in opposite directions. However, different particles (e.g., electrons and protons) rotate at different rates, causing one circularly polarized component to propagate through the medium at a slightly different velocity than that of the other. The net result is a gradual rotation of the plane associated with the linear polarization. Placing a limit on the rotation of the polarization vector, therefore, also delimits the intensity of the magnetic field, whose strength determines the velocity differential between the two circularly polarized components.

[11]However, at the time of this writing, preliminary results from the Pierre Auger Observatory are already ruling out BL Lac objects, a subclass of blazars, as sources of EHE cosmic rays. With 6 times more events with energy above 10 EeV than all previous searches, the new catalog shows no excess correlation between the arrival directions of EHE cosmic rays and the known BL Lac locations.

[12]See, e.g., Elbert and Sommers (1995).

in the early universe.[13] These events would occur homogeneously throughout the cosmos and generate supermassive particles that would in turn decay into quarks and leptons. These events would therefore lead to secondary EHE protons and photons with an energy spectrum and relative abundances characteristic of the underlying hadronization process. As they propagate through the intergalactic medium, these cascade products would themselves continue to produce a shower of leptons and pions via the GZK effect.

In the second class of top-down models, the supermassive particles responsible for the observed EHE cosmic rays are produced directly in the early universe,[14] with a lifetime longer than a Hubble time. The motivation for such a scenario is the possible identification of these particles with dark matter, which would then mean that most of the EHE cosmic rays observed at Earth would be produced in the halo of our Galaxy.

The GZK effect would have no bearing on such a population of energetic particles. There is therefore a clear distinction between these two classes of top-down models in terms of their predicted energy spectrum, photon fraction limit, and the neutrino flux limit. Again, we will eventually know whether either, or none, of these categories are compatible with the data as the Pierre Auger Observatory continues to improve the EHE cosmic-ray statistics.

13.2 GALAXY CLUSTERS

Galaxy clusters are the most massive gravitational structures in the universe, with a hot diffuse plasma ($k_B T_e \sim 10\,\text{keV}$, where $k_B = 1.38 \times 10^{-16}$ ergs K^{-1} is Boltzmann's constant) that fills the intergalactic medium (see plate 26). Aside from its characteristic thermal bremsstrahlung emission, this hot tenuous plasma reveals itself through a distortion to the CMB spectrum, known as the Sunyaev-Zel'dovich effect.[15] Cosmic microwave background (CMB) photons passing through this hot, intracluster medium (ICM) have a $\sim 1\%$ chance of inverse Compton scattering (see section 5.7) off the energetic electrons, causing a small (~ 1 mK) fluctuation in the CMB spectrum.

An image of the X-ray surface brightness of the Abell 1689 cluster is shown in plate 26, along with contours (showing the Sunyaev-Zel'dovich Effect) overlaid for direct comparison. Using spatially resolved spectroscopy, one may even determine the temperature and metal abundance of the hot plasma (see figure 13.2). These properties are based on fits to an optically thin emission model, with an absorbing column density N_H fixed at the galactic value.

But galaxy clusters are far more than mere large-scale sources of diffuse X-rays. They receive attention over the entire range of spectral energies, though observations in each energy regime come with their own quirks, and no single, simple model is adequate in explaining the whole spectrum. Nonetheless, it is clear now that cosmic

---

[13] See, e.g., Hill (1983).
[14] See, e.g., Berezinsky, Kachelriess, and Vilenkin (1997).
[15] See Sunyaev and Zel'dovich (1970).

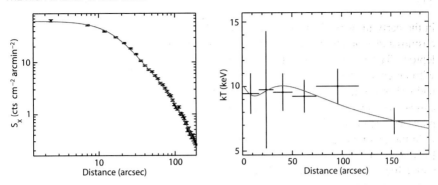

Figure 13.2 For the same Galaxy cluster (Abell 1689) shown in plate 26, we see here the radial profile of the background subtracted X-ray surface brightness (left panel) and the radial profiles of the (inferred) *Chandra* temperatures. In both cases, the solid curves indicate the best fits to the data. (Image from Bonamente et al. 2006)

rays must play an important role energetically within the ICM, even to the point of possibly providing a significant source of heat for the 10-keV plasma. The latest observations, from radio, to ultraviolet, to X-ray, and finally to γ-rays, point to the fact that Coulomb collisions between the cluster's high-energy proton population and its thermal electrons may also be the root of nonthermal excesses sometimes seen in the extreme ultraviolet and X-ray regimes.

The presence of a relativistic population of particles in the ICM, albeit electrons rather than protons, is suggested by the highly polarized radio luminosities ($L_r \sim 10^{40}$–10^{42} ergs s^{-1}) measured from emitting regions extending over Mpc scales, thus far observed in over 30 galaxy clusters.[16] This is synchrotron emission, requiring both a large magnetic field ($B > 0.1 \ \mu$G) and highly relativistic ($\gamma > 100$), nonthermal electrons not associated directly with any particular galaxies.

Since such electrons may diffuse only a few kpc away from a source before losing their energy via radiative losses (see section 5.6), to account for the Mpc extent of cluster radio emission one must propose a mechanism for constantly replacing the radiated energy. The most likely scenario appears to be a "secondary" model,[17] which resolves this difficulty by having a constant injection of (nonthermal) electrons produced when high-energy protons collide with hydrogen in the ICM, creating a pionic cascade and subsequent decays into muons and then electrons and positrons (see discussion at the end of section 12.2.1).

It should be pointed out, however, that the secondary model for the nonthermal-synchrotron emitting electrons in galaxy clusters is not universally accepted. Conclusive observations of the secondary process have been few. In Hercules A and Hydra A[18] and in MS 0735,[19] vacated areas of X-ray bremsstrahlung indicate that the background of thermal electrons is being pushed aside by relativistic electrons

[16] See, e.g., Kim et al. (1990), Giovannini et al. (1993), and Kempner and Sarazin (2001).
[17] See, e.g., Dennison (1980).
[18] See Nulsen et al. (2005).
[19] McNamara et al. (2005).

injected from the axes of a central source, implying a phenomenally large power, $L_p \sim 10^{44}$ erg s^{-1}. These bubbles coincide with sources of radio emission as the protons produce a cascade of synchrotron-emitting electrons, confirming that the secondary decay process does occur in clusters. The protons released during such an episode are confined and built up within the cluster over cosmic times, totaling some 10^{61} erg of energy in nonthermal particles—very near energy equipartition with the thermal background.[20]

The energy liberated during the infall of the galaxy group NGC 4839 toward the center of the Coma cluster, observed by XMM-Newton,[21] is an example of another potential source of high-energy protons. NGC 4839 achieves a velocity of $\sim$1400 km s^{-1}, and since the sound speed corresponding to Coma's gas temperature of $\sim$8 keV is $\sim$1000 km s^{-1}, the subcluster's supersonic motion is expected to produce shocks (which may have already been seen in the X-ray images of NGC 4839). It is likely that the shock compression in these regions energizes a fraction of the thermal particles within the intracluster medium (ICM) by first-order Fermi acceleration. However, how the protons are actually energized is not directly relevant to this analysis. For example, since the cosmic rays remain trapped within the cluster for over a Hubble time, it is likely that second-order acceleration by a turbulent distribution of Alfvén waves may be even more effective at helping the protons attain their highest energies (see section 4.3). Cluster mergers provide a viable mechanism for energizing the cosmic rays, regardless of how the individual particles are ultimately accelerated.

At any rate, the prospects for providing a definitive confirmation or rejection of the secondary model are bright indeed, due to the fact that in the Coma cluster and a dozen others, the diffuse radio emission exhibits a spectral break, with a steeper index at higher energy. This has been interpreted as the signature of electron escape,[22] of a reacceleration of already relativistic electrons—possibly energized by a cluster merger event[23]—or of energy-dependent radiative losses acting to impose a maximum electron energy during shock acceleration itself.[24]

But whereas the associated γ-ray emission in nonsecondary models is therefore expected to be undetectable (due to this downturn in the relativistic electron population), the secondary model itself results in a γ-ray flux from the decay of π^0's produced in proton–proton collisions, and the relativistic hadrons are not expected to have a downturn similar to that of the electrons. Thus, upcoming missions, such as GLAST (see section 1.5 and plate 4), should be able to determine which of these two camps is correct and provide conclusive evidence for the energetic dominance of cosmic rays in the ICM.

The X-ray luminosity ($L_X \sim 10^{45}$ ergs s^{-1}) from galaxy clusters (plate 26) is generally interpreted as due to thermal bremsstrahlung by a tenuous thermal plasma characterized by a temperature in the range $\sim$2–10 keV (figure 13.2), and a central

---

[20] See, Völk, Aharonian, and Breitschwerdt (1996), Berezinsky, Blasi, and Ptuskin (1997), and Hinton and Domainko (2007).

[21] Neumann et al. (2001).

[22] See, e.g., Rephaeli (1979).

[23] See, e.g., Brunetti et al. (2004).

[24] See, e.g., Webb, Drury, and Biermann (1984).

density $\sim 10^{-3}$ cm^{-3}. The thermal X-ray-emitting leptons coexist with the nonthermal radio-emitting electrons. What has drawn significant attention in recent years is the slight excess seen above the thermal component starting at $\sim$20–25 keV and extending out to energies greater than $\sim$45 keV. Though still somewhat controversial, this additional nonthermal component has thus far been reported for the Coma cluster, Abell 2256, and Abell 754.[25] Hard X-ray detections have been reported in several other Abell clusters (2199, 2319, and 3667) in the redshift range 0.023 < z < 0.056, though apparently with weaker signals. Most of these excesses above the thermal emission can be fitted with a photon, power-law spectrum and index $\sim$2.

The most common interpretation of the hard X-radiation is inverse Compton scattering between the radio-emitting electrons and the cosmic microwave background radiation. However, if the same electrons are responsible for both the radio synchrotron and nonthermal X-ray emission, then the implied magnetic field B must be an order of magnitude smaller than that observed; otherwise, the number density of electrons required for the inverse Compton emission would overproduce synchrotron emission in the radio. In the Coma cluster, for example, Faraday rotation measurements (see discussion in section 13.1) suggest that $B \sim 6 \, \mu$G,[26] in sharp contrast with the value derived from *Beppo*-SAX observations within the context of the inverse Compton scattering scenario, which instead requires a field $B \sim 0.16 \, \mu$G.

It is very likely, therefore, that the correct explanation for the hard X-ray emission in galaxy clusters will incorporate processes other than inverse Compton scattering. Almost certainly, this will involve Coulomb interactions between the ICM cosmic rays and the thermal electrons (i.e., those responsible for the thermal bremsstrahlung emission), which produce a quasi-thermal lepton component distinct from the power-law electrons created in the pionic cascades.[27]

Interactions such as these may also be relevant to other long-standing problems with galaxy clusters, such as the deficit of soft X-rays observed from their core. Models of the high-energy emission throughout the ICM predict that the tenuous, hot gas cools as it gradually radiates away its energy as it falls towards the center of the cluster. Low-resolution X-ray imaging[28] largely confirms the existence of these so-called "cooling flows."

However, newer high-resolution X-ray spectra of several clusters obtained with the Reflection Grating Spectrometer on XMM-*Newton* show a soft X-ray spectrum inconsistent with a simple cooling flow. Though emission lines from gas at $\sim$2–3 keV are observed, those expected from gas cooling below 1–2 keV are missing, indicating that the plasma is cooling down to about 2–3 keV, but not below this temperature.

Several remedies have been proposed, but it is not yet clear which, if any, of them is the explanation for this odd behavior. Most of the tentative solutions attempt to inject additional energy into the accreting gas, via heating, mixing, or differential absorption. Other models account for the absent cooling flow at the core of the

---

[25] For an example of how these observations were made and interpreted, see Fusco-Femiano et al. (2003).

[26] See, e.g., Feretti et al. (1995).

[27] See Wolfe and Melia (2007).

[28] See, e.g., Fabian et al. (2001).

cluster by invoking an inhomogeneous metallicity. But continuous or sporadic heating generates additional questions, such as why the heat should be targeted solely on the cooler gas. In addition, there is no clear observational evidence yet for the required widespread heating, or the presence of shocks, except in some radio lobes associated with AGNs, which occupy only a portion of the volume.

Solutions based on metallicity adopt the view that metals in the ICM may not be spread uniformly; if they are clumped, then little line emission is expected from the gas cooling below ~ 1 keV. In this scenario, the low-metallicity portion of the plasma cools without line emission, whereas the strength of the soft X-ray lines from the metal-rich gas depends on the mass fraction of that gas and not on the overall abundance, since soft X-ray line emission dominates the cooling function below ~ 2 keV.

Interactions between the cosmic rays and the infalling thermal plasma may also inhibit excessive cooling if the cosmic-ray distribution is inhomogeneous. Clearly, if they pervade the ICM without gradients, they could not preferentially heat the cooler gas, which is found predominantly near the cluster's core. Whatever the case, the cooling-flow problem shapes up as one of the major areas of investigation for the coming years.

The high-energy emission from galaxy clusters has been one of the most difficult to interpret, and we are only now able to probe the physical conditions in the ICM, thanks to our vastly improved observational capability in X-rays with *Chandra* and XMM-*Newton* and, soon to be, in γ-rays with GLAST. This situation is due in part to the remoteness of the sources we are trying to study, but also to the evident complexities that obviate a simple, self-consistent solution.

It is rather clear now that cosmic rays must be present in the intergalactic medium, with sufficient numbers to effect the energetics of the ICM gas. But we don't know yet where they originate. It is likely that several mechanisms all contribute to their overall distribution, including Fermi acceleration at shocks produced during cluster mergers, and deposition into the intergalactic medium by the relativistic jets in AGNs. These relativistic hadrons apparently produce—via proton–proton collisions and subsequent pion decay—the nonthermal leptons responsible for the synchrotron emission seen on Mpc scales. And their Coulomb collisions with the ~ 10-keV thermal plasma in the ICM accelerate a fraction of these particles to energies as high as ~ 45 keV, where they produce the nonthermal, hard X-ray signature only now starting to emerge as a distinct spectral component. The study of galaxy clusters is work in progress, and will continue to be one of the most active branches of high-energy astrophysics for many years to come.

13.3 DIFFUSE EMISSION

We have reached full circle—an extensive, arching journey that has taken us from the high-energy, full-sky maps of chapter 2 to a study (through the rest of this book) of the many point sources that appear in them. We have learned, however, that the high-energy sky is filled with far more than white dwarfs, neutron stars, and black holes. Already in this chapter, we have concluded that cosmic rays, particularly at the highest energies, present quite a challenge to theorists attempting to divine their origin, and the largest structures in the universe—galaxy clusters—also happen to be among the biggest emitters of X-ray (and possibly γ-ray) photons.

But we are not finished yet, because all of the sources we have talked about—galaxy clusters and the various classes of compact objects—are sometimes too far away (or simply too faint) for us to resolve individually, and cosmic rays are themselves sources of high-energy radiation in the interstellar medium. Not surprisingly, therefore, maps such as plates 5, 7, 9, and 11, reveal high-energy emission throughout the sky. A major challenge is to discern between radiation produced within our Galaxy and that originating elsewhere. Only then can we produce an image of the true cosmic background, which appears to be dominated by distant, unresolved AGNs, with an important contribution from clusters and the so-called warm–hot intergalactic medium at lower energy.

13.3.1 The X-ray Background

Below $\sim$10 keV, one of the largest extended features in the X-ray sky is the galactic X-ray background, often referred to as the galactic ridge X-ray emission, discovered in the late 1960s.[29] This emission extends over more than 100° along the galactic plane, but only a few degrees in the lateral direction. Its spectrum appears to be due to optically thin bremsstrahlung processes in a thermal plasma with a temperature $\sim$5–10 keV,[30] but such a hot plasma would not be bound to the Galaxy.

This inconsistency has led to an alternative interpretation—that the galactic ridge X-ray background is instead the superposition of weak, unresolved X-ray sources.[31] This hypothesis has been largely confirmed by recent observations with RXTE and INTEGRAL, which indicate that the ridge X-rays trace the near-infrared emission and, therefore, the stellar mass distribution in the Galaxy.[32] The fact that the observed X-ray flux to stellar mass ratio is compatible with the X-ray luminosity function of cataclysmic variables (section 9.3) and stars with active coronae near the Sun only strengthens this conclusion.

Away from the galactic plane, the bulk of the extragalactic X-ray background in the 0.1- to 10-keV band has been resolved into discrete sources with the deepest ROSAT, *Chandra* (see plate 6), and XMM-*Newton* observations. Follow-up studies[33] with ground-based optical telescopes have led to the identification of these objects as either unobscured (with X-ray flux $S_X > 10^{-14}$ ergs cm^{-2} s^{-1}) or obscured ($10^{-15.5} < S_X < 10^{-14}$ ergs cm^{-2} s^{-1}) AGNs, with ever fainter and redder optical counterparts.

The newly discovered[34] *Chandra* and XMM-*Newton* sources are predominantly Seyfert galaxies (see section 12.3.1) at a median luminosity of $\sim$10^{43} ergs s^{-1} and a

---

[29] See, e.g., Cooke, Griffiths, and Pounds (1969).
[30] See Koyama et al. (1989).
[31] See, e.g., Worrall et al. (1982).
[32] See Revnivtsev et al. (2006).
[33] See, e.g., Barger et al. (2003).
[34] See Hasinger et al. (2005).

median redshift around 0.7. The higher-luminosity AGNs, on the other hand, have a space density peaking around $z \sim 2$, similar to that of optically selected quasars. An important consequence of this kind of work is a clarified view of the luminosity-dependent evolution in source density, which ultimately will help us understand the correlations displayed in figures 12.5, 12.6, and 12.7.

However, the cosmic X-ray background is not *entirely* due to unresolved point sources. The best estimates are that the *Chandra* Deep Fields North (plate 6) and South, the most sensitive observations of the X-ray sky to date, resolve about 80% of the extragalactic X-ray background at 1–2 keV[35] and $\sim$50% at 7 keV.[36]

Some of the unresolved flux is due to the warm–hot intergalactic medium with a temperature 10^5–10^7 K, that is thought to comprise most of the baryons in the local universe.[37] Theoretical models of this plasma predict a signal dominated by line emission from oxygen and iron, with prominent features at 0.5–0.8 keV, with measurable flux up to 1.5 keV. This hot gas has been difficult to observe directly, though some X-ray absorption lines have been attributed to it.[38]

13.3.2 The Gamma-Ray Background

Above $\sim$1 MeV, the galactic diffuse continuum is produced primarily (via neutral pion decays) in energetic interactions of relativistic nucleons with gas in the interstellar medium and (via inverse Compton scattering of the ambient low-energy radiation field) by electrons, which also radiate via bremsstrahlung. These various processes dominate in different parts of the spectrum, so deciphering the many components can provide useful information, e.g., about the cosmic-ray distribution at large distances from Earth. Knowing about the role of these relativistic particles in producing a high-energy background is also essential for identifying the contributions to the diffuse emission from other sources, both galactic and extragalactic.

The sole exception to the general character of the diffuse emission above 10 keV is evident in the older OSSE image shown in plate 7, with an updated version (produced with the Spectrometer, SPI, aboard the INTEGRAL satellite) shown in figure 13.3.

We learned in section 2.2 that the dominant emission we see here, at $\sim$511 keV, is due to the annihilation (within the bulge) of approximately 1.5×10^{43} positrons per second. The line width is just under 3 keV (FWHM), and the emission appears to be spatially extended over a scale of about $8°$, which at the distance to the galactic center corresponds to approximately 1.2 kpc. A single point source, such as the supermassive black hole in the middle, has been excluded with a high level of confidence.[39]

To this day, the origin of the annihilation radiation is a mystery. Although the physical processes responsible for producing positrons are quite well understood and

[35] See, e.g., Moretti et al. (2003).
[36] See, e.g., Worsley et al. (2005).
[37] See, e.g., Cen and Ostriker (1999).
[38] See Nicastro et al. (2005).
[39] See Jean et al. (2006).

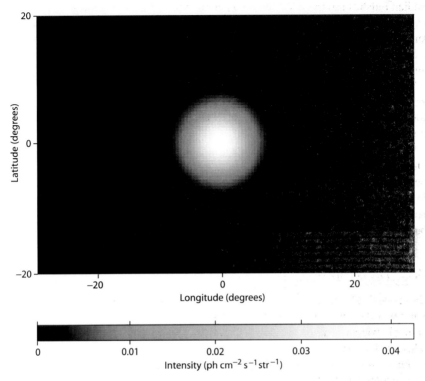

Figure 13.3 Cosmic electron–positron annihilation (511-keV) radiation was first detected in balloon observations of the galactic center in the 1970s. In the most recent measurements, made with SPI on the INTEGRAL satellite, the 511-keV sky distribution appears to be adequately described by a spherical Gaussian radial profile with a full-width half-maximum of 8° and a ∼3° uncertainty. The data hint at a possible elongation in longitude, which may suggest a faint contribution from a disk component. Assuming that the centroid of the 511-keV emission is at the galactic center, one infers a flux from the bulge of $(0.96^{+0.21}_{-0.14}) \times 10^{-3}$ ph cm^{-2} s^{-1}. There may also be a slight hint of a displacement of the centroid away from the galactic center, to the position (in galactic coordinates) $l = -1.0° \pm 0.7°$ and $b = 0.3° \pm 0.7°$, which would put it close to the hard X-ray source 1E 1740.2–2942. This object has been mentioned as a possible source of positrons in the galactic bulge (see Misra and Melia 1993). (Image from Weidenspointner et al. 2004)

several possible classes of source for them have been identified, no self-consistent picture exists to account for (1) such a high rate of annihilation and (2) the observed positron distribution, which is unlike that of any disk-dominated population we know.

In recent years, significant progress has been made in understanding positron annihilation spectroscopy, though, in the end, the more we've learned, the more perplexing the diffuse 511-keV emission has become. We now know that positrons cannot be produced in the galactic bulge with an energy greater than about 3 MeV—an incredibly delimiting constraint. This is based on two compelling arguments, both having to do with a comparison between the inferred annihilation rate and the overall diffuse γ-ray emission observed from this region with COMPTEL and EGRET.

Any process that produces electron–positron pairs ($\chi\chi \to e^+e^-$) is necessarily accompanied by the channel $\chi\chi \to e^+e^-\gamma$ as well, arising from electromagnetic radiative corrections.[40] This associated γ-ray emission is known as *internal bremsstrahlung*, arising from the self-interaction of the two newly created charged particles, not from propagation effects in the surrounding medium. The compelling nature of the constraint one may derive from this interaction is due to the fact that it does not depend on the specific particle physics associated with χ itself. This applies regardless of whether χ is a high-energy photon, or some other yet unidentified dark matter particle.

From the Feynman diagrams for these two interactions ($\chi\chi \to e^+e^-$ and $\chi\chi \to e^+e^-\gamma$), one can calculate exactly the γ-ray flux for a given positron creation rate. It turns out that to account for an annihilation rate of $\sim 1.5 \times 10^{43}$ s^{-1} (assuming that the positron distribution is in steady state), the associated γ-ray emission due to internal bremsstrahlung violates the observed COMPTEL/EGRET diffuse limits if the positrons are produced with an energy above ~ 20 MeV.

Once produced, these positrons continue to generate correlated high-energy emission as they interact with the medium through which they propagate. They lose most of their energy by Coulomb collisions, and then either form positronium "in flight" or continue to lose energy until they thermalize with the ambient medium. At that point, they undergo several processes, including charge exchange or direct annihilation with neutral atoms (H or He) or molecules (H$_2$), radiative recombination or direct annihilation with free electrons, or capture onto grains. The direct annihilation events produce two photons at 511 keV (plus whatever kinetic energy is left at that point), whereas the other eventualities lead again to the formation of positronium, which, depending on its spin state—para-positronium with anti-aligned spins, occurring 25% of the time, or ortho-positronium, with parallel spins—decays into 2 or 3 photons, respectively.

In the two-photon final state, the emerging radiation has an energy of 511 keV, whereas, for ortho-positronium, the photons fill a continuous distribution between 0 and 511 keV. However, a problem arises with the radiation produced when energetic positrons annihilate with electrons in flight while the former are undergoing ionization energy loss in matter. As noted long ago, up to $\sim 20\%$ of these particles annihilate before they reach a sufficiently low-energy to form positronium.[41] A recent calculation of the γ-ray spectrum produced by the positrons annihilating while they are still relativistic has shown that the emerging photon flux would exceed the COMPTEL/EGRET limit unless the positrons are injected into the interstellar medium with an energy below ~ 3 MeV.[42]

Together, these two constraints on the energy with which positrons may be created in the galactic bulge leave theorists with very little room to maneuver. For example, the particle cascades resulting from a collision between a relativistic cosmic-ray

---

[40] See Beacom, Bell, and Bertone (2005).
[41] See Heitler (1954).
[42] See Beacom and Yüksel (2006).

proton and hydrogen in the interstellar medium (see section 12.2.1) would severely conflict these limits and are therefore apparently ruled out as the origin of the 511-keV annihilation radiation. The decay of dark matter with a mass in excess of ~ 3 MeV is also inconsistent with the data, effectively eliminating all but a tiny patch of phase space where these particles may exist. However, the injection limits are still consistent with positrons originating from the radioactive decay of nuclei, such as ^{56}Co, ^{26}Al, and ^{44}Ti, ejected into the interstellar medium by supernovae and massive stars, and ^{22}Na produced in certain novae.

Of these, ^{26}Al is the only isotope whose contribution to the positron rate may be reasonably quantified, thanks to measurements of the diffuse 1.8-MeV line intensity, which contributes to the overall 1- to 30-MeV map shown in plate 9. The radioactive ^{26}Al has a half-life of almost one million years, decaying into an excited state of ^{26}Mg 82% of the time by emission of a positron and 18% of the time by electron capture. Subsequently, ^{26}Mg decays to its ground state 99.7% of the time by the emission of a 1.8-MeV γ-ray.

Based on the measured 1.8-MeV flux integrated over a (generous) radius of $15°$, one infers from this decay chain a 511-keV annihilation flux only 7% of the measured bulge value. In addition, the 1.8-MeV distribution is clearly disk dominated, which is morphologically inconsistent with the SPI-image of the 511-keV emission shown in figure 13.3.

The positron injection rate due to the decay of the other radioactive isotopes is uncertain due to the fact that they undergo beta decay while still embedded within the supernova ejecta, making it very difficult for the positrons to escape through the thick medium. Nonetheless, even if all of these particles could eventually find their way out of the expanding remnant, there is still the question of how one should reconcile the spheroidal shape of the annihilation region with the (expected) disk-dominated distribution of supernova explosions.

Finally, the observed nova rate points to a positron contribution from ^{22}Na of less than one percent of the measured value. We are therefore led to the conclusion that none of the known distributions of radioactive isotopes can account for the positron annihilation rate observed from the Galaxy's central bulge.

The nature of galactic positrons—their origin, propagation, and annihilation—is clearly still a mystery. It is almost as if we have too much observational information now, since no self-consistent picture seems possible. Perhaps some of the assumptions we take for granted in our analysis are incorrect. The positrons may not be in steady state, for example, or the magnetic fields through which they diffuse may not be as turbulent as we think, so our estimate of their propagation time is wrong. There is an answer—we just have to find it.

As we look to even higher energies, our understanding of the high-energy background becomes more secure. We can see in figure 13.4, for example, that the galactic diffuse γ-ray emission, from ~ 1 MeV to ~ 1 Tev, is a blend of components arising from interactions between cosmic rays and the interstellar medium, inverse Compton scattering effects, and bremsstrahlung emission. The extragalactic background spectrum (thin solid line) is significantly less intense than its galactic counterpart, and is shown here (in the COMPTEL and EGRET range) for comparison; a more detailed view (extending from 1 keV to 1 TeV) is displayed in figure 13.5.

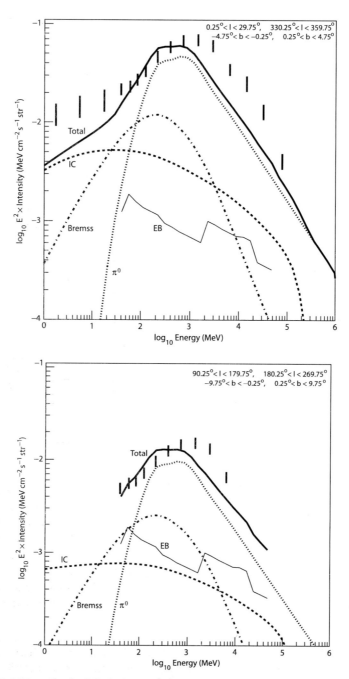

Figure 13.4 The galactic diffuse high-energy emission is produced primarily by energetic interactions of nucleons with gas via neutral pion production and decay, and by electrons via inverse Compton (IC) scattering and bremsstrahlung. These processes dominate in different portions of the spectrum and vary with line-of-sight angle through the Galaxy. The top panel

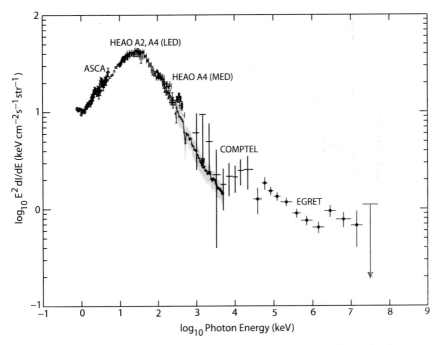

Figure 13.5 The extragalactic diffuse X-ray and gamma-ray background emission is a super-position of all unresolved high-energy sources. Active galactic nuclei (see figure 12.15) are the dominant class of gamma-ray emitters up to the highest energies. The spectrum is not consistent with a simple power law, and exhibits some positive curvature consistent with a blazar origin. However, important contributions are made by other extragalactic sources, including (unresolved) galaxy clusters and gamma-ray burst events. (Image from Strong, Moskalenko, and Reimer 2004b)

However, the cosmic-ray distribution used for the fits in figure 13.4 is not exactly the same as that deduced from observations near Earth (see section 13.1), which, without modification, does not produce a good match to the data. In these fits, the electron distribution is renormalized upward by a factor $\sim$4 relative to the local density, while the proton distribution is renormalized upward by a factor $\sim$1.8. In addition, both populations apparently must steepen, with a break energy at $\sim$20 GeV for electrons and $\sim$10 GeV for protons. These variations from local conditions have been attributed to energy losses and stochastic propagation and acceleration effects

---

Figure 13.4 (*continued*) shows the galactic diffuse emission in the inner Galaxy (i.e., for galactic longitude $330° < l < 30°$ and latitude $0° < b < 5°$); the bottom panel is a representation of the diffuse emission from the outer Galaxy (i.e., for $90° < l < 270°$ and $0° < b < 10°$). The model contributions to the total emission (thick, solid line) are as follows: (dots) π^0-decay, (dash) IC, (dash-dot) bremsstrahlung, and the extragalactic background (labeled EB and shown with a thin solid line). EGRET and COMPTEL data are shown as dark bars. See also figure 13.5. (Images from Strong, Moskalenko, and Reimer 2004a)

across the Galaxy.[43] If these variations are real, they would provide *prima facie* evidence in support of the view that cosmic rays up to the knee originate within the Milky Way itself.

Given its relative faintness compared to the galactic diffuse emission, the extragalactic γ-ray background is difficult to identify in any model-independent fashion. The spectrum shown in figure 13.5 is a superposition of all unresolved sources of high-energy γ-radiation in the universe, which is dominated by AGNs, particularly at the highest energies sampled in this survey (see figure 12.15). Nonetheless, given the inherent difficulty in measuring the γ-ray signal through the murky galactic field, there is no consensus yet on what fraction of the background is due to unresolved supermassive black holes; predictions range from 25% all the way up to 100%. Other plausible contributions may come from galaxy clusters[44] (section 13.2), energetic particles accelerated in shocks (section 4.3) associated with large-scale cosmological structure formation,[45] and distant γ-ray burst events (section 11.2).

Other, more speculative proposals for the origin of this extragalactic γ-ray spectrum perhaps reveal the edge in our knowledge of the high-energy activity occurring at large redshifts. Many of these ideas bring to a focus the interface between astrophysics and particle physics, which finds a natural setting in the early Universe. If reliably derived, the extragalactic background may potentially provide important information about the phase of baryon–antibaryon annihilation,[46] the evaporation of primordial black holes,[47] and the annihilation of so-called weakly interacting massive particles (WIMPs).[48]

High-energy astrophysics is a maturing science, though by no means is the pace of discovery slowing down. If anything, we are entering an age in which cross-disciplinary activity is becoming a necessity, as we noted in the previous paragraph. More and more, the boundary between modern astrophysics and particle physics is dissolving, and nowhere is this in greater evidence than in the high-energy domain.

We have learned, for example, that a dominant source of γ-rays is neutral pion decays in the cascades produced by proton–proton scatterings. But these interactions create charged pions as well, which decay into muons, and then electrons and positrons. All along the chain, energetic neutrinos are produced, and some of these reach Earth in sufficient numbers to be detected with the km^3 arrays now being built around the globe. Thus, although neutrino astronomy is becoming a field in its own right, the neutrino and γ-ray signals are intimately connected, and it is inconceivable to imagine modeling a source producing either of these particles without a consideration of the other.

Needless to say, the greatest discoveries lie ahead of us. Our future detectors and satellites will have a probative power to uncloak the high-energy sky far beyond the exquisite views we have produced thus far. Larger collecting areas will push

---

[43] See, e.g., Strong, Moskalenko, and Reimer (2004a).
[44] Ensslin et al. (1997).
[45] Loeb and Waxman (2000).
[46] See, e.g., Gao et al. (1990) and Dolgov and Silk (1993).
[47] See Hawking (1974).
[48] See, e.g., Jungman et al. (1996).

timing resolution well below the millisecond range, and the development of X-ray polarimetry will provide us with valuable information concerning the geometry of the source. Even X-ray interferometry will be possible in space, with the promise of providing us with an actual high-energy image of an event horizon, e.g., of the supermassive black hole at the center of our Galaxy and in the nucleus of M87. May the excitement continue.

SUGGESTED READING

The origin of cosmic rays continues to be somewhat of a mystery almost a century after their discovery, though tangible progress is now being made. Almost certainly, shocks are responsible for producing energetic particles within the Galaxy, at least below the knee. For a review, see Blandford and Eichler (1987).

But with the unprecedented observations made with the Pierre Auger Observatory, it is becoming increasingly clear that the highest energy particles originate from outside the Galaxy. Evidence is growing that at least some of these particles are accelerated within the giant radio lobes of AGNs. See Mollerach et al. (2007).

Though now aging, the following review article captures many of the physics issues associated with cosmic-ray acceleration, propagation, and detection: Hillas (1984).

The issues pertaining to the high-energy emission from galaxy clusters continue to shift as new observations and theoretical work refine our view of these enormous structures. For example, a recent set of observations with *Chandra* illustrates how complex the emission region can be. See Fabian et al. (2000, 2006).

The properties of the intracluster gas in clusters are significantly affected by nongravitational processes, including star formation and the energy injection by AGNs. These effects contribute to a redshift evolution in the cluster emissivity. For a review, see Rosati, Borgani, and Norman (2002).

These influences notwithstanding, a lingering problem with galaxy clusters is the absence of significant cooling in their cores. Given the observed level of X-ray emission from these regions, the inferred radiative cooling timescale is shorter than the age of the cluster, yet the temperature drops by no more than a factor 3 toward the center. The gas should be cooling rapidly, but instead something keeps it warm. This disparity is sometimes referred to as the "cooling flow problem." For a review, see Peterson and Fabian (2006).

As X-ray and gamma-ray instruments continue to improve in sensitivity and spatial resolution, it becomes increasingly clear that (at least most of) the high-energy background can be resolved into known classes of sources. As an example of how dramatic this process can be, read about the impact *Chandra* had with regard to the X-ray background in Mushotzky et al. (2000).

Bibliography

Abraham, J., and the Auger Collaboration, "Properties and Performance of the Prototype Instrument for the Pierre Auger Observatory," *Nuclear Instruments and Methods in Physics Research* A523 (2004), pp. 50–95.

Abramowicz, M. A., M. Calvani, and L. Nobili, "Thick Accretion Disks with Super-Eddington Luminosities," *The Astrophysical Journal* 242 (1980), pp. 772–788.

Abramowicz, M. A., and W. Kluźniak, "A Precise Determination of Black Hole Spin in GRO J1655–40," *Astronomy and Astrophysics* 374 (2001), pp. L19–L20.

Adelman-McCarthy, J. K., M. A. Agüeros, S. A. Allam, et al., "The Fifth Data Release of the Sloan Digital Sky Survey," *The Astrophysical Journal Supplements* 172 (2007), pp. 634–644.

Aharonian, F., A.G. Akhperjanian, K.-M. Aye, et al., "Very High Energy Gamma Rays from the Direction of Sagittarius A*," *Astronomy and Astrophysics* 425 (2004), pp. L13–L17.

Aharonian, F., A.G. Akhperjanian, J.A. Barris, et al., "Limits on the TeV Flux of Diffuse Gamma Rays as Measured with He HEGRA Air Shower Array," *Astroparticle physics* 17 (2002), pp. 459–475.

Aharonian, F., A.G. Akhperjanian, A.R. Bazer-Bachi, et al., "The H.E.S.S. Survey of the Inner Galaxy in Very High-Energy Gamma Rays," *The Astrophysical Journal* 636 (2006), pp. 777–797.

Alfvén, H., "Magnetohydrodynamic Waves, and the Heating of the Solar Corona," *Monthly Notices of the Royal Astronomical Society* 107 (1947), pp. 211–219.

Alfvén, H., and N. Herlofsen, "Cosmic Radiation and Radio Stars," *Physical Review* 78 (1950), p. 616.

Alpar, M. A., A. F. Cheng, M. A. Ruderman, and J. Shaham, "A New Class of Radio Pulsars," *Nature* 300 (1982), pp. 728–730.

Anantharamaiah, K. R., A. Pedlar, R. D. Ekers, and W. M. Goss, "Radio Studies of the Galactic Center, II: The Arc, Threads, and Related Features at 90 cm (330 MHz)," *Monthly Notices of the Royal Astronomical Society* 249 (1991), pp. 262–281.

Antonucci, R., "Unified Models for Active Galactic Nuclei and Quasars," *Annual Reviews of Astronomy and Astrophysics* 31 (1993), pp. 473–521.

Angel, J.R.P., E. F. Borra, and J. D. Landstreet, "The Magnetic Fields of White Dwarfs," *The Astrophysical Journal* 45 (1981), pp. 457–474.

Arguado, E. and J. E. Burt, *Understanding Weather and Climate,* Upper Saddle River, NJ: Prentice Hall (1999).

Avni, Y., A., Soltan, H. Tananbaum, G. Zamorani, "A Method for Determining Luminosity Functions Incorporating both Flux Measurements and Flux Upper Limits, with Applications to the Average X-ray to Optical Luminosity Ratio for Quasars," *The Astrophysical Journal* 238 (1980), pp. 800–807.

Ayasli, S., and P. C. Joss, "Thermonuclear Processes on Accreting Neutron Stars—A Systematic Study," *The Astrophysical Journal* 256 (1982), pp. 637–665.

Baganoff, F. K., M. W. Bautz, W. N. Brandt, G. Chartas, et al., "Rapid X-ray Flaring from the Direction of the Supermassive Black Hole at the Galactic Centre," *Nature* 413 (2001), pp. 45–48.

Baganoff, F. K., Y. Maeda, M. Morris, et al., "Chandra X-Ray Spectroscopic Imaging of Sagittarius A* and the Central Parsec of the Galaxy," *The Astrophysical Journal* 591 (2003), pp. 891–915.

Balberg, S., and S. L. Shapiro, "Gravothermal Collapse of Self-Interacting Dark Matter Halos and the Origin of Massive Black Holes," *Physical Review Letters* 88 (2002), pp. 101301.1–101301.4.

Balbus, S. A., C. F. Gammie, and J. F. Hawley, "Fluctuations, Dissipation and Turbulence in Accretion Discs," *Monthly Notices of the Royal Astronomical Society* 271 (1994), pp. 197–201.

Balbus, S. A., and J. F. Hawley, "A Powerful Local Shear Instability in Weakly Magnetized Disks, I: Linear Analysis; II: Nonlinear Evolution," *The Astrophysical Journal* 376 (1991), pp. 214–233.

Band, D., J. Matteson, L. Ford, et al., "BATSE Observations of Gamma Ray Burst Spectra, I: Spectral Diversity," *The Astrophysical Journal* 413 (1993), pp. 281–292.

Baptista, R., K. Horne, R. A. Wade, et al., "HST Spatially-Resolved Spectra of the Accretion Disk and Gas Stream of the Nova-like variable UX Ursae Majoris," *Monthly Notices of the Royal Astronomical Society* 298 (1998), pp. 1079–1091.

Baptista, R., and J. E. Steiner, "Improving the Eclipse Mapping Method," *Astronomy and Astrophysics* 277 (1993), pp. 331–344.

Bardeen, J. M., W. H. Press, and S. A. Teukolsky, "Rotating Black Holes: Locally Nonrotating Frames, Energy Extraction, and Scalar Synchrotron Radiation," *The Astrophysical Journal* 178 (1972), pp. 347–369.

Barger, A. J., L. L. Cowie, M. W. Bautz, et al., "Supermassive Black Hole Accretion History Inferred from a Large Sample of Chandra Hard X-ray Sources," *Astronomical Journal* 122 (2001), pp. 2177–2194.

Barger, A. J., L. L. Cowie, P. Capak, et al., "Optical and Infrared Properties of vthe 2 Ms Chandra Deep Field North X-Ray Sources," *Astronomical Journal* 126 (2003), pp. 632–665.

Barish, B. C., and R. Weiss, "LIGO and the Detection of Gravitational Waves," *Physics Today* October (1999), pp. 44–50.

Beacom, J. F., N. F. Bell, and G. Bertone, "Gamma-Ray Constraint on Galactic Positron Production by MeV Dark Matter," *Physical Review Letters* 94 (2005), pp. 171301.1–171301.4.

Beacom, J. F., and H. Yüksel, "Stringent Constraint on Galactic Positron Production," *Physical Review Letters* 97 (2006), pp. 071102.1–071102.4.

Bélanger, G., A. Goldwurm, M. Renaud, et al., "A Persistent High-Energy Flux from the Heart of the Milky Way: INTEGRAL's View of the Galactic Center," *The Astrophysical Journal* 636 (2006a), pp. 275–289.

Bélanger, G., R. Terrier, O. C. de Jager, A. Goldwurm, and F. Melia, "Periodic Modulations in an X-ray Flare from Sagittarius A*," *Journal of Physics* 54 (2006b), pp. 420–426.

Belian, R. D., J. P. Conner, and W. D. Evans, "The Discovery of X-ray Bursts from a Region in the Constellation Norma," *The Astrophysical Journal Letters* 206 (1976), pp. L135–L138.

Bell, S. J., "Little Green Men, White Dwarfs, or Pulsars," *Annals of the New York Academy of Science* 302 (1977), pp. 685–689.

Bender, R., J. Kormendy, G. Bower, et al., "HST STIS Spectroscopy of the Triple Nucleus of M31: Two Nested Disks in Keplerian Rotation around a Supermassive Black Hole," *The Astrophysical Journal* 631 (2005), pp. 280–300.

Berezinsky, V. S., and A. Z. Gazizov, "Production of High-Energy Cosmic Neutrinos in $p\gamma$ and $n\gamma$ Scattering, I: Neutrino Yields for Power-Law Spectra of Protons and Neutrons," *Physical Review D* 47 (1993), pp. 4206–4216.

Berezinsky, V., M. Kachelriess, and A. Vilenkin, "Ultrahigh Energy Cosmic Rays without a Greisen-Zatsepin-Kuzmin Cutoff," *Physical Review Letters* 79 (1997), pp. 4302–4305.

Berezinsky, V. S., P. Blasi, and V. S. Ptuskin, "Clusters of Galaxies as Storage Room for Cosmic Rays," *The Astrophysical Journal* 487 (1997), pp. 529–535.

Bhattacharjee, P., and G. Sigl, "Origin and Propagation of Extremely High Energy Cosmic Rays," *Physics Reports* 327 (2000), pp. 109–247.

Bicknell, G. V., and M. C. Begelman, "Understanding the Kiloparsec-Scale Structure of M87," *The Astrophysical Journal* 467 (1996), pp. 597–621.

Bicknell, G. V., and D. B. Melrose, "In Situ Acceleration in Extragalactic Radio Jets," *The Astrophysical Journal* 262 (1982), pp. 511–528.

Bignami, G. F., and P. A. Caraveo, "Geminga: Its Phenomenology, Its Fraternity, and Its Physics," *Annual Reviews of Astronomy and Astrophysics* 34 (1996), pp. 331–382.

Bildsten, L., "Propagation of Nuclear Burning Fronts on Accreting Neutron Stars: X-ray Bursts and Sub-hertz Noise," *The Astrophysical Journal* 438 (1995), pp. 852–875.

Bisnovatyi-Kogan, G. S., and B. V. Komberg, "Pulsars and Close Binary Systems," *Soviet Astronomy* 18 (1974), pp. 217–221.

Blackman, E. G., "Distinguishing Solar Flare Types by Differences in Reconnection Regions," *The Astrophysical Journal Letters* 484 (1997), pp. L79–L82.

Blaes, O. M., "Stabilization of Non-axisymmetric Instabilities in a Rotating Flow by Accretion on to a Central Black Hole," *Monthly Notices of the Royal Astronomical Society* 227 (1987), pp. 975–992.

Blandford, R. D., and M. C. Begelman, "On the Fate of Gas Accreting at a Low Rate Onto a Black Hole," *Monthly Notices of the Royal Astronomical Society* 303 (1999), pp. L1–L5.

Blandford, R. D., and D. Eichler, "Particle Acceleration at Astrophysical Shocks—A Theory of Cosmic-Ray Origin," *Physics Reports* 154 (1987), pp. 1–75.

Blandford, R. D., and A. Königl, "Relativistic Jets as Compact Radio Sources," *The Astrophysical Journal* 232 (1979), pp. 34–48.

Blandford, R. D., and M. J. Rees, "A 'Twin-Exhaust' Model for Double Radio Sources," *Monthly Notices of the Royal Astronomical Society* 169 (1974), pp. 395–415.

Blasi, P., S. Burles, and A. V. Olinto, "Cosmological Magnetic Field Limits in an Inhomogeneous Universe," *The Astrophysical Journal Letters* 514 (1999), pp. L79–L82.

Bolton, C. T., "Identifications of CYG X-1 with HDE 226868," *Nature* 235 (1972), pp. 271–273.

Bonamente, M., M. K. Joy, S. J. LaRoque, et al., "Determination of the Cosmic Distance Scale from Sunyaev-Zel'dovich Effect and Chandra X-ray Measurements of High-Redshift Galaxy Clusters," *The Astrophysical Journal* 647 (2006), pp. 25–54.

Borione, A., M. C. Chantell, C. E. Covault, et al., "High Statistics Search for Ultra-high Energy γ-ray Emission from Cygnus X-3 and Hercules X-1," *Physical Review D* 55 (1997), pp. 1714–1731.

Bouchet, L., P. Mandrou, J. P. Roques, et al., "Sigma Discovery of Variable $e^+ - e^-$ Annihilation Radiation Near the Galactic Center Variable Compact Source 1E 1740.7–2942," *The Astrophysical Journal Letters* 383 (1991), pp. L45–L48.

Bowyer, S., E. T., Byram, T. A. Chubb, et al., "Cosmic X-ray Sources," *Science* 147 (1965), pp. 394–398.

Boyer, R. H., and R. W. Lindquist, "Maximal Analytic Extension of the Kerr Metric," *Journal of Mathematical Physics* 8 (1967), pp. 265–281.

Bradt, H. V., S. Rappaport, W. Mayer, et al., "X-Ray and Optical Observations of the Pulsar NP 0532 in the Crab Nebula," *Nature* 222 (1969), pp. 728–730.

Brandenburg, A., A. Nordlund, R. Stein, and U. Torkelsson, "Dynamo-Generated Turbulence and Large-Scale Magnetic Fields in a Keplerian Shear Flow," *The Astrophysical Journal* 446 (1995), pp. 741–754.

Bromley, B., F. Melia, and S. Liu, "Polarimetric Imaging of the Massive Black Hole at the Galactic Center," *The Astrophysical Journal Letters* 555 (2001), pp. L83–L87.

Brown, G. E., and H. A. Bethe, "A Scenario for a Large Number of Low-Mass Black Holes in the Galaxy," *The Astrophysical Journal* 423 (1994), pp. 659-664.

Brown, R. L., "Precessing Jets in Sagittarius A: Gas Dynamics in the Central Parsec of the Galaxy," *The Astrophysical Journal* 262 (1982), pp. 110–119.

Brunetti, G., P. Blasi, R. Cassano, et al., "Alfvénic Reacceleration of Relativistic Particles in Galaxy Clusters: MHD Waves, Leptons and Hadrons," *Monthly Notices of the Royal Astronomical Society* 350 (2004), pp. 1174–1194.

Burbidge, G. R., "Estimates of the Total Energy in Particles and Magnetic Field in the Non-thermal Radio Sources," *The Astrophysical Journal* 129 (1959), pp. 849–851.

Bussard, R. W., R. Ramaty, and R. J. Drachman, "The Annihilation of Galactic Positrons," *The Astrophysical Journal* 228 (1979), pp. 928–934.

Cannizzo, J. K., "The Accretion Disk Limit Cycle Model: Toward an Understanding of the Long-Term Behavior of SS Cygni," *The Astrophysical Journal* 419 (1993), pp. 318–336.

Carter, B., "Global Structure of the Kerr Family of Gravitational Fields," *Physical Review* 174 (1968), pp. 1559–1571.

Casares, J. P., "X-ray Binaries and Black Hole Candidates: A Review of Optical Properties," in *Lecture Notes in Physics* 563 (2001), pp. 277–327.

Casares, J., P. A. Charles, and T. Naylor, "A 6.5-day Periodicity in the Recurrent Nova V404 Cygni Implying the Presence of a Black Hole," *Nature* 355 (1992), pp. 614–617.

Cavallo, G., and M. J. Rees, "A Qualitative Study of Cosmic Fireballs and Gamma Ray Bursts," *Monthly Notices of the Royal Astronomical Society* 183 (1978), pp. 359–365.

Cen, R., and J. P. Ostriker, "Where Are the Baryons?" *The Astrophysical Journal* 514 (1999), pp. 1–6.

Chandrasekhar, S., *Stellar Structure,* Chicago: University of Chicago Press (1939).

———, *Truth and Beauty,* Chicago: University of Chicago Press (1987).

Charles, P.A., and M. J. Coe, "Optical, Ultraviolet and Infrared Observations of X-ray Binaries," in *Compact Stellar X-ray Sources,* ed. W.H.G. Lewin and M. van der Klis, Cambridge, UK: Cambridge University Press (2006), pp. 215–266.

Chen, W., N. Gherels, and M. Leventhal, "On the Optical Counterparts, Long-term Variabilities, Radio Jets, and Accretion Sources in 1E 1740.7–2942 and GRS 1758–258," *The Astrophysical Journal* 426 (1994), pp. 586–598.

Chen, W., and N. Gehrels, "The Progenitor of the New COMPTEL/ROSAT Supernova Remnant in VELA," *The Astrophysical Journal Letters* 514 (1999), pp. L103–L106.

Churazov, E., R. Sunyaev, S. Sazonov, M. Revnivtsev, and D. Varshalovich, "Positron Annihilation Spectrum from the Galactic Centre region Observed by SPI/INTEGRAL," *Monthly Notices of the Royal Astronomical Society* 357 (2005), pp. 1377–1386.

Clery, D., "Telescopes Break New Ground in Quest for Cosmic Rays," *Science* 305 (2004), pp. 1393–1394.

Coe, M. J., A. R. Engel, and J. J. Quenby, "Anti-correlated Hard and Soft X-ray Intensity Variations of the Black-Hole Candidates CYG X-1 and A0620-00," *Nature* 259 (1976), pp. 544–545.

Cooke, B. A., R. E. Griffiths, and K. A. Pounds, "Evidence for a Galactic Component of the Diffuse X-Ray Background," *Nature* 224 (1969), pp. 134–137.

Costa, E., F. Frontera, J. Heise, et al., "Discovery of an X-ray Afterglow Associated with the Gamma Ray Burst of 28 February 1997," *Nature* 387 (1997), pp. 783–785.

Cottam, J., F. Paerels, and M. Mendez, "Gravitationally Redshifted Absorption Lines in the X-ray Burst Spectra of a Neutron Star," *Nature* 420 (2002), pp. 51–54.

Crane, P. C., J. R. Dickel, and J. J. Cowan, "Detection of an Unresolved Nuclear Radio Source in M31," *The Astrophysical Journal Letters* 390 (1992), pp. L9–L12.

Crocker, R. M., M. Fatuzzo, R. Jokipii, F. Melia, and R. R. Volkas, "The AGASA/SUGAR Anisotropies and TeV Gamma Rays from the Galactic Center: A Possible Signature of Extremely High-Energy Neutrons," *The Astrophysical Journal* 622 (2005), pp. 892–909.

Cunningham, C. T., "The Effects of Redshifts and Focusing on the Spectrum of an Accretion Disk Around a Kerr Black Hole," *The Astrophysical Journal* 202 (1975), pp. 788–802.

Dennison, B., "Formation of Radio Halos in Clusters of Galaxies from Cosmic-Ray Protons," *The Astrophysical Journal Letters* 239 (1980), pp. L93–L96.

Di Matteo, T., S. W. Allen, A. C. Fabian, et al., "Accretion onto the Supermassive Black Hole in M87," *The Astrophysical Journal* 582 (2003), pp. 133–140.

Dingus, B. L., "Observations of the Highest Energy Gamma Rays from Gamma Ray Bursts," in *Gamma Ray Burst and Afterglow Astronomy 2001,* ed. G. R. Ricker and R. K. Vanderspek, New York: AIP (2003), pp. 240–243.

Dolgov, A., and J. Silk, "Baryon Isocurvature Fluctuations at Small Scales and Baryonic Dark Matter," *Physical Review D* 47 (1993), pp. 4244–4255.

Dubus, G., P.A. Charles, K. S. Long, et al., "The Eclipsing X-ray Pulsar X-7 in M33," *Monthly Notices of the Royal Astronomical Society* 302 (1999), pp. 731–734.

Eichler, D., M. Livio, T. Piran, et al., "Nucleosynthesis, Neutrino Bursts and Gamma Rays from Coalescing Neutron Stars," *Nature* 340 (1989), pp. 126–128.

Eisenhauer, F., R. Genzel, T. Alexander, et al., "SINFONI in the Galactic Center: Young Stars and Infrared Flares in the Central Light-Month," *The Astrophysical Journal* 628 (2005), pp. 246–259.

Elbert, J. W., and P. Sommers, "In Search of a Source for the 320 EeV Fly's Eye Cosmic Ray," *The Astrophysical Journal* 441 (1995), pp. 151–161.

Ensslin, T. A., P. L. Biermann, P. P. Kronberg, and X.-P. Wu, "Cosmic-Ray Protons and Magnetic Fields in Clusters of Galaxies and Their Cosmological Consequences," *The Astrophysical Journal* 477 (1997), pp. 560–567.

Evans, R. D., *The Atomic Nucleus,* New York: McGraw-Hill, 1955.

Evans, P. A., and C. Hellier, "What Can XMM-*Newton* Tell Us About the Spin Periods of Intermediate Polars," in *The Astrophysics of Cataclysmic Variables and Related Objects,* ed. by J. M. Hameury and J. P. Lasota, San Francisco: Astronomical Society of the Pacific (2005), pp. 165–171.

Fabian, A. C., R. F. Mushotzky, P.E.J. Nulsen, and J. R. Peterson, "On the Soft X-ray Spectrum of Cooling Flows," *Monthly Notices of the Royal Astronomical Society Letters* 321 (2001), pp. L20–L24.

Fabian, A. C., M. J. Rees, L. Stella, and N. E. White, "X-ray Fluorescence from the Inner Disc in Cygnus X-1," *Monthly Notices of the Royal Astronomical Society* 238 (1989), pp. 729–736.

Fabian, A. C., J. S. Sanders, S. Ettori, et al., "Chandra Imaging of the Complex X-ray Core of the Perseus Cluster," *Monthly Notices of the Royal Astronomical Society Letters* 318 (2007), pp. L65–L68.

Fabian, A. C., J. S. Sanders, G. B. Taylor, et al., "A Very Deep Chandra Observation of the Perseus Cluster: Shocks, Ripples and Conduction," *Monthly Notices of the Royal Astronomical Society* 366 (2006), pp. 417–428.

Fabian, A. C., I. Smail, K. Iwasawa, et al., "Testing the Connection Between the X-ray and Submillimeter Source Populations Using *Chandra,*" *Monthly Notices of the Royal Astronomical Society* 315 (2000), pp. L8–L12.

Fabian, A. C., S. Vaughan, K. Nandra, et al., "A Long Hard Look at MCG-6-30-15 with XMM-Newton," *Monthly Notices of the Royal Astronomical Society* 335 (2002), pp. L1–L5.

Fan, X., V. K. Narayanan, R. H. Lupton, et al., "A Survey of $z > 5.8$ Quasars in the Sloan Digital Sky Survey, I: Discovery of Three New Quasars and the Spatial Density of Luminous Quasars at $z \sim 6$," *The Astronomical Journal* 122 (2001), pp. 2833–2849.

Fanaroff, B. L., and J. M. Riley, "The Morphology of Extragalactic Radio Sources of High and Low Luminosity," *Monthly Notices of the Royal Astronomical Society* 167 (1974), pp. 31P–36P.

Feigelson, E. D., H. Bradt, J. McClintock, R. Remillard, C. M. Urry, et al., "H 0323+022—a New BL Lacertae Object with Extremely Rapid Variability," *The Astrophysical Journal* 302 (1986), pp. 337–351.

Fender, R., S. Corbel, T. Tzioumis, et al., "Quenching of the Radio Jet During the X-Ray High State of GX 339-4," *The Astrophysical Journal Letters* 519 (1999), pp. L165–L168.

Feretti, L., D. Dallacasa, G. Giovannini, et al., "The Magnetic Field in the Coma Cluster," *Astronomy and Astrophysics* 302 (1995), pp. 680–690.

Ferland, G. J., "Hazy, A Brief Introduction to CLOUDY 94.00," Lexington: University of Kentucky, Department of Physics, Astronomy Internal Report, 1996.

Ferrarese, L., and D. Merritt, "A Fundamental Relation Between Supermassive Black Holes and Their Host Galaxies," *The Astrophysical Journal Letters* 539, (2000), pp. L9–L12.

Fermi, E., "On the Origin of the Cosmic Radiation," *Physical Review Letters* 75 (1949), pp. 1169–1174.

Filippenko, A. V., "Optical Spectra of Supernovae," *Annual Reviews of Astronomy and Astrophysics* 35 (1997), pp. 309–355.

Filippenko, A. V., T. Matheson, and L. C. Ho, "The Mass of the Probable Black Hole in the X-Ray Nova GRO J0422+32," *The Astrophysical Journal* 455 (1995), pp. 614–622.

Fleysher, R., and the Milagro Collaboration, "Search for Diffuse TeV Gamma-Ray Emission from the Galactic Plane, Using the Milagro Gamma-Ray Telescope," *Bulletin of the American Astronomical Society* 34 (2002), p. 676.

Ford, H. C., R. J. Harms, Z. I. Tsvetanov, et al., "Narrowband HST Images of M87: Evidence for a Disk of Ionized Gas Around a Massive Black Hole," *The Astrophysical Journal Letters* 435 (1994), pp. L27–L30.

Fox, D. B., D. A. Frail, P. A. Price, et al., "The Afterglow of GRB 050709 and the Nature of the Short-Hard γ-ray Bursts," *Nature* 437 (2005), pp. 845–850.

Fox, J. G., "Evidence Against Emission Theories," *American Journal of Physics* 33 (1965), pp. 1–17.

———, "Constancy of the Velocity of Light," *Journal of the Optical Society* 57 (1967), pp. 967–968.

Frail, D., S. R. Kulkarni, S. R. Nicastro, et al., "The Radio Afterglow from the Gamma Ray Burst of 8 May 1997," *Nature* 389 (1997), pp. 261–263.

Frank, J., A. King, and D. Raine, *Accretion Power in Astrophysics,* Cambridge, UK: Cambridge University Press (2002).

Friedman, H., S. W. Lichtman, and E. T. Byram, "Photon Counter Measurements of Solar X-rays and Extreme Ultraviolet Light," *Physical Review* 83 (1951), pp. 1025–1030.

Fritz, G., R. C. Henry, J. F. Meekins, et al., "X-ray Pulsar in the Crab Nebula," *Science* 164 (1969), pp. 709–712.

Fryxell, B. A., K. Olson, P. Ricker, et al., "FLASH: An Adaptive Mesh Hydrody-namics Code for Modeling Astrophysical Flashes," *The Astrophysical Journal Supplements* 131 (2000), pp. 273–334.

Fryxell, B. A., and S. E. Woosley, "Finite Propagation Time in Multidimensional Thermonuclear Runaways," *The Astrophysical Journal* 261 (1982), pp. 332–336.

Fusco-Femiano, R., M. Orlandini, S. De Grandi, et al., "Hard X-ray and Radio Observations of Abell 754," *Astronomy and Astrophysics* 398 (2003), pp. 441–446.

Gaensler, B. M., N. M. McClure-Griffiths, M. S. Oey, et al., "A Stellar Wind Bubble Coincident with the Anomalous X-Ray Pulsar 1E 1048.1–5937: Are Magne-tars Formed from Massive Progenitors?" *The Astrophysical Journal Letters* 620 (2005), pp. L95–L98.

Galama, T. J., P. M. Vreeswijk, J. van Paradijs et al., "An Unusual Supernova in the Error Box of the Gamma Ray Burst of 25 April 1998," *Nature* 395 (1998), pp. 670–672.

Galloway, D. K., M. P. Muno, J. M. Hartman, et al., "Thermonuclear (Type I) X-ray Bursts Observed by the Rossi X-ray Timing Explorer," *The Astrophysical Journal Supplements* (2006), in press.

Gänsicke, B. T., "Observational Population Studies of Cataclysmic Variables— The Golden Era of Surveys," in *The Astrophysics of Cataclysmic Variables and Related Objects,* ed. by J.-M. Hameury and J.-P. Lasota, San Francisco: Astronomical Society of the Pacific (2005), pp. 3–17.

Gao, Y.-T., F. W. Stecker, M. Gleiser, and D. B. Cline, "Large-scale Anisotropy in the Extragalactic γ-ray Background as a Probe for Cosmological Antimatter," *The Astrophysical Journal Letters* 361 (1990), pp. L37–L40.

Garcia, M. R., B. F. Williams, F. Yuang, et al., "A Possible Detection of M31* with *Chandra*," *The Astrophysical Journal* 632 (2005), pp. 1042–1047.

Gehrels, N., G. Chincarini, P. Giommi et al., "The Swift Gamma-Ray Burst Mission," *The Astrophysical Journal* 611 (2004), pp. 1005–1020.

Gehrels, N., C. L. Sarazin, P. T. O'Brien, et al., "A Short γ-ray Burst Apparently Associated with an Elliptical Galaxy at Redshift $z = 0.225$," *Nature* 437 (2005), pp. 851–854.

Gehrels, N., and E. D. Williams, "Temperatures of Enhanced Stability in Hot Thin Plasmas," *The Astrophysical Journal Letters* 418 (1993), pp. L25–L28.

Gebhardt, K., J. Kormendy, L. C. Ho, et al., "A Relationship Between Nuclear Black Hole Mass and Galaxy Velocity Dispersion," *The Astrophysical Journal Letters* 539 (2000), pp. L13–L16.

Genzel, R., R. Schödel, T. Ott, A. Eckart, T. Alexander, F. Lacombe, D. Rouan, and B. Aschenbach, "Near-IR Flares from Accreting Gas Around the Supermassive Black Hole in the Galactic Center," *Nature* 425 (2003), pp. 934–937.

Giacconi, R., H. Gursky, F. Paolini, and B. Rossi, "Evidence for X-Rays from Sources Outside the Solar System," *Physical Review Letters* 9 (1962), pp. 439–443.

Gierliński, M., A. A. Zdziarski, J. Poutanen, et al., "Radiation Mechanisms and Geometry of Cygnus X-1 in the Soft State," *Monthly Notices of the Royal Astronomical Society* 309 (1999), pp. 496–512.

Gilfanov, M., M. Revnivtsev, and S. Molkov, "Boundary Layer, Accretion Disk and X-ray Variability in the Luminous LMXBs," *Astronomy and Astrophysics* 410 (2003), pp. 217–230.

Giovannini, G., L. Feretti, T. Venturi, et al., "The Halo Radio Source Coma C and the Origin of Halo Sources," *The Astrophysical Journal* 406 (1993), pp. 399–406.

Gold, T., "Rotating Neutron Stars as the Origin of the Pulsating Radio Sources," *Nature* 218 (1968), pp. 731–732.

Goldwurm, A., J. Ballet, B. Cordier, et al., "SIGMA/GRANAT Soft Gamma Ray Observations of the X-ray Nova in Musca: Discovery of Positron Annihilation Emission Line," *The Astrophysical Journal Letters* 389 (1992), pp. L79–L82.

Goodman, J., "Are Gamma-Ray Bursts Optically Thick?" *The Astrophysical Journal Letters* 308 (1986), pp. L47–L50.

———, "Radio Scintillation of Gamma Ray-Burst Afterglows," *New Astronomy* 2 (1997), pp. 449–460.

Greisen, K., "End to the Cosmic-Ray Spectrum?" *Physical Review Letters* 16 (1966), pp. 748–750.

Grindlay, J., H. Gursky, H. Schnopper, et al., "Discovery of Intense X-ray Bursts from the Globular Cluster NGC 6624," *The Astrophysical Journal Letters* 205 (1976), pp. L127–L130.

Guetta, D., and T. Piran, "The Luminosity and Redshift Distributions of Short-Duration GRBs," *Astronomy and Astrophysics* 435 (2005), pp. 421–426.

Haardt, F., and L. Maraschi, "A Two-Phase Model for the X-ray Emission from Seyfert Galaxies," *The Astrophysical Journal Letters* 380 (1991), pp. L51–L54.

Hachisu, I., M. Kato, and T. Kato, "Detection of Two-Armed Spiral Shocks on the Accretion Disk of the Eclipsing Fast Nova V1494 Aql," *The Astrophysical Journal Letters* 606 (2004), pp. L139–L142.

Haehnelt, M. G., and M. J. Rees, "The Formation of Nuclei in Newly Formed Galaxies and the Evolution of the Quasar Population, *Monthly Notices of the Royal Astronomical Society* 263 (1993), pp. 168–178.

Haislip, J. B., M. C. Nysewander, D. E. Reichart, et al., "A Photometric Redshift of $z = 6.39 \pm 0.12$ for GRB 050904," *Nature* 440 (2006), pp. 181–183.

Halpern, J. P., and F. Y. H. Wang, "A Broadband X-ray Study of the Geminga Pulsar," *The Astrophysical Journal* 477 (1997), pp. 905–915.

Hansen, C. J., and H. M. van Horn, "Steady-State Nuclear Fusion in Accreting Neutron-Star Envelopes," *The Astrophysical Journal* 195 (1975), pp. 735–741.

Harding, A. K., and D. Lai, "Physics of Strongly Magnetized Neutron Stars," *Reports on Progress in Physics* 69 (2006), pp. 2631–2708.

Hartman, R. C., D. L. Bertsch, S. D. Bloom, A. W. Chen, et al., "The Third EGRET Catalog of High-Energy Gamma-Ray Sources," *The Astrophysical Journal Supplements* 123 (1999), pp. 79–202.

Hasinger, G., T. Miyaji, and M. Schmidt, "Luminosity-Dependent Evolution of Soft X-ray Selected AGN," *Astronomy and Astrophysics* 441 (2005), pp. 417–434.

Hawking, S. W., "Black Hole Explosions?" *Nature* 248 (1974), pp. 30–31.

Hawley, J. F., "Global Magnetohydrodynamical Simulations of Accretion Tori," *The Astrophysical Journal* 528 (2000), pp. 462–479.

Hawley, J. F., and S. A. Balbus, "A Powerful Local Shear Instability in Weakly Magnetized Disks, II: Nonlinear Evolution," *The Astrophysical Journal* 376 (1991), pp. 223–233.

———, "The Dynamical Structure of Nonradiative Black Hole Accretion Flows," *The Astrophysical Journal* 573 (2002), pp. 738–748.

Hawley, J. F., C. F. Gammie, and S. A. Balbus, "Local Three-Dimensional Magnetohydrodynamic Simulations of Accretion Disks," *The Astrophysical Journal* 440 (1995), pp. 742–763.

Hay, H. J., J. P. Schiffer, T. E. Cranshaw, and P. A. Egelstaff, "Measurement in the Red Shift in an Accelerated System Using the Mössbauer Effect in Fe^{57}," *Physical Review Letters* 4 (1960), p. 165.

Heger, A., A. Cumming, D. K. Galloway, and S. E. Woosley, "Models of Type I X-ray Bursts from GS 1826-24: A Probe of rp-Process Hydrogen Burning," *The Astrophysical Journal Letters* 671 (2007), pp. L141–L144.

Heitler, W., *Quantum Theory of Radiation*, Oxford, UK: Oxford University Press, 1954.

Henke, B. L., E. M. Gullikson, and J. C. Davis, "X-ray Interactions: Photoabsorption, Scattering, Transmission, and Reflection at $E = 50$ eV to 30 keV," *Atomic Data and Nuclear Data Tables* 54 (1993), pp. 181–342.

Henriksen, R. N., A. H. Bridle, and K. L. Chan, "Synchrotron Brightness Distribution of Turbulent Radio Jets," *The Astrophysical Journal* 257 (1982), pp. 63–74.

Hess, V., "Observation of Penetrating Radiation in Seven Balloon Flights," *Physik Zeitschrift* 13 (1912), pp. 1084–1091.

Hewish, A., S. J. Bell, J. D. Pilkington, P. F. Scott, and R. A. Collins, "Observation of a Rapidly Pulsating Radio Source," *Nature* 217 (1968), pp. 709–713.

Hill, C. T., "Monopolonium," *Nuclear Physics B* 224 (1983), pp. 469–490.

Hillas, A. M., "The Origin of Ultra-High-Energy Cosmic Rays," *Annual Review of Astronomy and Astrophysics* 22 (1984), pp. 425–444.

Hinton, J., W. Domainko, and E.C.D. Pope, "Gamma Ray Emission Associated with Cluster-Scale AGN Outbursts," *Monthly Notices of the Royal Astronomical Society* 382 (2007), pp. 466–472.

Hjorth, J., D. Watson, J.P.U. Fynbo, et al., "The Optical Afterglow of the Short γ-Ray Burst GRB 050709," *Nature* 437 (2005), pp. 859–861.

Hoyle, R. A., J. Narlikar, and J. A. Wheeler, "Electromagnetic Waves from Very Dense Stars," *Nature* 203 (1964), pp. 914–916.

Hsu, J. L., J. Arons, and R. I. Klein, "Numerical Studies of Photon Bubble Instability in a Magnetized, Radiation-Dominated Atmosphere," *The Astrophysical Journal* 478 (1967), pp. 663–677.

Hunter, S.D., D. L. Bertsch, J. R. Catelli, T. M. Dame, et al., "EGRET Observations of the Diffuse Gamma-Ray Emission from the Galactic Plane," *The Astrophysical Journal* 481 (1997), pp. 205–240.

Hurley, K., R. Sari, and S. G. Djorgovski, "Cosmic Gamma Ray Bursts, Their Afterglows, and Their Host Galaxies," in *Compact Stellar X-ray Sources*, ed. W. Lewin and M. van der Klis Cambridge, UK: Cambridge University Press (2006), pp. 587–622.

Igumenshchev, I. V., X. Chen, and M. A. Abramowicz, "Accretion Discs Around Black Holes: Two-Dimensional, Advection-Cooled Flows," *Monthly Notices of the Royal Astronomical Society* 278 (1996), pp. 236–250.

Igumenshchev, I. V., and R. Narayan, "Three-Dimensional Magnetohydrodynamic Simulations of Spherical Accretion," *The Astrophysical Journal* 566 (2002), pp. 137–147.

Jaroszyński, M., M. A. Abramowicz, and B. Paczyński, "Supercritical Accretion Disks Around Black Holes," *Acta Astronomica* 30 (1980), pp. 1–34.

Jean, P., J. Knödlseder, W. Gillard, et al., "Spectral Analysis of the Galactic e^+e^- Annihilation Emission," *Astronomy and Astrophysics* 445 (2006), pp. 579–589.

Jenkins, C. J., and P.A.G. Scheuer, "What Docks the Tails of Radio Source Components?" *Monthly Notices of the Royal Astronomical Society* 174 (1976), pp. 327–333.

Johnson, W. N., F. R., Harnden, and R. C. Haymes, "The Spectrum of Low-Energy Gamma Radiation from the Galactic-Center Region," *The Astrophysical Journal Letters* 172 (1972), pp. L1–L7.

Jones, F. C., "A Theoretical Review of Diffusive Shock Acceleration," *The Astrophysical Journal Supplements* 90 (1994), pp. 561–565.

Joss, P. S., "X-ray Bursts and Neutron-Star Thermonuclear Flashes," *Nature* 270 (1977), pp. 310–314.

Joss, P. C., and S. A. Rappaport, "Neutron Stars in Interacting Binary Systems," *Annual Reviews of Astronomy and Astrophysics* 22 (1984), pp. 537–592.

Jungman, G., M. Kamionkowski, and K. Griest, "Supersymmetric Dark Matter," *Physics Reports* 267 (1996), pp. 195–373.

Kalogera, V., and G. Baym, "The Maximum Mass of a Neutron Star," *The Astrophysical Journal Letters* 470 (1996), pp. L61–L64.

Kaper, L., E.P.J. van den Heuvel, and P. A. Woudt, *Black Holes in Binaries and Galactic Nuclei,* New York: Springer (2001).

Kaspi, V. M., "Recent Progress on Anomalous X-ray Pulsars," *Astrophysics and Space Science* 308 (2007), pp. 1–11.

Kato, S., "Trapped One-Armed Corrugation Waves and QPOs," *Publications of the Astronomical Society of Japan* 42 (1990), pp. 99–113.

Kellermann, K. I., M. L. Lister, D. C. Homan, et al., "Superluminal Motion and Relativistic Beaming in Blazar Jets," in *High Energy Blazar Astronomy, ASP Conference Proceedings, Vol. 299,* eds. L. O. Takalo and E. Valtaoja, San Francisco: Astronomical Society of the Pacific (2003), pp. 117–124.

Kellermann, K. I., R. Sramek, M. Schmidt, et al., "VLA Observations of Objects in the Palomar Bright Quasar Survey," *The Astronomical Journal* 98 (1989), pp. 1195–1207.

Kempner, J. C., and C. L. Sarazin, "Radio Halo and Relic Candidates from the Westerbork Northern Sky Survey," *The Astrophysical Journal* 548 (2001), pp. 639–651.

Kent, S. M., "Sloan Digital Sky Survey," *Astrophysics and Space Science* 217 (1994), pp. 27–30.

Kerr, R. P., "Gravitational Field of a Spinning Mass as an Example of Algebraically Special Metrics," *Physical Review Letters* 11 (1963), pp. 237–238.

Kim, K.-T., P. P. Kronberg, P. E. Dewdney, et al., "The Halo and Magnetic Field of the Coma Cluster of Galaxies," *The Astrophysical Journal* 355 (1990), pp. 29–37.

Kitamoto, S., H. Tsunemi, S. Miyamoto, and K. Hayashida, "Discovery and X-ray Properties of GS 1124–683 (a.k.a. Nova Muscae)," *The Astrophysical Journal* 394 (1992), pp. 609–614.

Klebesadel, R., I. Strong, and R. Olson, "Observations of Gamma Ray Bursts of Cosmic Origin," *The Astrophysical Journal Letters* 182 (1973), pp. L85–L88.

Klein, O., and Y. Nishina, "Über die Streuung von Strahlung Durch Freie Elektronen Nach Der Neuen Relativistischen Quantendynamik von Dirac," *Zeitschrift für Physik* 52 (1929), pp. 853–868.

Klotz, A., M. Boër, and J. L. Atteia, "Observational Constraints on the Afterglow of GRB 020531," *Astronomy and Astrophysics* 404 (2003), pp. 815–818.

Kniffen, D. A., D. L. Bertsch, C. E. Fichtel, et al., "Time Variability in the Gamma ray Emission of 3C 279," *The Astrophysical Journal* 411 (1993), pp. 133–136.

Kniffen, D. A., R. C. Hartman, D. J. Thompson, et al., "Gamma Radiation from the Crab Nebula Above 35 MeV," *Nature* 251 (1974), pp. 397–399.

Kosack, K., H. M. Badran, I. H. Bond, P. J. Boyle, S. M. Bradbury, J. H. Buckley, D. A. Carter-Lewis, et al., "TeV Gamma-Ray Observations of the Galactic Center," *The Astrophysical Journal Letters* 608 (2004), pp. L97–L100.

Kouveliotou, C., C. A. Meegan, G. J. Fishman, et al., "Identification of Two Classes of Gamma Ray Bursts," *The Astrophysical Journal Letters* 413 (1993), pp. L101–L104.

Kowalenko, V., and F. Melia, "Towards Incorporating a Turbulent Magnetic Field in an Accreting Black Hole Model," *Monthly Notices of the Royal Astronomical Society* 310 (1999), pp. 1053–1061.

Koyama, K., H. Awaki, H. Kunieda, et al., "Intense 6.7-keV Iron Line Emission from the Galactic Centre," *Nature* 339 (1989), pp. 603–605.

Kraft, R. P., J. Mathews, and J. L. Greenstein, "Binary Stars Among Cataclysmic Variables, II: Nova WZ Sagittae: A Possible Radiator of Gravitational Waves," *The Astrophysical Journal* 136 (1962), pp. 312–314.

Krolik, J., and J. Hawley, "Where Is the Inner Edge of an Accretion Disk Around a Black Hole?" *The Astrophysical Journal* 573 (2002), pp. 754–763.

Kronberg, P. P., "Extragalactic Magnetic Fields," *Reports on Progress in Physics* 57 (1994), pp. 324–382.

Kurfess, J. D., "Key Results from the Oriented Scintillation Spectrometer Experiment," *Astronomy and Astrophysics* 120 (1996), pp. 5–12.

Lamb, D. Q., and F. K. Lamb, "Nuclear Burning in Accreting Neutron Stars and X-ray Bursts," *The Astrophysical Journal* 220 (1978), pp. 291–302.

Lamb, D. Q., and F. Melia, "Synchronization-Induced Period Gaps and Ultra-short Periods in Magnetic Cataclysmic Binaries," *The Astrophysical Journal Letters* 321 (1987), pp. L133–L137.

Lamb, D. Q., and D. E. Reichart, "Gamma Ray Bursts as a Probe of the Very High Redshift Universe," *The Astrophysical Journal* 536 (2000), pp. 1–18.

Landau, L., and E. M. Lifshitz, *Quantum Electrodynamics,* Oxford, UK: Butterworth-Heinemann (1982).

———, *Fluid Mechanics*, Oxford, UK: Butterworth-Heinemann (1987).

Laor, A., "Line Profiles from a Disk Around a Rotating Black Hole," *The Astrophysical Journal* 376 (1991), pp. 90–94.

Large, M. I., A. E. Voughan, and B. Y. Mills, "A Pulsar Supernova Association?" *Nature* 220 (1968), pp. 340–341.

LaRosa, T. N., N. E. Kassim, and T.J.W. Lazio, "A Wide-Field 90 Centimeter VLA Image of the Galactic Center Region," *The Astronomical Journal* 119 (2000), pp. 207–240.

Lasota, J.-P., J.-M. Hameury, and J.-M. Huré, "Dwarf Novae at Low Mass Transfer Rates," *Astronomy and Astrophysics* 302 (1995), pp. L29–L32.

Lehmann, I., G., Hasinger, M. Schmidt, et al., "The ROSAT Deep Survey, VI: X-ray sources and Optical Identifications of the Ultra Deep Survey," *Astronomy and Astrophysics* 371 (2001), pp. 833–857.

Leventhal, M., C. J. MacCallum, and P. D. Stang, "Detection of 511 keV Positron Annihilation Radiation from the Galactic Center Direction," *The Astrophysical Journal Letters* 225 (1978), pp. L11–L14.

Lewin, W.H.G., and P. C. Joss, "X-ray Bursters and the X-ray Sources of the Galactic Bulge," *Space Science Reviews* 28 (1981), pp. 3–87.

Lewin, W.H.G., J. van Paradijs, and R. E. Taam, "X-Ray Bursts," *Space Science Reviews* 62 (1993), pp. 223–389.

Lewin, W.H.G., J. van Paradijs, and E.P.J. van den Heuvel, *X-ray Binaries,* Cambridge, UK: Cambridge University Press (1997).

Liang, E. P., "Multiwavelength Signatures of Galactic Black Holes: Observations and Theory," *Physics Reports* 302 (1998), pp. 67–142.

Liang, E. P., and D. P. Dermer, "Interpretation of the Gamma-Ray Bump from Cygnus X-1," *The Astrophysical Journal Letters* 325 (1988), pp. L39–L42.

Liebert, J., and H. S. Stockman, "The AM Herculis Magnetic Variables," in *Cataclysmic Variables and Low-Mass X-Ray Sources*, ed. D. Q. Lamb and J. Patterson, Dordrecht: Reidel (1985), pp. 151–177.

Lifshitz, E., *Statistical Physics,* Oxford, UK: Pergamon (1980).

Lightman, A. P., and D. M. Eardley, "Black Holes in Binary Systems: Instability of Disk Accretion," *The Astrophysical Journal* 187 (1974), pp. L1–L3.

Liu, S., M. Fromerth, and F. Melia, "Line Emission from Cooling Emission Flows in the Nucleus of M31," *The Astrophysical Journal* 565 (2002), pp. 952–958.

Loeb, A., and E. Waxman, "Cosmic γ-Ray Background from Structure Formation in the Intergalactic Medium," *Nature* 405 (2000), pp. 156–158.

Longair, M. S., M. Ryle, and P.A.G. Scheuer, "Models of Extended Radio Sources," *Monthly Notices of the Royal Astronomical Society* 164 (1973), pp. 243–270.

Lorimer, D. R., "Binary and Millisecond Pulsars," *Living Reviews in Relativity* 8 (2005), pp. 7–87.

Lutovinov, A. A., S. A., Grebenev, M. N. Pavlinsky, and R. A. Sunyaev, "X-ray Bursts from the Source A1742–294 in the Galactic-Center Region," *Astronomy Letters* 27 (2001), pp. 501–506.

Lynden-Bell, D., "Galactic Nuclei as Collapsed Old Quasars," *Nature* 223 (1969), pp. 690–694.

Malzac, J. et al., "Bimodal Spectral Variability of Cygnus X-1 in an Intermediate State," *Astronomy and Astrophysics*, 448 (2006), pp. 1125–1137.

Manchester, R. N., G. B. Hobbs, A. Teoh, et al., "The Australia Telescope National Facility Pulsar Catalogue," *The Astronomical Journal* 129 (2005), pp. 1993–2006.

Markoff, S., M. A. Nowak, and J. Wilms, "Going with the Flow: Can the Base of Jets Subsume the Role of Compact Accretion Disk Coronae?" *The Astrophysical Journal* 635 (2005), pp. 1203–1216.

Markowitz, A., and R. Edelson, "An RXTE Survey of Long-Term X-ray Variability in Seyfert 1 Galaxies," *The Astrophysical Journal* 547 (2001), pp. 684–692.

Mazets, E. P., S. V. Golenetskii, and V. N. Ilinskii, "A Burst of Cosmic Gamma-Ray Flare Observed with Kosmos-461," *Journal of Experimental and Theoretical Physics Letters* 19 (1974), pp. 77–81.

McClintock, J. E., K. Horne, and R. A. Remillard, "The DIM Inner Accretion Disk of the Quiescent Black Hole A0620-00," *The Astrophysical Journal* 442 (1995), pp. 358–365.

McClintock, J. E., and R. A. Remillard, "Black Hole Binaries," in *Compact Stellar X-ray Sources*, ed. W.H.G. Lewin and M. van der Klis, Cambridge, UK: Cambridge University Press (2006), pp. 157–214.

McNamara, B. R., P.E.J. Nulsen, M. W. Wise, et al., "The Heating of Gas in a Galaxy Cluster by X-ray Cavities and Large-Scale Shock Fronts," *Nature* 433 (2005), pp. 45–47.

Meegan, C., G. Fishman, R. Wilson, et al., "Spatial Distribution of Gamma Ray Bursts Observed by BATSE," *Nature* 355 (1992), pp. 143–145.

Melia, F., "An Accreting Black Hole Model for Sagittarius A*," *The Astrophysical Journal Letters* 387 (1992a), pp. L25–L28.

———, "The Nucleus of M31," *The Astrophysical Journal Letters* 398 (1992b), pp. L95–L98.

———, *Electrodynamics,* Chicago: University of Chicago Press (2001a).

———, "X-rays from the Edge of Infinity," *Nature* 413 (2001b), pp. 25–26.

———, *The Edge of Infinity,* Cambridge, UK: Cambridge University Press (2003).

———, *The Galactic Supermassive Black Hole,* Princeton, NJ: Princeton University Press (2007).

———, *Cracking the Einstein Code,* Chicago: University of Chicago Press (2009).

Melia, F., B. C. Bromley, S. Liu, and C. K. Walker, "Measuring the Black Hole Spin in Sagittarius A*," *The Astrophysical Journal Letters* 554 (2001), pp. L37–L40.

Melia, F., and A. Königl, "The Radiative Deceleration of Ultrarelativistic Jets in Active Galactic Nuclei," *The Astrophysical Journal* 340 (1989), pp. 162–180.

Melia, F., and V. Kowalenko, "Magnetic Field Dissipation in Converging Flows," *Monthly Notices of the Royal Astronomical Society* 327 (2001), pp. 1279–1287.

Menou, K., A. A. Esin, R. Narayan et al., "Black Hole and Neutron Star Transients in Quiescence," *The Astrophysical Journal* 520 (1999), pp. 276–291.

Mereghetti, S., "The Zoo of X-ray Pulsars," in *Frontier Objects in Astrophysics and Particle Physics*, ed. F. Giovannelli and G. Mannocchi, Bologna: Italian Physical Society (2001), pp. 239–249.

Merloni, A., and A. Fabian, "Coronal Outflow Dominated Accretion Discs: A New Possibility for Low-Luminosity Black Holes?" *Monthly Notices of the Royal Astronomical Society* 332 (2002), pp. 165–175.

Mészáros, P., "Theories of Gamma Ray Bursts," *Annual Review of Astronomy and Astrophysics* 40 (2002), pp. 137–169.

Mészáros, P., and M. J. Rees, "Optical and Long-Wavelength Afterglow from Gamma Ray Bursts," *The Astrophysical Journal* 476 (1997), pp. 232–237.

———, "Poynting Jets from Black Holes and Cosmological Gamma Ray Bursts," *The Astrophysical Journal Letters* 482 (1997), pp. L29–L32.

———, "Collapsar Jets, Bubbles, and Fe Lines," *The Astrophysical Journal Letters* 556 (2001), pp. L37–L40.

Meyer, F., and E. Meyer-Hofmeister, "On the Elusive Cause of Cataclysmic Variable Outbursts," *Astronomy and Astrophysics* 104 (1981), pp. L10–L12.

Mihalas, D., and B. W. Mihalas, *Foundations of Radiation Hydrodynamics*, New York: Dover (1999).

Milne, P. A., "Distribution of Positron Annihilation Radiation," *New Astronomy Reviews* 50 (2006), pp. 548–552.

Mineshige, S., and J. C. Wheeler, "Disk-Instability Model for Soft-X-ray Transients Containing Black Holes," *The Astrophysical Journal* 343 (1989), pp. 241–253.

Mirabel, I. F., "Microquasars: Open Questions and Future Perspectives," *Astrophysics and Space Science Supplement* 276 (2001), pp. 319–327.

Mirabel, I. F., L. F. Rodriguez, B. Cordier, et al., "A Double-Sided Radio Jet from the Compact Galactic Centre Annihilator 1E 140.7–2942," *Nature* 358 (1992), pp. 215–217.

Misra, R., and F. Melia, "Formation of a Jet in the Galactic Center Black Hole Candidate 1E 1740.7–2942," *The Astrophysical Journal Letters* 419 (1993), pp. L25–L29.

———, "Hot Accretion Disks with Electron–Positron Pair Winds," *The Astrophysical Journal* 449 (1995), pp. 813–825.

Miyoshi, M., J. Moran, J. Hernstein, et al., "Evidence for a Black Hole from High Rotation Velocities in a Sub-Parsec Region of NGC 4258," *Nature* 373 (1995), pp. 127–129.

Moffet, A. T., J. Gubbay, D. S. Robertson, and A. J. Legg, "High Resolution Observations of Variable Radio Sources," *IAU Symposium 44, External Galaxies and Quasi-Stellar Objects,* ed. D. S. Evans, Dordrecht: Reidel (1972), pp. 228–229.

Mollerach, S., et al., "Studies of Clustering in the Arrival Directions of Cosmic Rays Detected at the Pierre Auger Observatory Above 10 EeV," *Contribution to the 30th International Cosmic-ray Conference, Merida, Mexico* (2007), eprint arXiv:0706.1749.

Moretti, A., S. Campana, D. Lazzati, and G. Tagliaferri, "The Resolved Fraction of the Cosmic X-Ray Background," *The Astrophysical Journal* 588 (2003), pp. 696–703.

Muno, M. P., F. K. Baganoff, M. W. Bautz, et al., "A Deep Chandra Catalog of X-ray Point Sources Toward the Galactic Center," *The Astrophysical Journal* 589 (2003), pp. 225–241.

Muno, M. P., R. A. Remillard, E. H. Morgan, et al., "Radio Emission and the Timing Properties of the Hard X-ray State of GRS 1915+105," *The Astrophysical Journal* 556 (2001), pp. 515–532.

Mushotzky, R. F., L. L. Cowie, A. J. Barger, and K. A. Arnaud, "Resolving the Extragalactic Hard X-ray Background," *Nature* 404 (2000), pp. 459–464.

Nakar, E., and T. Piran, "Time-scales in Long Gamma Ray Bursts," *Monthly Notices of the Royal Astronomical Society,* 331 (2002), pp. 40–44.

Nandra, K., and K. A. Pounds, "GINGA Observations of the X-Ray Spectra of Seyfert Galaxies," *Monthly Notices of the Royal Astronomical Society* 268 (1994), pp. 405–429.

Nayakshin, S., and F. Melia, "Magnetic Flares and the Observed $\tau_T \approx 1$ in Seyfert Galaxies," *The Astrophysical Journal Letters* 490 (1997), pp. L13–L16.

Neumann, D. M., M. Arnaud, and R. Gastaud, "The NGC 4839 Group Falling into the Coma Cluster Observed by XMM-*Newton*," *Astronomy and Astrophysics* 365 (2001), pp. L74–L79.

Nicastro, F., S. Mathur, E. Martin, et al., "The Mass of the Missing Baryons in the X-ray Forest of the Warm-Hot Intergalactic Medium," *Nature* 433 (2005), pp. 495–498.

Norman, C. A., D. B. Melrose, and A. Achterberg, "The Origin of Cosmic Rays above $10^{18.5}$ eV," *The Astrophysical Journal* 454 (1995), pp. 60–68.

Norton, A. J., G. A. Wynn, and R. V. Somerscales, "The Spin Periods and Magnetic Moments of White Dwarfs in Magnetic Cataclysmic Variables," *The Astrophysical Journal* 614 (2004), pp. 349–357.

Nulsen, P.E.J., D. C. Hambrick, B. R. McNamara, et al., "The Powerful Outburst in Hercules A," *The Astrophysical Journal Letters* 625 (2005), pp. L9–L12.

Ogle, P. M., T. Brookings, C. R. Canizares, et al., "Testing the Seyfert Unification Theory: Chandra HETGS Observations of NGC 1068," *Astronomy and Astrophysics* 402 (2003), pp. 849–864.

Orosz, J. A., "Inventory of Black Hole Binaries," in *Proceedings of IAU Symposium No. 212,* ed. K. van der Hucht, A. Herrero, and C. Esteban, San Francisco: Astronomical Society of the pacific (2003), pp. 365–371.

Pacholczyk, A. G., *Radio Astrophysics,* New York: Freeman (1970).

Pacini, F., "Energy Emission from a Neutron Star," *Nature* 216 (1967), pp. 567–568.

Paczyński, B., "Evolutionary Processes in Close Binary Systems," *Annual Reviews of Astronomy and Astrophysics* 9 (1971), pp. 183–208.

————, "Gamma-Ray Bursters at Cosmological Distances," *The Astrophysical Journal Letters* 308 (1986), pp. L43–L46.

————, "Advection Dominated Accretion Flows: A Toy Disk Model," *Acta Astronomica* 48 (1998), pp. 667–676.

Paczyński, B., and P. J. Wiita, "Thick Accretion Disks and Supercritical Luminosities," *Astronomy and Astrophysics* 88 (1980), pp. 23–31.

Page, D. N., and K. S. Thorne, "Disk-Accretion onto a Black Hole. Time-Averaged Structure of Accretion Disk," *The Astrophysical Journal* 191 (1974), pp. 499–506.

Papaloizou, J.C.B., and D.N.C. Lin, "Theory of Accretion Disks, I: Angular Momentum Transport Processes," *Annual Review of Astronomy and Astrophysics* 33 (1995), pp. 505–540.

————, "Theory of Accretion Disks, II: Application to Observed Systems," *Annual Review of Astronomy and Astrophysics* 34 (1996), pp. 703–747.

Parker, E. N., "The Hydrodynamic Theory of Solar Corpuscular Radiation and Stellar Winds," *The Astrophysical Journal* 132 (1960), pp. 821–866.

———, *Cosmical Magnetic Fields: Their Origin and Their Activity,* Oxford, UK: Clarendon Press (1979).

Patterson, J., "The Evolution of Cataclysmic and Low-Mass X-ray Binaries," *The Astrophysical Journal Supplements* 54 (1984) pp. 443–493.

Pearson, T. J., and A.C.S. Readhead, "The Milliarcsecond Structure of a Complete Sample of Radio Sources, II: First-Epoch Maps at 5 GHz," *The Astrophysical Journal* 328 (1988), pp. 114–142.

Pedlar, A., K. R. Anantharamaiah, R. D. Ekers, W. M. Goss, J. H. van Gorkom, U. J. Schwarz, and J.-H. Zhao, "Radio Studies of the Galactic Center, I: The Sagittarius A Complex," *The Astrophysical Journal* 342 (1989), pp. 769–784.

Peterson, J. R., and A. C. Fabian, "X-ray Spectroscopy of Cooling Clusters," *Physics Reports* 427 (2006), pp. 1–39.

Petschek, H. E., "Magnetic Field Annihilation," *The Physics of Solar Flares, Proceedings of the AAS-NASA Symposium* (1964), p. 425.

Phinney, E. S., and S. R. Kulkarni, "Binary and Millisecond Pulsars," *Annual Review of Astronomy and Astrophysics* 32 (1994), pp. 591–639.

Piran, T., "The Physics of Gamma Ray Bursts," *Reviews of Modern Physics* 76 (2005), pp. 1143–1210.

Preece, R. D., M. S. Briggs, R. S. Mallozzi, et al., "The BATSE Gamma-Ray Burst Spectral Catalog, I: High Time Resolution Spectroscopy of Bright Bursts Using High Energy Resolution Data," *The Astrophysical Journal Supplements* 126 (2000), pp. 19–36.

Pringle, J. E., "Thermal Instabilities in Accretion Discs," *Monthly Notices of the Royal Astronomical Society* 177 (1976), pp. 65–71.

Qin, Y.-P., G.-Z. Xie, S.-J. Xue, et al., "The Hardness–Duration Correlation in the Two Classes of Gamma-Ray Bursts," *Publications of the Royal Astronomical Society of Japan* 52 (2000), pp. 759–761.

Rappaport, S. A., F. Verbunt, and P. C. Joss, " A New Technique for Calculations of Binary Stellar Evolution, with Application to Magnetic Braking," *The Astrophysical Journal* 275 (1983), pp. 713–731.

Rees, M. J., "Appearance of Relativistically Expanding Radio Sources," *Nature* 211 (1966), pp. 468–470.

———, "Studies in Radio Source Structure," *Monthly Notices of the Royal Astronomical Society* 135 (1967), pp. 345–360.

———, "New Interpretation of Extragalactic Radio Sources," *Nature* 229 (1971), pp. 312–317.

Rees, M. J., and P. Mészáros, "Relativistic Fireballs—Energy Conversion and Time-scales," *Monthly Notices of the Royal Astronomical Society* 258 (1992), pp. P41–P43.

———, "Unsteady Outflow Models for Cosmological Gamma Ray Bursts," *The Astrophysical Journal Letters* 430 (1994), pp. L93–L96.

Remillard, R. A., and J. E. McClintock, "X-ray Properties of Black-Hole Binaries," *Annual Reviews of Astronomy and Astrophysics* 44 (2006), pp. 49–92.

Rephaeli, Y., "Relativistic Electrons in the Intracluster Space of Clusters of Galaxies—The Hard X-ray Spectra and Heating of the Gas," *The Astrophysical Journal* 227 (1979), pp. 364–369.

Revnivtsev, M., S. Sazonov, M. Gilfanov, et al., "Origin of the Galactic Ridge X-ray Emission," *Astronomy and Astrophysics* 452 (2006), pp. 169–178.

Reynolds, C. S., and M. A. Nowak, "Fluorescent Iron Lines as a Probe of Astrophysical Black Hole Systems," *Physics Reports* 377 (2003), pp. 389–466.

Reynolds, C. S., M. J. Ward, A. C. Fabian, and A. Celotti, "A Multiwavelength Study of the Seyfert 1 Galaxy MCG-6-30-15," *Monthly Notices of the Royal Astronomical Society* 291 (1997), pp. 403–417.

Rhoads, J. E., "The Dynamics and Light Curves of Beamed Gamma Ray Burst Afterglows," *The Astrophysical Journal* 525 (1999), pp. 737–749.

Rhoades, C. E., and R. Ruffini, "Maximum Mass of a Neutron Star," *Physical Review Letters* 32 (1974), pp. 324–327.

Ritter, H., "Catalogue of Cataclysmic Binaries, Low-Mass X-ray Binaries, and Related Objects," *Astronomy and Astrophysics Supplements* 57 (1984), pp. 385–418.

Robinson, E. L., E. S. Barker, A. L. Cochran, et al., "MV Lyrae—Spectrophotometric Properties of Minimum Light," *The Astrophysical Journal* 251 (1981), pp. 611–619.

Rockefeller, G., C. L. Fryer, F. Melia, and M. S. Warren, "Diffuse X-rays from the Inner 3 Parsecs of the Galaxy," *The Astrophysical Journal* 604 (2004), pp. 662–670.

Romani, R. W., "A Census of Low Mass Black Hole Binaries," *Astronomy and Astrophysics* 333 (1998), pp. 583–590.

Rosati, P., S. Borgani, and C. Norman, "The Evolution of X-ray Clusters of Galaxies," *Annual Review of Astronomy and Astrophysics* 40 (2002), pp. 539–577.

Ruggles, C. L. N., and N. J. Saunders, *Astronomies and Cultures,* Boulder, co: University Press of Colorado (1993).

Rybicki, G. B., and A. P. Lightman, *Radiative Processes in Astrophysics,* Wiley-Interscience (1985).

Salpeter, E. E., "Accretion of Interstellar Matter by Massive Objects," *The Astrophysical Journal* 140 (1964), pp. 796–800.

Sari, R., T. Piran, and R. Narayan, "Spectra and Light Curves of Gamma Ray Burst Afterglows," *The Astrophysical Journal Letters* 497 (1998), pp. L17–L20.

Sarma, A. P., T. H. Troland, R. M. Crutcher, et al., "Magnetic Fields in Shocked Regions: Very Large Array Observations of H_2O Masers," *The Astrophysical Journal* 580 (2002), pp. 928–937.

Scheuer, P.A.G., "Models of Extragalactic Radio Sources with a Continuous Energy Supply from a Central Object," *Monthly Notices of the Royal Astronomical Society* 166 (1974), pp. 513–528.

Schmidt, M., "3C 273: A Star-like Object with Large Red-shift," *Nature* 197 (1963), p. 1040.

Schmidt, M., J. C. Higdon, and G. Hueter, "Application of the V/V(max) Test to Gamma Ray Bursts," *The Astrophysical Journal Letters* 329 (1988), pp. L85–L87.

Schoenfelder, V., et al., "Instrument Description and Performance of the Imaging Gamma-Ray Telescope COMPTEL Aboard the Compton Gamma-Ray Observatory," *The Astrophysical Journal Supplements* 86 (1993), pp. 657–692.

Schwarzschild, M., *Structure and Evolution of the Stars,* Princeton, NJ: Princeton University Press (1958).

Seyfert, C. K., "Nuclear Emission in Spiral Nebulae," *The Astrophysical Journal* 97 (1943), pp. 28–40.

Shakura, N. I., and R. A. Sunyaev, "Black Holes in Binary Systems. Observational Appearance," *Astronomy and Astrophysics* 24 (1973), pp. 337–355.

Shapiro, S. L., A. P. Lightman, and D. M. Eardley, "A Two-Temperature Accretion Disk Model for Cygnus X-1—Structure and Spectrum," *The Astrophysical Journal* 204 (1976), pp. 187–199.

Shapiro, S. L., and S. A. Teukolsky, *Black Holes, White Dwarfs and Neutron Stars: The Physics of Compact Objects,* New York: Wiley-Interscience (1983).

Shklovsky, I. S., "On the Nature of the Crab Nebula's Optical Emission," in *A Source Book in Astronomy & Astrophysics, 1900–1975,* ed. K. R. Lang and O. Gingerich, Cambridge, MA Harvard University Press (1979), pp. 488–493.

Shu, P. H., and S. H. Lubow, "Mass, Angular Momentum, and Energy Transfer in Close Binary Systems," *Annual Review of Astronomy and Astrophysics* 19 (1981), pp. 277–293.

Shvartsman, V. F., "Halos Around "Black Holes," *Soviet Astronomy* 15 (1971), p. 37.

Siegmund, O.H.W., and K. A. Flanagan, *EUV, X-Ray, and Gamma-Ray Instrumentation for Astronomy X: Proceedings of SPIE 21–23 July 1999, Denver, Colorado,* Bellingham, WA: SPIE–International Society for Optical Engineers (1999).

Silk, J., and M. J. Rees, "Quasars and Galaxy Formation," *Astronomy and Astrophysics* 331 (1998), pp. L1–L4.

Soltan, A., "Masses of Quasars," *Monthly Notices of the Royal Astronomical Society* 200 (1982), pp. 115–122.

Sreekumar, P., D. L. Bertsch, B. L. Dingus, et al., "Constraints on the Cosmic Rays in the Small Magellanic Cloud," *Physical Review Letters* 70 (1993), pp. 127–129.

Staelin, D. H., and E. C. Reifenstein, "Pulsating Radio Sources Near the Crab Nebula," *Science* 162 (1968), pp. 1481–1483.

Stanek, K. Z., T. Matheson, P. M. Garnavich, et al., "Spectroscopic Discovery of the Supernova 2003dh Associated with GRB 030329," *The Astrophysical Journal Letters* 591 (2003), pp. L17–L20.

Stella, L, Vietri, M., and S. M. Morsink, "Correlations in the Quasi-periodic Oscillation Frequencies of Low-Mass X-ray Binaries and the Relativistic Precession Model," *The Astrophysical Journal Letters* 524 (1999), pp. L63–L66.

Stirling, A. M., R. E. Spencer, C. J. de la Force, et al., "A Relativistic Jet from Cygnus X-1 in the Low/Hard X-ray State," *Monthly Notices of the Royal Astronomical Society* 327 (2001), pp. 1273–1278.

Stocke, J. T., S. L. Morris, I. M. Gioia, et al., "The Einstein Observatory Extended Medium-Sensitivity Survey, II: The Optical Identifications," *The Astrophysical Journal Supplements* 76 (1991), pp. 813–874.

Stone, J. M., Hawley, J. F., Gammie, C. F., and Balbus, S. A., "Three-Dimensional Magnetohydrodynamical Simulations of Vertically Stratified Accretion Disks," *The Astrophysical Journal* 463 (1996), pp. 656–673.

Stone, J. M., and J. E. Pringle, "Magnetohydrodynamical Non-radiative Accretion Flows in Two Dimensions," *Monthly Notices of the Royal Astronomical Society* 322 (2001), pp. 461–472.

Strohmayer, T. E., J. H. Swank, and W. Zhang, "The Periods Discovered by RXTE in Thermonuclear Flash Bursts," *Nuclear Physics B* 69 (1998), pp. 129–134.

Strong, A. W., I. V. Moskalenko, and O. Reimer, "Diffuse Galactic Continuum Gamma Rays: A Model Compatible with EGRET Data and Cosmic Ray Measurements," *The Astrophysical Journal* 613 (2004a), pp. 962–976.

———, "A New Determination of the Extragalactic Diffuse Gamma Ray Background From EGRET Data," *The Astrophysical Journal* 613 (2004b), pp. 956–961.

Stubbs, P., "Red Shift Without Reason," *New Scientist* 50 (1971), pp. 254–255.

Sunyaev, R., E. Churazov, M. Gilfanov, et al., "Three Spectral States of 1E 1740.7-2942—From Standard Cygnus X-1 Type Spectrum to the Evidence of Electron–Positron Annihilation Feature," *The Astrophysical Journal* 383 (1991), pp. L49–L52.

Sunyaev, R., and Ya. B. Zel'dovich, "Small-Scale Fluctuations of Relic Radiation," *Astrophysics and Space Science* 7 (1970), pp. 3–19.

Swordy, S. P., D. Mueller, P. Meyer, et al., "Relative Abundances of Secondary and Primary Cosmic Rays at High Energies," *The Astrophysical Journal* 349 (1990), pp. 625–633.

Taam, R., "X-ray Bursts from Thermonuclear Runaways on Accreting Neutron Stars," *The Astrophysical Journal* 241 (1980), pp. 358–366.

———, "Angular Momentum Loss and the Evolution of Cataclysmic Binaries," *The Astrophysical Journal* 268 (1983), pp. 361–367.

Tagger, M., and F. Melia, "A Possible Rossby Wave Instability Origin for the Flares in Sagittarius A*," *The Astrophysical Journal Letters* 636 (2006), pp. L33–L36.

Tagger, M., and R. Pellat, "An Accretion–Ejection Instability in Magnetized Disks," *Astronomy and Astrophysics* 349 (1999), pp. 1003–1016.

Tanaka, Y., and N. Shibazaki, "X-ray Novae," *Annual Review of Astronomy and Astrophysics* 34 (1996), pp. 607–644.

Tananbaum, H., H. Gursky, E. M. Kellogg, R. Giacconi, and C. Jones, "Observation of a Correlated X-ray Transition in Cygnus X-1," *The Astrophysical Journal Letters* 177 (1972a), pp. L5–L10.

Tananbaum, H., H. Gursky, E. M. Kellogg, R. Levinson, et al., "Discovery of a Periodic Pulsating Binary X-Ray Source in Hercules from UHURU," *The Astrophysical Journal Letters* 174 (1972b), pp. L143–L149.

Taylor, J. H., R. N. Manchester, and A. G. Lyne, "Catalog of 558 Pulsars," *The Astrophysical Journal Supplements* 88 (1993), pp. 529–568.

Taylor, J. H., and D. R. Stinebring, "Recent Progress in Understanding Pulsars," *Annual Reviews of Astronomy and Astrophysics* 24 (1986), pp. 285–327.

Thomas, H. C., "Consequences of Mass Transfer in Close Binary Systems," *Annual Review of Astronomy and Astrophysics* 15 (1977), pp. 127–151.

Thompson, C., and R. C. Duncan, "The Soft Gamma Repeaters as Very Strongly Magnetized Neutron Stars, I: Radiative Mechanism for Outbursts," *Monthly Notices of the Royal Astronomical Society* 275 (1995), pp. 255–300.

Thompson, D. J., Fichtel, C. E., Kniffen, D. A. et al., "SAS-2 High-Energy Gamma-Ray Observations of the VELA Pulsar," *The Astrophysical Journal Letters* 200 (1975), pp. L79–L82.

Tremaine, S., "An Eccentric-Disk Model for the Nucleus of M31," *The Astronomical Journal* 110 (1995), pp. 628–633.

Trinchieri, G., and G. Fabbiano, "The Discrete X-ray Source Population in M31," *The Astrophysical Journal* 382 (1991), pp. 82–99.

Tsuchiya, K., R. Enomoto, L. T. Ksenofontov, M. Mori, T. Naito, A. Asahara, G. V. Bicknell, et al., "Detection of Sub-TeV Gamma Rays from the Galactic Center Direction by CANGAROO-II," *The Astrophysical Journal Letters* 606 (2004), pp. L115–L118.

Ulrich, M.-H., L. Maraschi, and C. M. Urry, "Variability of Active Galactic Nuclei," *Annual Review of Astronomy and Astrophysics* 35 (1997), pp. 445–502.

van der Klis, M., "Millisecond Oscillations in X-ray Binaries," *Annual Review of Astronomy and Astrophysics* 38 (2000), pp. 717–760.

Van Hoven, G., "Solar Flares and Plasma Instabilities—Observations, Mechanisms, and Experiments," *Solar Physics* 49 (1976), pp. 95–116.

van Paradijs, J., "Catalogue of X-ray Binaries," in *X-ray Binaries*, ed. W.H.G. Lewin, J. van Paradijs, and E.P.J. van den Heuvel, Cambridge, UK: Cambridge University Press (1995), pp. 536–577.

van Paradijs, J., P. J. Groot, T. Galama et al., "Transient Optical Emission from the Error Box of the Gamma Ray Burst of 28 February 1997." *Nature* 386 (1997), pp. 686–689.

Verbunt, F., and C. Zwaan, "Magnetic Braking in Low-Mass X-ray Binaries," *Astronomy and Astrophysics* 100 (1981), pp. L7–L9.

Villasenor, J. S., D. Q. Lamb, G. R. Ricker, et al., "Discovery of the Short γ-ray Burst GRB 050709," *Nature* 437 (2005), pp. 855–858.

Voges, W., B. Aschenbach, and T. Boller, "The ROSAT All-Sky Survey Bright Source Catalogue," *Astronomy and Astrophysics* 349 (1999), pp. 389–405.

Völk, H. J., F. A. Aharonian, and D. Breitschwerdt, "The Nonthermal Energy Content and Gamma Ray Emission of Starburst Galaxies and Clusters of Galaxies," *Space Science Reviews* 75 (1996), pp. 279–297.

Wald, R. M., *General Relativity,* Chicago: University of Chicago Press (1984).

Wang, Q. D., E. V. Gotthelf, and C. C. Lang, "A Faint Discrete Source Origin for the Highly Ionized Iron Emission from the Galactic Centre," *Nature* 415 (2002), pp. 148–150.

Warner, B., "Intermediate Polars," in *Cataclysmic Variables and Low-Mass X-ray Binaries*, ed. D. Q. Lamb and J. Patterson, Dordrecht: Reidel (1985), pp. 269–279.

Warren, M. S., and J. K. Salmon, *A Parallel Hashed Oct-tree N-body Algorithm*, Washington, DC: IEEE Computer Society Press (1993), pp. 12–21.

———, "A Portable Parallel Particle Program," *Computer Physics Communications,"* 87 (1995), pp. 266–290.

Webb, G. M., L. Drury, and P. Biermann, "Diffusive Shock Acceleration of Energetic Electrons Subject to Synchrotron Losses," *Astronomy and Astrophysics* 137 (1984), pp. 185–201.

Webbink, R. F., S. A. Rappaport, and G. J. Savonije, "On the Evolutionary Status of Bright, Low-Mass X-ray Sources," *The Astrophysical Journal* 270 (1983), pp. 678–693.

Webbink, R. F., and D. T. Wickramasinghe, "Cataclysmic Variable Evolution: AM Her Binaries and the Period Gap," *Monthly Notices of the Royal Astronomical Society* 335 (2002), pp. 1–9.

Webster, B. L., and P. Murdin, "Cygnus X-1—A Spectroscopic Binary with a Heavy Companion?" *Nature* 235 (1972), pp. 37–38.

Wehrle, A. E., E. Pian, C. M. Urry, et al., "Multiwavelength Observations of a Dramatic High-Energy Flare in the Blazar 3C 279," *The Astrophysical Journal* 497 (1998), pp. 178–187.

Weidenspointner, G., V. Lonjou, J. Knödlseder, et al., "SPI Observations of Positron Annihilation Radiation from the 4th Galactic Quadrant: Sky Distribution," in *Proceedings of the 5th INTEGRAL Workshop on the INTEGRAL Universe,* ed. V. Schönfelder, G. Lichti, and C. Winkler, Paris: ESA Publications (2004), pp. 133–137.

Weinberg, S., *Gravitation and Cosmology: Principles and Applications of the General Theory of Relativity,* New York: Wiley, 1972.

White, N.E., J.H. Swank, and S. S. Holt, "Accretion Powered X-ray Pulsars," *The Astrophysical Journal* 270 (1983), pp. 711–734.

White, R. L., R. H. Becker, X. Fan, and M. A. Strauss, "*Hubble* Space Telescope Advanced Camera for Surveys Observations of the $z = 6.42$ Quasar SDSS J1148+5251: A Leak in the Gunn-Peterson Trough," *The Astronomical Journal* 129 (2005), pp. 2102–2107.

Wickramasinghe, D. T., and L. Ferrario, "Magnetism in Isolated and Binary White Dwarfs," *Publications of the Astronomical Society of the Pacific* 112 (2000), pp. 873–924.

Wittels, J. J., I. I. Shapiro, and W. D. Cotton, "Confirmation of a Conspiracy— Dual-Band VLBI Maps of the Flat-Spectrum Radio Source 2021+614," *The Astrophysical Journal Letters* 262 (1982), pp. L27–L30.

Wolfe, B., and F. Melia, "The Broadband Spectrum of Galaxy Clusters," *The Astrophysical Journal* 675 (2007), pp. 156–162.

Woosley, S. E., "Gamma Ray Bursts from Stellar Mass Accretion Disks around Black Holes," *The Astrophysical Journal* 405 (1993), pp. 273–277.

Woosley, S. E., and J. S. Bloom, "The Supernova Gamma Ray Burst Connection," *Annual Review of Astronomy and Astrophysics* 44 (2006), pp. 507–556.

Worrall, D. M., F. E. Marshall, E. A. Boldt, and J. H. Swank, "HEAO-1 Measurements of the Galactic Ridge," *The Astrophysical Journal* 255 (1982), pp. 111–121.

Worsley, M. A., A. C. Fabian, F. E. Bauer, et al., "The Unresolved Hard X-ray Background: The Missing Source Population Implied by the *Chandra* and XMM-*Newton* Deep Fields," *Monthly Notices of the Royal Astronomical Society* 257 (2005), pp. 1281–1287.

Wu, K., et al., "Accretion Flow in Magnetic Cataclysmic Variables," *Chinese Journal of Astronomy and Astrophysics* 3 (2003), pp. 235–244.

Yakovlev, D. G., and C. J. Pethick, "Neutron Star Cooling," *Annual Review of Astronomy and Astrophysics* 42 (2004), pp. 169–210.

Zahn, J. P., "Tidal Friction in Close Binary Stars," *Astronomy and Astrophysics* 57 (1977), pp. 383–394.

Zamorani, G., J. P. Henry, T. Maccacaro, et al."X-ray Studies of Quasars with the Einstein Observatory, II," *The Astrophysical Journal* 245 (1981), pp. 357–374.

Zatsepin, G. T., and V. A. Kuzmin, "Upper Limit of the Spectrum of Cosmic Rays," *Journal of Experimental and Theoretical Physics Letters* 4 (1966), pp. 78–80.

Zdziarski, A. A., A. C. Fabian, K. Nandra, et al., "Physical Processes in the X-Ray/Gamma-Ray Source of IC4329A," *Monthly Notices of the Royal Astronomical Society* 269 (1994), pp. L55–L60.

Zdziarski, A. A., J. Poutanen, W. S. Paciesas, and L. Wen, "Understanding the Long-Term Spectral Variability of Cygnus X-1 with Burst and Transient Source Experiment and All-Sky Monitor Observations," *The Astrophysical Journal* 578 (2002), pp. 357–373.

Zel'dovich, Ya. B., and I. D. Novikov, "Relativistic Astrophysics, Part I," *Soviet Physics—Uspekhi* 7 (1965), pp. 763–788.

Zhang, B., "Gamma Ray Bursts in the Swift Era," *Chinese Journal of Astronomy and Astrophysics* 7 (2007), pp. 1–50.

Zhang, B., Fan, Y.Z., Dyks, J., et al., "Physical Processes Shaping GRB X-ray Afterglow Lightcurves: Theoretical Implications from the SWIFT XRT Observations," *The Astrophysical Journal* 642 (2006), pp. 354–370.

Zhang, W., S. E. Woosley, and A. Heger, "The Propagation and Eruption of Relativistic Jets from the Stellar Progenitors of Gamma-Ray Bursts," *The Astrophysical Journal* 608 (2004), pp. 365–377.

Zingale, M., F. X. Timmes, B. Fryxell, et al., "Helium Detonations on Neutron Stars," *The Astrophysical Journal Supplements* 133 (2001), pp. 195–220.

Życki, P. T., C. Done, and D. A. Smith, "The 1989 May Outburst of the Soft X-ray Transient GS 2023+338 (V404 Cyg)," *Monthly Notices of the Royal Astronomical Society* 309 (1999), pp. 561–575.

Index